Decontamination of Warfare Agents

Edited by
André Richardt and
Marc-Michael Blum

Further Reading

N. Khardori (Ed.)

Bioterrorism Preparedness
Medicine – Public Health – Policy

2006
ISBN: 978-3-527-31235-1

H.-J. Jördening, J. Winter (Eds.)

Environmental Biotechnology
Concepts and Applications

2005
ISBN: 978-3-527-30585-8

Decontamination of Warfare Agents

Enzymatic Methods for the Removal of B/C Weapons

Edited by
André Richardt and Marc-Michael Blum

WILEY-VCH Verlag GmbH & Co. KGaA

The Editors

Dr. André Richardt
German Armed Forces Institute for Protection Technology–NBC Protection
Business Area of Biological and Chemical Decontamination
Humboldtstr. 100
29633 Munster
Germany

Dr. Marc-Michael Blum
University of Frankfurt
Institute of Biophysical Chemistry
Marie-Curie-Str. 9
60439 Frankfurt
Germany

Library of Congress Card No.:
applied for

British Library Cataloguing-in-Publication Data
A catalogue record for this book is available from the British Library.

Bibliographic information published by the Deutsche Nationalbibliothek
The Deutsche Nationalbibliothek lists this publication in the Deutsche Nationalbibliografie; detailed bibliographic data are available in the Internet at <http://dnb.d-nb.de>.

Typesetting SNP Best-set Typesetter Ltd., Hong Kong
Printing betz-druck GmbH, Darmstadt
Binding Litges & Dopf Buchbindese, GmbH, Heppenheim

Printed in the Federal Republic of Germany
Printed on acid-free paper

ISBN: 978-3-527-31756-1

Foreword

The present book is intended to give a comprehensive overview of the history and state-of-the-art of a novel experimental technique: the enzymatic decontamination of biological and chemical warfare (C/BW) agents. By this, it not only intends to point out the actual relevance and need of modern technologies, but also tries to contribute efforts to a better international coordination of promising research and development work in this field. Decontamination in general defines a technical process used to reduce or ideally remove or destroy contaminants, and moreover, to prevent the spread of contaminants from persons and equipment. Contamination with C/BW agents can occur without warning, as a result of attacks and collateral damage during military operations, as well as by deliberate release of highly toxic materials and infectious agents, or industrial accidents and natural outbreaks of diseases. Historically, military communities have the highest requirement for powerful decontamination technologies, not at least presumably because they are potentially confronted with a higher abundance and severity of C/BW agent attacks in comparison with possible terrorist or criminal scenarios with localized release of minor quantities.

The main reason to search for new technologies is that the current decontamination procedures for C/BW agents depend on higher temperature treatment and harsh chemical reactive components e. g. strong oxidants and chemical disinfectants. These techniques and compounds are often not suitable for different materials, cause logistic problems and are themselves hazardous, corrosive and environmentally critical. In many Armed Forces, aggressive and toxic substances like hypochlorite and formaldehyde are still the pivotal reactive components for C/BW-decontamination. The common goal of the various decontamination measures is to rapidly destroy known CW agents as well as to disinfect bacteria, fungi and protozoa, inactivate viruses and detoxify toxins.

For all decontamination procedures of clothing, material and vehicles, the military user would rather prefer a modern technique that is equally effective for CW and BW agents. However, this approach is technically very ambitious. The decontamination of BW agents including bacterial spores, which are known to be highly resistant to temperature and disinfectants, represents a particular challenge with respect to optimizing the proportions of efficiency and logistic burden, as well as

Decontamination of Warfare Agents.
Edited by André Richardt and Marc-Michael Blum

ISBN: 978-3-527-31756-1

its environmental impact. On the other hand, with the search for novel reactive components, biotechnology-based decontaminants became increasingly important with respect to these requirements. As will be outlined in this book, biotechnological approaches based on biochemical catalysis using enzymes on a technical scale may be competent alternatives or at least additives for existing protection measures against C/BW agents. They allow decontamination reactions under ambient (mild) conditions and follow a highly effective catalytic, rather than a reagent-consuming, stoichiometric principle. The data, as presented here, outline a good prognosis for such reagents. However, there is also strong evidence that more research and development efforts in this field – ideally on the basis of internationally coordinated cooperations – are indispensable to efficiently exploit this valuable biotechnology potential for future decontamination techniques, be it for military or civil defense.

Acting diretor of Bundeswehr
Research Institute of Protection
Technologies – NBC-Protection

Contents

Decontamination of Warefare Agents.
Edited by André Richardt and Marc-Michael Blum

ISBN: 978-3-527-31756-1

Preface

Since the end of the Cold War, the threat scenarios arising from the potential use of chemical and biological (CB) warfare agents have changed fundamentally. During the Cold War, both blocks stockpiled tens of thousands of tons of chemical warfare agents and pursued biological warfare programmes. In case of a military conflict between NATO and the Warsaw Pact Central, Europe was the most likely battlefield and large-area attacks with CB agents were among the most realistic scenarios. Decontamination systems were designed to be able to return contaminated equipment back into combat service as quickly as possible. Environmental concerns and potential long-term damage to equipment due to the use of aggressive and harmful chemicals were not an issue. Human casualties and fatalities were likely and although military planning tried to minimize these casualties, large-scale attacks would have resulted in substantial numbers of affected personnel. As the civilian population had only rudimentary protection against these agents, the numbers of affected persons would be even higher. After the end of the Cold War, the scenario shifted towards the use of CB agents, both finally banned by international treaty regimes, in smaller conflicts where terrorists, religious fanatics or other insurgents could use these weapons in asymmetric scenarios to cause fear, panic and uncertainty, especially affecting the civilian population. In addition to this, the last twenty years saw revolutionary advancements in the life sciences and the phenomenon of globalization went along with an enormous growth of worldwide traffic of both goods and people. Both developments could add to a growing threat, making the use of CB agents even more likely.

When monitoring public discussion, it is often striking that the fear of being faced with new dangerous substances is based on unrealistic assumptions and insufficient information. Media, scientists and other actors sometimes abuse the veil of mystery that has rested on the field of chemical and biological warfare agents since their first use to raise attention and to justify the requests for money. But even a realistic and unemotional assessment of the current threats reveals that the use of CB agents, especially against civilian population, is possible and becomes even more likely as we see a growing technical sophistication among insurgents and terrorist movements. Also, several countries remain outside the treaty regimes

Decontamination of Warefare Agents.
Edited by André Richardt and Marc-Michael Blum

ISBN: 978-3-527-31756-1

banning CB agents. Several of them are thought to pursue active weapon programs and raise proliferation concerns.

New comprehensive concepts to counter the CB threat are needed. The changed scenarios call for a unified strategy that covers the military side as well as civil defense. In these concepts, decontamination plays a crucial role flanked by detection and protection technologies. Modern decontamination should be fast, safe, environmentally benign and useable in civilian environments. New and highly advanced decontamination technologies have been under investigation for many years and some of them are on the brink of being introduced into service in the very near future. Biotechnology-based methods using enzymes are among these technologies and are the main topic of this volume.

Our intention to write this book with the help of many specialists in the field was driven by the perception that scientists should not only know the basics of their field, but also be aware of the requirements for effective decontaminants and of general concepts if they plan to introduce new technologies in this security-relevant field. Therefore, this book is intended to serve as a bridge between basic science, on the one hand, and applied engineering and product development on the other. Biotechnology-based decontaminants could be used to fill capability gaps where other highly corrosive and harmful decontaminants fail. However, new technologies rarely replace old ones right away. In an interim phase, new and old technologies will coexist and acknowledging that public procurement decisions are often made for a decade or more, this phase might turn out be quite long. We also intend to add to the public debate by trying to demystify chemical and biological agents and the possible threat posed by them. Discussion should be based on scientific knowledge, which all the authors of this volume try to put forward in a way that should enable the reader to take part in the debate with the necessary background to either make valuable contributions themselves, or to be able to assess and judge arguments put forward by others.

We would like to thank our friends, colleagues, co-authors as well as the editorial staff at Wiley-VCH for their support, ideas and remarks. Special thanks go to our families for their patience during the endeavour of this book.

André Richardt and Marc-Michael Blum

List of Contributors

Hans-Jürgen Altmann
Armed Forces Scientific Institute for Protection Technologies–NBC-Protection (WIS)
Humboldtstr. 100
29633 Munster
Germany

Marc-Michael Blum
Scientific Services
Ledererstrasse 23
D-80331 Munich
Germany

Steffen Danielsen
Novozymes A/S
Molecular Biotechnology
Protein Design
Building 1U office 1.20
Krogshøjvej 36
DK-2880 Bagsværd
Denmark

Roland Dierstein
Armed Forces Scientific Institute for Protection Technologies–NBC-Protection (WIS)
Humboldtstr. 100
29633 Munster
Germany

Friedrich Frischknecht
Department of Parasitology
University of Heidelberg Medical School
Im Neuenheimer Feld 324
D-69120 Heidelberg
Germany

Alexander Grabowski
Wehrwissenschaftliches Institut für Schutztechnologien–ABC-Schutz (WIS) 400
Humboldtstr. 100
29633 Munster
Germany

Thomas Hellweg
TU Berlin
Stranski Laboratorium
Sekr. ER1
Strasse des 17.Juni 112
10623 BERLIN
Germany

Birgit Hülseweh
Armed Forces Scientific Institute for Protection Technologies–NBC-Protection (WIS)
Humboldtstr. 100
29633 Munster
Germany

Decontamination of Warefare Agents.
Edited by André Richardt and Marc-Michael Blum

ISBN: 978-3-527-31756-1

Kai Kehe
InstPharmTox–Toxikologische Epidemiologie
Neuherbergstr.11
80937 München
Germany

S.J. Mitchell
Decon team, C&D, Detection
Bldg 383, DSTL Porton Down
Salisbury
Wilts SP4 0JQ
United Kingdom

Bärbel Niederwöhrmeier
Armed Forces Scientific Institute for Protection Technologies–NBC-Protection (WIS)
Humboldtstr. 100
29633 Munster
Germany

Bernd Niemeyer
Helmut-Schmidt-University /
University of the Federal Armed Forces Hamburg
Holstenhofweg 85
D-22043 Hamburg
Germany

Lars Ostergaard
Novozymes A/S
Dept Protein Design, Krogshoejvej, 36
DK-2880 Bagsvaerd
Denmark

André Richardt
Armed Forces Scientific Institute for Protection Technologies–NBC-Protection (WIS)
Humboldtstrasse 1
29633 Munster
Geermany

Heiko Russmann
Wehrwissenschaftliches Institut für Schutztechnologien–ABC-Schutz (WIS) 220
Humboldtstr. 100
29633 Munster
Germany

Horst Thiemann
InstPharmTox–Toxikologische Epidemiologie
Neuherbergstr.11
80937 München
Germany

Stefan Wellert
TU Berlin
Stranski-Laboratorium für Physikalische und Theoretische Chemie
Straße des 17.Juni 112
D-10623 Berlin
Germany

Franz Worek
InstPharmTox–Toxikologische Epidemiologie
Neuherbergstr.11
80937 München
Germany

1 The History of Biological Warfare

Friedrich Frischknecht

1.1 Introduction

Poisons have been used for assassinations for as long as humans can remember. It is unclear when they were first used intentionally for the purpose of warfare. However, during the last century, several tens of thousands of people were killed when disease agents or toxins were used in warfare, mainly during Japanese attacks on China in World War II. Two international treaties were established, first in 1925 and then in 1972, to prohibit the use of biological weapons in war, but both failed to inhibit research and large-scale productions of biological and toxin weapons in several countries. Since the letters containing anthrax that were sent in the wake of the September 11, 2001 terrorist attacks on the United States, it has been feared that pathogens could again be used in large scale for either terrorist or warfare purposes and billions of dollars have been poured into research of questionable scientific merit. To put the potential of such future threats into perspective, I recount some historical examples of biological warfare and terrorism.

1.2 Pre-Twentieth Century Examples

When humans first brought war upon each other, more soldiers were incapacitated from diseases than from the hand of their human enemies. This observation might have lead to the early deployment of poisoning substances during war (see Table 1.1). To our current knowledge, poisons were administered to enemy water supplies as early as the sixth century BC. Animal cadavers substituted for poisons during the Greek and Roman eras and Emperor Barbarossa used human corpses to the same end, although it is likely that a simple spoiling of the water supply, rather than the spreading of disease, was intended [1–3]. Hannibal suggested a more active approach when advising the Bithynians to catapult jars filled with snakes towards enemy ships in 184 BC [4]. The panic created, rather than

Decontamination of Warefare Agents.
Edited by André Richardt and Marc-Michael Blum

ISBN: 978-3-527-31756-1

Table 1.1 Selection of possible events of biological warfare.

Year	*Event*	*Disease agents and outcomes*
<1000 BCE. (Trojan War)	Legend of Scythian archers using poison arrows.	Clostridium (?) causing gangrene and tetanus.
<500 BCE.	Assyrians poisoned enemy wells.	Rye ergot fungus causing hallucinations.
590 BCE.	Greeks poison water supply of Kirrha during the first Sacred War.	Hellebore root causing diarrhea. Kirrha falls and the population is slaughtered.
184 BCE.	Bithynians catapult jars filled with snakes towards enemy ships.	Snakes causing panic. Sea battle is won.
1155	Emperor Barbarossa poisons water wells.	Decomposing human bodies.
1346	Tartars catapult plague victims over the walls of Caffa.	Yersinia pestis (?) causing plague. City is abandoned.
1495	Spanish sell wine mixed with blood from leprosy patients to enemy.	
1763	British distribute blankets from smallpox patients to native Americans	Variola virus causing smallpox. Epidemic develops.
1797	Napoleon floods fields around Mantua to enhance malaria.	
1915–1918	Germans infect animals of Allies	Anthrax and Glanders.
1932–1945	Japanese conduct large-scale human experiments and biological warfare in China.	Many different pathogens killing tens of thousands.

Note that the understanding of disease causing agents does not predate the work by Pasteur and Koch and thus a clear distinction between chemical and biological warfare can often not be made prior to that time. Also cause and consequence as well as intention and outcome are not always clearly established for many events as discussed in the text.

poisonous bites, likely decided the battle, revealing human psychology as a second important dimension during biological attacks. Catapulting infected human bodies and excrement constituted a further step during the sieges of many towns, although it is not clear if these actions contributed much to the spread of disease. When the Tartars besieged the Crimean city of Caffa (now Feodosiya, Ukraine) in 1346, they catapulted victims of the bubonic plague into the city where the Black Death soon caused collapse. As the occupying Genovese fled and the victors moved on, both spread the disease that ended up killing some 50% of the European and Chinese population (Figure 1.1). This changed the course of human history forever [5, 6] (Figure 1.1). However, whether biological warfare was the beginning of this greatest of medieval disasters remains impossible to prove. The fleas that transmit the disease between humans leave dead bodies rather quickly, therefore calling into question if the corpses that flew into Caffa were flea infested [3, 7]. Rats moving in and out of the city walls might have spread the disease much more

A

B

Figure 1.1 Late medieval illustrations of the Black Death. (A) Suffering plague victims as illustrated in a German bible from 1411. (B) The 1349 burning of Jews blamed for causing the pandemic as an example of the social repercussions unexplained infectious diseases caused throughout history.

efficiently, although it is not clear if they would have moved far enough. Furthermore, it is not clear if the attackers intended to spread disease or simply wanted to get rid of the stinking bodies of their dead comrades. And lastly, it is even disputed if *Yersinia pestis* was the cause of the Black Death at all [8]. These uncertainties illustrate one of the biggest problems in biological warfare history. How can one be sure an attack actually leads to the spread of infections, while they could just as well result from coincidental natural infection in an "unnaturally" large aggregation of–or new encounters between–humans.

With time, humans became more inventive and thought of more elaborate ways of distributing disease agents. In 1495, Spanish forces supplied their French adversaries with wine contaminated with the blood of leprosy patients during battles in Southern Italy [9]. In the seventeenth century, Polish troops tried to fire saliva from rabid dogs towards their enemies. The first pledge against the use of poisoned weapons was made between France and Germany in the 1675 Strasbourg Agreement. And while Russian troops might still have catapulted plague victims into the Baltic city of Reval during a war with Sweden in 1710, and the Tunisians tried plague infected clothing 75 years later, Napoleon went a step further. He attempted to use swamp fever during the siege of Mantua in 1797, by flooding the fields around the city, hoping to thus induce the spread of the disease now known as the mosquito-transmitted malaria. Twelve years later he looked on happily at British troops dying from swamp fever in the marches of Holland, where they set up camp [10].

Accidental spreading of diseases like smallpox, influenza, measles and tuberculosis took the lives of most indigenous people of the Americas, making the Spanish conquest possible [6, 11, 12]. Later, in North America, the French and English used smallpox-drenched clothing that they distributed among Indians [2, 13]. However, similar to the situation in Caffa, it is not clear if the following smallpox epidemic was caused by this measure, or rather because of contacts from trading activities between Indians and colonists [7, 14]. During the civil war in the United States, the Confederate doctor Luke Blackburn attempted to infect federal troops using clothing carried by smallpox and yellow fever patients. However, yellow fever was soon discovered to be solely transmitted by mosquitoes, leaving claims that soldiers died from such an attack rather unbelievable. On the other hand, Union troops were forbidden by an army order to use poison in any manner.

1.3
From World War I to World War II

With the foundation of microbiology by Pasteur and Koch, bio-warfare could take new dimensions as weapons could be designed on a rational basis. Recognizing this, two international declarations (1874 in Brussels and 1899 in The Hague) tried to prohibit the use of poisoned arms [15]. Nevertheless, as with chemical weapons, the Germans pioneered the use of biological weapons during World War I, albeit on a miniscule scale. Covert operations were using both anthrax and glanders to infect animals directly or to contaminate animal feed in Romania, France, Mesopotamia, Argentina and the United States [7, 13]. Fearing each other and relying on misinformed intelligence reports, a number of European countries had biological warfare programmes established before the onset of World War II.

In North America, Sir Frederick Banting, Nobel Prize-winning discoverer of insulin, started what could be called the first private biological weapon research center [16]. The United States joined their British allies, who, like the French, feared a German biological attack [16]. However, unlike the Nazis, some Japanese were truly enthusiastic about the potential of biological weapons [16–20]. The radical nationalist Shiro Ishii started his research in 1930 and became the first head of Japan's bio-weapon program during World War II (Figure 1.2). At its peak, the program employed over 5000 people, killing as many as 600 prisoners a year in human experiments in the largest of its many centers. At least 25 different

Figure 1.2 Japanese biological warfare during World War II. (A) Shiro Ishii in 1932, photographed by Masao Takezawa. (B) Reconstructed building of Unit 731 at Harbin as photographed in 2002 by Markus Källander. Both images are public domain pictures from Wikipedia.

disease-causing agents were cruelly tested on prisoners and unsuspecting civilians. Water wells in Chinese villages were poisoned to study cholera and typhus outbreaks. Plague-infested fleas were dropped by plane over Chinese cities or distributed by saboteurs in rice fields and along roads, causing epidemics in areas where the plague was unknown, some persisting for years after the war ended. It is estimated that several tens of thousands of people died as a consequence of offensive biological research, including soldiers on both sides of Soviet-Japanese battles when suicide squads and artillery shells were employed to spread disease in 1939 [18, 19].

1.4 Secret Projects and Cold War Allegations

After the war, the Americans granted freedom to all researchers in exchange for information on their human experiments [17–20]. In this way, mass murderers and war criminals became respected citizens, including founders of pharmaceutical companies. Masaji Kitano, the successor of Ishii, published post-war papers on human experiments using "monkey" instead of "human" when referring to the experiments in wartime China [18]. A cover-up was almost successful when, 45 years after the war, numerous bones were found during construction work near Ishii's old laboratory in Tokyo [17]. In 1947, to profit from the secret Japanese knowledge, United States President Truman withdrew the Geneva Protocol from Senate, which still had not ratified it. Soon after, the US military conducted open-air tests releasing both pathogenic and non-pathogenic microbes on test animals, human volunteers and unsuspecting civilians [2, 16, 21]. *Bacillus globigii* and *Serratia marcensces,* two harmless bacteria, were released from naval vessels on the Virginia coast and off San Francisco, infecting around 800 000 people in San Francisco alone. Furthermore, bacterial aerosols were released at bus stations and airports [16]. The most infamous test was the 1966 contamination of the New York metro system with *Bacillus globigii* (a so-called anthrax simulant) to study the spread of a pathogen in a big city [16, 22]. However, with the opposition to the Vietnam War and his own wish for a Noble Peace Prize growing, US President Nixon decided to abandon offensive biological weapon research and signed the Biological Weapon and Toxin Convention (BWTC) in 1972. While most stocks were destroyed the same year, some quantities of toxins were kept aside by the CIA for several years [22].

Allegations that other countries produce biological weapons have served as excuses to develop these weapons and have been exploited as welcome means for propaganda. During the Korean War the Chinese, North Koreans and Soviets accused the US-led troops of deploying biological weapons of various kinds. Although it is now largely seen as a propaganda move, the secret deal between the US and Japanese bio-weapon researchers did not help diffuse the allegations [23]. In reverse, the United States accused their Vietnamese enemy of dropping fungal toxins onto the American Hmong allies in Laos. After some confusion it turned out that the yellow rain associated with the reported variety of syndromes was bee feces [24].

Closer to home, Cuba complained frequently about American biological warfare, and in 1997 was the first country to officially file a complaint under article 5 of the BWTC, accusing the US of dissipating a plant pathogen [15]. Although this allegation is probably not substantial, the US did invest in biological warfare against Fidel Castro and Frederik Lumumba of Congo [22]. However, not everything ridiculed by politicians or scientists turned out to be fiction. The outbreak of anthrax in the city of Sverdlovsk (now Ekaterinburg, Russia) in 1979 was seen by many in the West as a breach of the BWTC. The Soviets naturally refuted such accusations.

After signing the BWTC, the Soviet Union created a gigantic project employing at its height over 50 000 people directly working on the development and production of biological weapons [25]. While stockpiling large amounts of anthrax and smallpox, also for the use on missiles, an unknown number of people lost their lives. Most famously, when a filter was not replaced between shifts. The resulting outbreak of inhalation anthrax in Sverdlovsk killed at least 66 civilians [25, 26]. In another accident within a high security lab, a virologist injected himself with marburg virus [25]. After his death, his colleagues re-isolated a more virulent virus from his body and continued to work on this particular strain! Similarly, there are speculations about the association of a bio-weapon research center located on a small island in the Aral Sea with an outbreak of smallpox in the city of Aralsk, as well as the occasional deaths of fisherman and researchers [22, 27]. With the fall of the Soviet Union it is now generally feared that not all stocks of dangerous pathogens including smallpox (Figure 1.3) have been destroyed, and may have been taken abroad by scientists now working in other countries [22, 25, 28]. Furthermore, it is not clear to what extend Russia is continuing with secret programs.

1.5
Towards the Twenty-first Century: Madmen on the Run

Apart from state-sponsored bio-warfare programmes, a number of single person or non-governmental group-driven attacks have occurred or were alleged [29]. Here just three examples. Just after World War II a peculiar bank robbery occurred in Tokyo. A man identified himself as an official from the Ministry of Health and Welfare and made every employee of the branch drink a potion with the effect that 12 out of 16 people died after several minutes. An alleged high level cover-up possibly led to the imprisonment of an innocent person who was sentenced to death, but never killed, fuelling speculations that former members of the Japanese bio-warfare program were somehow involved [17].

A large outbreak of salmonella in a small town along the Columbia River in Oregon in 1984 caused the gastroenteritis of 751 people [30] when a religious sect wanted to poison a whole community to influence an upcoming local election [22]. The bacteria were found in a number of salad bars in various restaurants in a

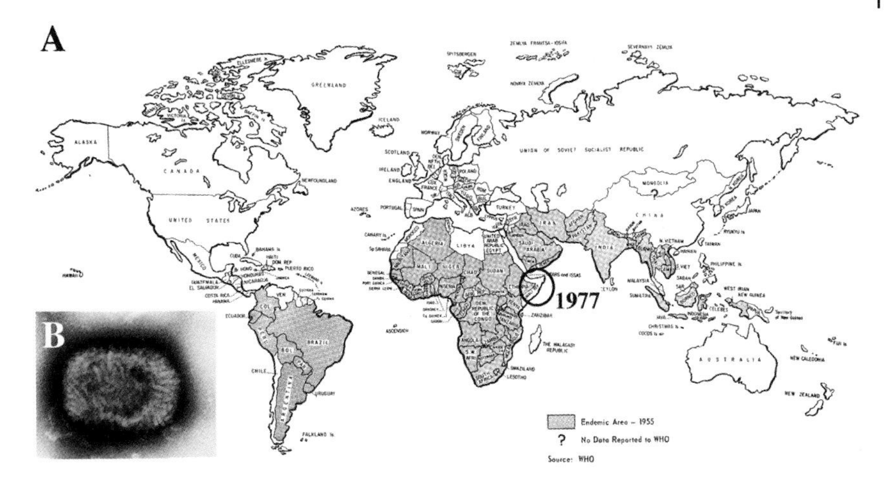

Figure 1.3 Smallpox and its eradication. (A) Countries with endemic smallpox in the year 1955 are highlighted. The circle indicates the last case of naturally acquired smallpox in Somalia 22 years later. (B) Electron micrograph of a variola virus particle. (C) Pre-1979 Asian poster promoting youth vaccination for Smallpox and Measles. (D) Ali Moaw Maalin at 23 years of age became the last person worldwide to contract smallpox (Variola minor) in Somalia in 1977. All images are from the Center of Disease Control Public Image Library and courtesy of WHO and CDC.

manner that could not be traced back to a naturally occurring outbreak and the sect was eventually identified as culprit due to the testimony of an insider [30]. The sect obtained the bacterial strain simply, from a commercial supplier, and produced it in a hospital setting in its own grounds. This clearly indicates how easy it might be for organized groups to collect a mix of reagents in order to put together a small bio-terrorism program. All it would seem to need are a few letters to "colleagues" at scientific institutions whose duty it is to send out published materials in order to share it with the rest of the community [31].

Another set of attacks occurred in 1995, when the Japanese Aum Shinrikyo cult in Tokyo killed 12 people and injured over 5000 by using the nerve gas sarin [32].

The investigations into the sect revealed that it had also tried to distribute anthrax within the city. However, this remained without success. The production of the spores was apparently possible, but the sect failed to produce the right kind of aerosol needed for successful dissemination [15, 33]. Sadly, the still unidentified perpetrator(s) of the 2001 anthrax attacks in the United States were more successful by sending contaminated letters that eventually killed 5 unsuspecting citizens. Together with copycat hoax attacks, the letters caused an estimated economic loss of several hundred million dollars [15, 22, 34].

1.6
Future Threats – Science or Fiction?

Public interest in the potential danger of biological warfare and terrorism has been sparked by the events on and following September 11, 2001 as well as the depiction of fictitious biological attacks in books and movies [35, 36]. While some experts predict major outbreaks with devastating effects, others assume that, with reasonable precautions, such epidemics could be avoided [15, 37]. Leitenberg (2001) interprets the example of the Aum Shinrikyo attacks as evidence that terrorists might not find it straightforward to develop an effective biological weapon. In contrast, state-sponsored biological warfare programs and their successful concealment indicate that complacency might be the wrong strategy in dealing with this fuzzy potential threat. However, when deciding on how to distribute research money, we should always compare the speculative nature of potential future biological attacks with the grim reality of daily life for millions of people, who continue to suffer and die often from already preventable infections.

1.7
Acknowledgements

My research is supported by grants from the German Federal Ministry for Education and Research (BioFuture Programme) and the German Research Foundation (Sonderforschungsbereich 544 "Control of Tropical Infectious Diseases" and Schwerpunktprogram 1128 "Optical Analysis of the Structure and Dynamics of Supra-molecular Biological Complexes"). I am grateful to Barbara Janssens, Markus Meissner, Sylvia Münter and Jörg Sacher for comments on the manuscript.

References

1 Clarke, R. (1968) *The Silent Weapon*, David McKay, New York, USA.
2 Poubard, J.A. and Miller, L.A. (1992) History of biological warfare: catapults to capsomeres. *Annals of the New York Academy of Sciences*, **666**, 9–20.
3 Wheelis, M. (1999) Biological warfare before 1914, in *Biological And Toxin*

Weapons: Research, Development And Use From The Middle Ages To 1945 (eds E. Geissler and J.E.v.C. Moon) Stockholm International Peace Research Institute, Oxford University Press, Oxford, UK.

4 Noah, D.L., Huebner, K.D., Darling, R.G. and Waeckerle, J.F. (2002) The history and threat of biological warfare and terrorism. *Emergency Medicine Clinics of North America*, **20**, 255–71.

5 Derbes, V.J. (1966) De Mussi and the great plague of 1348. *The Journal of the American Medical Association*, **196**, 59–62

6 Garrett, L. (1994) *The Coming Plague. Newly Emerging Diseases in a World out of Balance*, Penguin books, New York, USA.

7 Christopher, G.W., Cieslak, T.J., Pavlin, J.A. and Eitzen, E.M. (1997) Biological warfare: a historical perspective. *The Journal of the American Medical Association*, **278**, 412–7.

8 Cohn, S. (2003) *The Black Death Transformed: Disease and Culture in Early Renaissance Europe*, Hodder Arnold, London.

9 Iserson, K.V. (2002) *Demon Doctors*, Galen Press, Tuscon, AZ, USA.

10 Rocco, F. (2003) *The Miraculous Fever Tree: Malaria and the Quest for a Cure That Changed the World*, Harper Collins, New York, USA.

11 McNeill, W.H. (1976) *Plagues and People*, Doubleday, New York, USA.

12 Oldstone, M.B.A. (1999) *Viruses Plagues and History*, Oxford University Press, Oxford, UK.

13 Poubard, J.A., Miller, L.A. and Granshaw, L. (1989) The use of smallpox as a biological weapon in the French and Indian war of 1763. *ASM News*, **55**, 122–4.

14 Fenner, F., Henderson, D.A., Arita, I., Jezek, Z. and Ladnyi, I.D. (1988) *Smallpox and Its Eradication*, World Health Organization, Geneva, Switzerland. http://www.who.int/emc/diseases/smallpox/Smallpoxeradication.html

15 Leitenberg, M. (2001) Biological weapons in the twentieth century: a review and analysis. *Critical Reviews in Microbiology*, **27**, 267–320.

16 Regis, E. (1999) *The biology of doom. The History of America's Secret Germ Warfare Project*, Henry Holt and Co., New York, USA.

17 Gold, H. (1996) *Unit 731 Testimony. Japan's Wartime Human Experimentation Program*, Yenbooks, Tokyo, Japan.

18 Harris, S. (1992) Japanese biological warfare research on humans: a case study of microbiology and ethics. *Annals of the New York Academy of Sciences*, **666**, 21–52.

19 Harris, S.H. (1994) *Factories of Death*, Routledge, New York, USA.

20 Williams, P. and Wallace, D. (1989) *Unit 731: Japan's secret biological warfare in World War II*, Houghton and Staughton, London, UK.

21 Cole, L.A. (1988) *Clouds of Secrecy: The Army's Germ Warfare Tests over Populated Areas*, Rowman and Littlefield, Lanham, USA.

22 Miller, J., Engelsberg, S. and Broad, W. (2002) *Germs. Biological Weapons and America's Secret War*, Simon & Schuster, New York, USA.

23 Moon, J.E.v.C. (1992) Biological war allegations: the Korean War case. *Annals of the New York Academy of Sciences*, **666**, 53–83.

24 Seeley, T.D., Nowicke, J.W., Meselson, M., Guillemin, J. and Akratanakul, P. (1985) Yellow rain. *Scientific American*, **253** (3), 122–31.

25 Alibek, K. and Handelman, S. (1999) *Biohazard*, Random House, New York, USA.

26 Meselson, M., Guillemin, J., Hugh-Jones, M., Langmuir, A., Popova, I., Shelokov, A. and Yampolskaya, O. (1994) The Sverdlovsk anthrax outbreak of 1979. *Science*, **266**, 1202–8.

27 Enserink, M. (2002) Did bioweapons test cause a deadly smallpox outbreak? *Science*, **296**, 2116–7.

28 Domaradskij, I.V. and Orent, W. (2001) The memoirs of an inconvenient man: revelations about biological weapons research in the Soviet Union. *Critical Reviews in Microbiology*, **27**, 239–66.

29 Purver, R. (2002) Chemical and biological terrorism: the treat according to the open literature, Canadian Security and Intelligence Service. http://www.csis-scrs.gc.ca/en/publications/other/c_b_terrorism01.asp (Last accessed Oct, 9, 2007.).

30 Török, T.J., Tauxe, R.V., Wise, R.P., Livengood, J.R., Sokolow, R., Mauvais, S., Birkness, K.A., Skeels, M.R., Horan, J.M. and Foster, L.R. (1997) A large community outbreak of salmonellosis caused by intentional contamination of restaurant salad bars. *The Journal of the American Medical Association*, **278**, 389–95.

31 Breithaupt, H. (2000) Toxins for terrorists. *EMBO Reports*, **1**, 298–301.

32 Cole, L.A. (1996) The specter of biological weapons. *Scientific American*, (**12**), 30–5.

33 Atlas, R.A. (2001) Bioterrorism before and after September 11. *Critical Reviews in Microbiology*, **27**, 355–79.

34 Hughes, J.M., Gerberding, J.L., et al. (2002) Bioterrorism–related Anthrax. *Emerging Infectious Diseases*, **8** (*10*), 1013–1159.

35 Preston, R. (1997) *The Cobra Event*, Ballantine Books, New York, USA.

36 Morris, M.E. (1996) *Biostrike*, Avon Books, New York, USA.

37 O'Toole, T. (1999) Smallpox: an attack scenario. *Emerging Infectious Diseases*, **5**, 540–6.

Further Suggested Reading

Frischknecht, F. (2003) History of biological warfare. Human experimentation, modern nightmares and lone madmen in the twentieth century. *EMBO Reports*, **S4**, 47–52.

Geissler, E. and Moon, J.E.v.C. (1999) *Biological and Toxin Weapons: Research, Development and Use from the Middle Ages to 1945*, Stockholm International Peace Research Institute, Oxford University Press, Oxford, UK.

Guillemin, J. (2005) *Biological Weapons: From the Invention of State-Sponsored Programs to Contemporary Bioterrorism*. Columbia University Press, New York, USA.

Lederberg, J. and Cohen, W.S. (1999) *Biological Weapons: Limiting the Threat*, MIT Press, Boston, USA.

Wheelis, M., Rozsa, L. and Dando, M. (2006) *Deadly Cultures*, Harvard University Press, Cambridge, MA, USA.

Stockholm International Peace Research Insitute (1971) *The Problems of Chemical and Biological warfare*, Vol. I. Humanities Press, New York, USA.

2
History of Chemical Warfare

André Richardt

2.1
Introduction

Throughout human history there are numerous examples of chemicals or chemical weapons used or proposed during the course of a campaign or battle. The twentieth century saw the dawn of a new age in battlefield tactics and the abuse of chemicals as chemical warfare agents (CWAs). Although there have been many attempts to ban chemical warfare agents (CWAs), their devastating potential makes them attractive still. Therefore, it is likely that the emergence of chemical terrorism is going to be a significant threat in the twenty-first century. To understand the potential of current and future threats, this chapter provides a short but comprehensive history of chemical warfare, from its beginnings, up to the emergence of chemical terrorism. For information in detail, the interested reader is forwarded to additional literature and websites.

2.2
The Beginning

The use of chemical weapons dates from at least 1000 BC when the Chinese used arsenical smokes [1]. In 600 BC Solon of Athens put hellebore roots in the drinking water of Kirrha. By using noxious smoke and flame, allies of Sparta took an Athenian-held fort in the Peloponnesian War, between 420 and 430 BC. About 200 BC, the Carthaginians used mandrake root left in wine to sedate the enemy. Stink bombs of poisonous smoke and shrapnel were designed by the Chinese, along with a chemical mortar that fired cast-iron stink shells. Other conflicts during the succeeding centuries saw the use of smoke and flame [1, 2]. During the seventh century AD the Greeks invented Greek fire (Figure 2.1) [2]. This formulation that probably contained sulfur, rosin, naphtha, pitch, saltpeter and lime, floated on water and was designed for naval operations [2, 3]. However, the exact composition of the original Greek fire is still unknown. Toxic smoke projectiles were designed

Decontamination of Warefare Agents.
Edited by André Richardt and Marc-Michael Blum

ISBN: 978-3-527-31756-1

Figure 2.1 Greek fire. Credit: Adrienne Mayor, Greek Fire, Poison Arrows & Scorpion Bombs: Biological and Chemical Warfare in the Ancient World (Overlook, 2003); painting by Bob Lapsley/Aramco Services/PADIA.

and used during the Thirty Years War. Leonardo da Vinci proposed a powder of sulfide of arsenic and verdigris in the fifteenth century [4]. Venice employed unspecified poisons in hollow explosive mortar shells during the fifteenth and sixteenth centuries. It also sent poison chests to its enemy to poison wells, crops, and animals. During the Crimean War, the use of cyanide-filled shells was proposed to break the siege of Sevastopol [1].

2.3
The Rise of a New Age – From WW I to WW II

However, the birth of modern chemical warfare agents can be dated in the early twentieth century. Inventions in modern inorganic chemistry during the late eighteenth and early nineteenth centuries, and the flowering of organic chemistry worldwide during the late nineteenth and early twentieth centuries, generated a renewed interest in chemicals as military weapons.

The chemical agents first used in combat during World War I were eighteenth- and nineteenth-century discoveries. Carl Scheele, a Swedish chemist, discovered chlorine in 1774. He also determined the properties and composition of hydrogen cyanide in 1782. Comte Claude Louis Berthollet, a French chemist, synthesized cyanogen chloride in 1802. Sir Humphry Davy, a British chemist, synthesized phosgene in 1812. Dichloroethyl sulfide (mustard agent) was synthesized in 1822, again in 1854, and finally fully identified by Victor Meyer in 1886. John Stenhouse, a Scottish chemist and inventor, synthesized chloropicrin in 1848 [1]. In 1887, some considerations for the use of tear agents (lacrimators) for military purposes were brought up in Germany. A rudimentary chemical warfare program was started by the French with the development of a tear gas grenade containing ethyl bromoacetate. Also, in France there were some discussions regarding filling artillery shells with chloropicrin [5]. The French Gendarmerie had successfully employed riot-control agents for civilian crowd control. These agents were also used in small quantities in minor skirmishes against the Germans, but were largely inefficient. However, these riot-control agents were the first chemicals used on a modern battlefield, and the research for more effective agents continued throughout the war. In the early stages of World War I, the British examined their own chemical technology for battlefield use. Their first investigations also covered tear agents, but later they put their effort towards more toxic chemicals. Nevertheless, the first large-scale use of chemicals during World War I was initiated by heavily industrialized Germany. Three thousand 105mm shells filled with the lung irritant dianisidine chlorosulfate were fired at British troops by the Germans near Neuve Chapelle on October 27th 1914, but with no visible effect. However, the British were the victims of the first large-scale chemical projectile attack. The Germans continued firing modified chemical shells with equally unsuccessful results [6]. The shortage of artillery shells led to the concept of creating a toxic gas cloud directly from its storage cylinder. This concept was invented by Fritz Haber of the Kaiser Wilhelm Institute for Physical Chemistry of Berlin in 1914. German units placed a total number of between 2000 and 6000 cylinders opposite the Allied troops defending the city of Ypres in Belgium. The cylinders contained a total of around 160 tons of chlorine. Once the cylinders were in place, and due to the critical importance of the wind, the Germans waited for the winds to shift to a westerly direction towards the trenches of the Allied troops. During the afternoon of April 22nd 1915, the chlorine gas was released with devastating effects. This attack caused probably between 800 (realistic) and 5000 (mainly propaganda) deaths [7, 8]. The Figure 2.2 and Figure 2.3 give an impression of the horror on the battlefield. The psychological effect was demoralizing to the Allied troops and the Allied front line simply fell apart. Although the success of this attack was greater than expected, the Germans were not prepared to make significant gains. Also, the Allied troops quickly restored a new front line and it took only a short period of time for them to be able to use chlorine themselves. In September 1915, they launched their own chlorine attack against the Germans at Loos. The importance of the wind direction is shown in Figure 2.4. The expansion of the armamentarium with chloropicrin and phosgene was the just the beginning of a deadly competition between both sides. Better protective masks, more dangerous chemicals and improved delivery systems were invented and developed [1].

A further step in a devastating chemical war was the use of a new kind of chemical agent. On the 11th of July 1917, again near Ypres in Belgium, sulfur mustard was used by the Germans in an artillery attack. Compared to the first agents, mustard was a more persistent vesicant on the ground, and this caused new problems. Not only was the air poisoned, but also the ground and equipment could be contaminated. This new agent was effective in low doses and affected the lungs, the eyes and the skin. Although fewer than 5% of mustard exposed soldiers died, mustard injuries could easily overwhelm the medical system. The military and civil medical systems were not prepared for thousands of poisoned soldiers.

Figure 2.2 Dressing the Wounded during a Gas Attack, a 1918 painting by the British war artist Austin Osman Spare. Image is public domain picture from Wikipedia.

Figure 2.3 Amaurotic British soldiers waiting for their treatment. Image is public domain picture from Wikipedia.

Figure 2.4 A French cylinder attack on German trenches in Flanders. The critical importance of the wind is apparent. Condensation of water vapor caused the cloudlike appearance of the gas. Image is public domain picture from Wikipedia.

Table 2.1 Estimated chemical casualties in World War 1.

Country	*Nonfatal Chemical Casualties*	*Chemical fatalities*
Russia (Prentiss, 1937 [6])	420 000	56 000
Russia (Kohn, 1973 [9])	65 000	6 300
Germany	191 000	9 000
France	182 000	8 000
British Empire	180 000	8 100
United States	71 000	1 500

Mustard injuries were slow to heal and the average convalescent period was of over six to seven weeks. Therefore, the need for protective equipment, that was heavy and bulky at that time, for soldiers and horses led to more difficult and dangerous fighting. World War I was the dawn of a new military age with devastating effects [9] (Table 2.1).

Throughout the 1920s there was evidence that the military use of chemical agents continued after the end of World War I and rumors of chemical warfare attacks plagued the world. Germany worked with Russia, which had suffered half a million chemical casualties during World War I, to improve their chemical agent offensive and defensive programs from the late 1920s to the mid-1930s. During the Russian Civil War and Allied intervention in the early 1920s, both sides had chemical weapons, and there were reports of isolated chemical attacks. Later accounts accused the British, French, and Spanish of using chemical warfare at various times and places during the 1920s. For example there were rumors that Great Britain has employed chemicals against the Russians and mustard against the Afghans north of the Khyber Pass [10]. Spain has been accused of having used mustard shells and bombs against the Riff tribes of Morocco.

2.4 The Italian–Ethiopian War

The first major use of chemical weapons after World War I was reported during the Italian–Ethiopian War. On October 3rd 1935, Mussolini launched an invasion of Ethiopia from its neighbors Eritrea, and Italian Somaliland. The Italian troops dropped mustard bombs and sprayed mustard agent from airplane tanks. Mustard agent was used as a "dusty agent" to burn the unprotected feet of the Ethiopians with devastating effects. It was the first time that special sprayers were prepared onboard aircraft to vaporize a fine, deadly rain. These fearful tactics succeeded and by May 1936, the Ethiopian army was completely routed and Italy controlled most of Ethiopia until 1941 when British and other allied troops re-conquered the country. There are some conclusions that the Italians were clearly superior and that the use of chemical agents in the war was nothing more than an experiment. Still, there were thousands of victims of Italian mustard gas [11, 12].

2.5
Japanese Invasion of China

The Japanese had an extensive chemical weapons program. They were producing agent and munitions in large quantities by the late 1930s. During the invasion of China in 1937, it was reported that Japanese forces began using chemical shells, tear gas grenades, and lacrimatory candles. In 1939, there was an escalation by the Japanese that led to the use of mustard and lewisite to great effect. The Chinese troops retreated whenever they saw smoke, thinking it was a chemical attack [13].

2.6
The First Nerve Agents

Searching for more potent insecticides, German chemist Dr. Gerhart Schrader of the I. G. Farben Company discovered an extremely toxic organophosphorus insecticide in 1936. This new compound was reported to the Chemical Weapons Section of the German military prior to patenting as required by German law of that time. The substance had devastating effects on the nervous system and was therefore classified for further research. The substance was named tabun and after World War II it was designated GA, for "German" Agent "A". The research continued and in 1938, a similar agent, sarin (GB), was designated with a toxicity of five times higher than tabun. A pilot plant for production was built in 1939, the beginning of World War II [14]. Germany produced and weaponized approximately 78 000 of tons of chemical warfare agents. The key nerve agent, in terms of production, was mustard. The Germans filled artillery shells, bombs, rockets, and spray tanks with the agent. Why these deadly agents were not used on the battlefield is a continuing mystery. There were rumors that Hitler, a mustard casualty during World War I, did not favor their use. Thus, the German nerve agent program remained as a secret until its discovery by the Allies. There are reports that Japan produced about 8000 tons of chemical agents during the war. The favored agents were mustard agent, a mustard–lewisite mixture, phosgene and hydrogen cyanide. They gained experience in their use during attacks on China. However, the greatest producer of chemical warfare agents was the United States of America. Ready for retaliation, if Germany had been using chemical warfare agents, the United States produced proximately 146 000 tons of chemical agents between 1940 and 1945. With the possible exception of Japan during attacks in China, no nation used chemical agents on the battlefield during World War II. However, the positioning of chemical weapons near the front line in the case of need resulted in one major disaster. In 1943 the Germans bombed an American ship, the John Harvey, in Bari Harbor, Italy. This was a ship loaded with two thousand 100-pound M47A1 mustard bombs. Over 600 military casualties and an unknown number of civilian victims resulted from the raid, when they were poisoned by ingestion, skin exposure to mustard-contaminated water and inhalation of mustard-laden smoke. The harbor cleanup took more than three weeks [15].

2.7 Living with the Danger of Chemical Warfare – From WW II to 2000

The end of World War II did not stop the development, stockpiling, or use of chemical weapons. During the Yemen War of 1963 through 1967, Egypt in all probability used mustard and nerve agents in support of South Yemen against royalist troops in North Yemen [16]. Attacks occurred on the town of Gahar and on the villages of Gabas, Hofal, Gadr, and Gadafa. Shortly after these attacks, the International Red Cross examined victims, soil samples, and bomb fragments, and officially declared that chemical weapons, identified as mustard agent and possibly nerve agents, had been used in Yemen. Prior to this, no country had ever used nerve agents in the combat. The combination of the use of nerve agents by the Egyptians in early 1967 and the outbreak of war between Egypt and Israel during the Six-Day War in June, finally attracted world attention to the events in Yemen [16].

The US, which used napalm, defoliants and riot-control agents in Vietnam and Laos, finally ratified the Geneva Protocol in 1975, but with the stated reservation that the treaty did not apply either to defoliants or riot-control agents (Figure 2.5). During the late 1970s and early 1980s, reports of the use of chemical weapons against the Cambodian refugees and against the Hmong tribesmen of central Laos surfaced, and the Soviet Union was accused of using chemical agents in Afghanistan. Widely publicized reports of Iraqi use of chemical agents against Iran during the 1980s led to a United Nations investigation that confirmed the use of the vesicant mustard and the nerve agent GA. Later during the war, Iraq apparently also began to use the more volatile nerve agent GB, and Iran may have used chemical agents to a limited extent in an attempt to retaliate for Iraqi attacks. After the conflict with Iran, Iraq's Saddam Hussein used his chemical weapons to deal with rebellious Iraqi Kurds who had been assisted by the Iranians. The Iraqis used

Figure 2.5 The use of Napalm by US troops in Vietnam. Image is public domain picture from Wikipedia.

mustard, possibly combined with nerve gases, against the Kurdish town of Halabjah in March 1988, killing thousands of people.

Other countries that have stockpiled chemical agents include countries of the former Soviet Union, Libya (the Rapta chemical plant, part of which may still be operational), and France. Over two dozen other nations may also have the capability to manufacture offensive chemical weapons. The development of chemical warfare programs in these countries is difficult to verify because the substances used in the production of chemical warfare agents are, in many cases, the same substances that are used to produce pesticides and other legitimate civilian products.

2.8
The Running Madmen – Emergence of Chemical Terrorism

Although terrorism was not unknown in the world through the twentieth century, it wasn't really until the 1980s that the issue began to acquire a higher profile. Although there had been some domestic terrorism from the left in the United States during the late 1960s and into the 1970s, most notably in the form of the Weather Underground group [17], by the end of that decade the focus had turned towards the right, first in the form of the Survivalists movement and then the rightist / white supremacist "militias" that followed them [18]. Religious cult organizations also presented a potential domestic terrorist threat. In 1994 a Japanese religious cult, Aum Shinrikyo, released nerve agents in a residual area of Matsumoto, Japan, and in 1995 they used sarin in a crowded Tokyo subway [1, 19, 20]. This is an example of how the use of chemical warfare agents by terrorists could be a significant threat to the civilian population. Although the public interest in and fear of the danger of chemical warfare terrorism was high after the Aum Shinrikyo attacks, we should be aware that in reality more people die worldwide in industrial accidents with toxic chemicals, especially in the third world, than in chemical attacks. However, the possibility of an attack with chemical warfare agents by terrorists cannot be denied, and therefore, modern precautions to protect and to help not only military persons but also civilians are necessary.

References

1 Smart, J.K. (1997) History of chemical and biological warfare: an American perspective, in *Textbook of Military Medicine: Medical Aspects of Chemical und Biological Warfare* (eds R. Zajtchuk and R.F. Bellamy), Office of the Surgeon General, US Dept of the Army, Washington, DC, pp. 9–86.

2 Mayor, A. (2003) *Greek Fire, Poison Arrows & Scorpion Bombs: Biological and Chemical Warfare in the Ancient World*, Overlook Duckworth, ISBN: 1585677348X.

3 Schecter, W.P. and Donald, E.F. (2005) The surgeon and acts of civilian terrorism: chemical agents. *Journal of the American College of Surgeons*, **200**, 128–35.

4 Dogaroiu, C. (2003) Chemical warfare agents. *Romanian Journal of Legal Medicine*, **11**, 132–40.

5 Haber, L.F. (1986) *The Poisonous Cloud: Chemical Warfare in the First World War*, Clarendon Press, Oxford, England, pp. 15–40.
6 Prentiss, A.M. (1937) *Chemicals in War: A Treatise on Chemical Warfare*, McGraw-Hill, New York, NY, pp. 343–689.
7 Hanslian, R. (1936) The gas attack at Ypres: a study in military history, I. *Chemical Warfare Bulletin*, **22** (*1*), 5.
8 Trumpener, U. (1975) The road to Ypres: the beginning of gas warfare in World War I. *The Journal of Modern History*, **47**, 460–80.
9 Kohn, S. (1973) *The Cost of the War to Russia*, Howard Fertig, New York, NY.
10 Barker, A.J. (1968) *The Civilizing Mission: The Italo–Ethiopian War*, Dial Press, New York, NY, 241–4.
11 Murphy, P. (1937) Gas in the Italo-Abyssinian campaign. *Chemical Warfare Bulletin*, **22** (*1*), 1–8.
12 Kohn, G.C. (1986) *Directionary of Wars*, Facts on File Publication, Vol. 226, New York, NY, pp. 433–4.
13 Williams, P. and Wallace, D. (1989) *Unit 731: Japan's Secret Biological Warfare in World War II*, The Free Press (Macmillan), New York, NY, pp. 65–70.
14 Wiseman, D.J.C. (1951) *Special Weapons and Types of Warfare*, The War Office, London, England, p. 150.
15 Infield, G. (1988) *Disaster at Bari*, Vol. 209, Bantam Books, New York, NY, pp. 230–1.
16 Shoham, D. (1998) Chemical and biological weapons in Egypt. *The Nonproliferation Review*, **Spring–Summer**, 48–58. http://cns.miis.edu/pubs/npr/vol05/53/shoham53.pdf (Last accessed Oct, 12, 2007.).
17 Jacobs, R. (1997) *The Way the Wind Blew: A History of the Weather Underground*, Verso Date, Verso, ISBN: 1-85984-167-8.
18 Lamy, P. (1996) *Millennial Rage: Survivalists, White Supremalists, and the Doomsday Prophesy*, Plenum Books, New York, ISBN 0306454092, pp. 1521–401.
19 Lifton, R.J. (2000) *Destroying the World to Save it: Aum Shinrikyo, Apocalyptic Violence and the New Global Terrorism*, Henry Holt & Co, ISBN 0805065113.
20 Tucker, J.B. (1999) Historical trends related to bioterrorism: an empirical analysis. *Emerging Infectious Diseases*, **5**, 498–504.

Additional Web Links

http://de.wikipedia.org/wiki/Chemische_Waffe
http://en.wikipedia.org/wiki/Chemical_warfare#Chemical_warfare_in_World_War_I
http://www.npr.org/templates/story/story.php?storyId=5390710
http://www.history.navy.mil/faqs/faq104-4.htm
http://www.globalsecurity.org/wmd/library/report/1997/cwbw/
http://www.armageddononline.org/chemical_warfare.php
http://sun.tzu.s.art.of.war.en.em-leading.info/en/poison+gas

3
Monitoring and New Threats of Chemical/Biological Weapons

André Richardt and Roland Dierstein

3.1
Introduction

The devastating potential of biological and chemical warfare agents is well known over a long period of time. Therefore, the question remains, is it possible to ban the use of biological and chemical agents as weapons in military conflict between nations? This chapter gives an overview of different international conventions to control chemical and biological weapons in warfare. It also gives an impression of the difficulties in enforcing these conventions in the context of the emergence of international terrorism.

3.2
International Conventions

In ancient times, the use of poison in warfare was already frowned upon and this found expression in the Latin saying "Armis bella, non venis geri" (warfare should be done by weapons, not by poison). Most of the early attempts to control the use of biological and chemical weapons were bilateral agreements. For example, in 1675 an agreement between France and the Holy Roman Empire of Germany was signed to prohibit the use of poisoned missiles [1]. In 1874 the first international convention to forbid poisons and poisoned weapons was signed in Brussels. However, international law with the intention of controlling chemical and biological weapons in modern warfare began as early as 1899 with the Hague Land Warfare Convention. Therein, the contracting nations committed themselves to use no projectiles whose only purpose was to spread suffocating and poisonous gases. In 1907 in an annex, the use of poison or poisoned weapons, as well as the use of weapons, projectiles or materials, which were suitable to cause unnecessary suffering were additionally prohibited. During the conference of arms traffic of the League of Nations the "Protocol for the Prohibition of the Use in War of Asphyxiating, Poisonous or other Gases, and of Bacteriological Methods of

Decontamination of Warfare Agents.
Edited by André Richardt and Marc-Michael Blum

ISBN: 978-3-527-31756-1

Warfare," the so-called Geneva Protocol, was finally signed on July 17, 1925 [2] (Figure 3.1). It forbids the employment of chemical and biological weapons in war. At present, 133 countries have agreed to this protocol.

The Protocol provided the basis for the 1972 Biological and Toxin Weapons Convention (BTWC) and the 1993 Chemical Weapons Convention (CWC) as well as some regional arms control agreements. Upon ratification or accession to the protocol, some nations declared that it would cease to be binding on them if their enemies, or the allies of their enemies, failed to respect the prohibitions of the protocol. In recent years, however, many of the reservations have been withdrawn, especially following the coming into force of the BTWC and the CWC. Countries that continued to hold reservations toward the protocol were: Algeria, Angola, Bahrain, Bangladesh, China, Fiji, India, Iraq, Israel, Jordan, Democratic People's Republic of Korea, Republic of Korea, Kuwait, Libya, Nigeria, Papua New Guinea, Portugal, Syria, the United States of America, Vietnam, and the former Yugoslavia.

In the second half of the past century, the General Assembly of the United Nations repeatedly had to condemn numerous actions that were contrary to the principles and objectives of the Geneva Protocol. New efforts became necessary to reaffirm the signatory nations' adherence to the principles and objectives of that protocol, and to call upon all to comply strictly with them. The United Nations recalled the initial aims of the Geneva Protocol to contribute to the strengthening of confidence between peoples, and the general improvement of the international atmosphere. They were convinced of "the importance and urgency of eliminating

PROTOCOLL FOR THE PROHIBITION OF THE USE IN WAR OF ASPHYXIATING; POISONOUS OP OTHER GASE; AND OF BACTERIOLOGICAL METHODS OF WARFARE

Opened for signature 17 June 1925, entered into force 8 February 1928

The undersigned Plenipotentiaries, in the name of their respective governments:

Whereas the use in war of asphyxiating, poisonous or other gases, and of all analogous liquids, materials or devices, has been justly condemned by the general opinion of the civilised world; and

Whereas the prohibition of such use has been declared in Treaties to which the majority of Powers of the world are Parties; and

To the end that this prohibition shall be universally accepted as a part of International Law, binding alike the conscience and the practice of nations;

Declare:

That the High Contracting Parties, so far as they are not already Parties to Treaties prohibiting such use, accept this prohibition, agree to extend this prohibition to the use of bacteriological methods of warfare and agree to be bound as between themselves according to the terms of this declaration.

The High Contracting Parties will exert every effort to induce other States to accede to the present Protocol. Such accession will be notified to the Government of the French Republic, and by the latter to all signatories and acceding Powers, and will take effect on the date of the notification by the Government of the French Republic.

The present Protocol, of which the English and French texts are both authentic, shall be ratified as soon as possible. It shall bear to-day's date.

The ratifications of the present Protocol shall be addressed to the Government of the French Republic, which will at once notify the deposit of such ratification to each of the signatory and acceding Powers.

The instruments of ratification of and accession to the present Protocol will remain deposited in the archives of the Government of the French Republic.

The present Protocol will come into force for each signatory Power as from the date of deposit of its ratification, and, from that moment, each Power will be bound as regards other Powers which have already deposited their ratifications.

In witness whereof the Plenipotentiaries have signed the present Protocol.

Done at Geneva in a single copy, the seventeenth day of June, One Thousand Nine Hundred and Twenty-Five.

Figure 3.1 Text of the Geneva Protocol (League of Nations 1925).

from the arsenals of nations, through effective measures, such dangerous weapons of mass destruction using chemical or bacteriological (biological) agents" and recognized that an agreement on the prohibition of bacteriological (biological) and toxin weapons would represent a possible first step towards the achievement of agreement on effective measures for the prohibition of the development, production and stockpiling of chemical weapons, and determined to continue negotiations to that end. Moreover, they were determined to exclude completely the possibility of bacteriological (biological) and chemical agents being used as weapons, convinced that "no effort should be spared to minimize this risk."

As a consequence, the "Convention on the Prohibition of the Development, Production and Stockpiling of Bacteriological (Biological) and Toxin Weapons and on their Destruction" – the BTWC – was signed in London, Moscow and Washington on April 10th 1972, and finally came into force on March 26th 1975. Depositary nations – governments that admitted the presence of stockpiles in their countries – were the United States of America, the United Kingdom, and the Union of Soviet Socialist Republics [3–5]. In the meantime, 146 nations ratified the agreement and a further 15 countries signed the contract. The signatory nations of this convention were determined to achieve effective progress towards a general and complete disarmament including the prohibition and elimination of all types of chemical weapons of mass destruction. They were convinced that the prohibition of the development, production and stockpiling of bacteriological (biological) weapons and their elimination, "through effective measures, would facilitate the achievement of general and complete disarmament" under strict and effective international control. Nonetheless, they recognized the significance of the Geneva Protocol as a basis for this convention, being conscious of the contribution that this effort had already made, and continues to make.

At its core, the BTWC demands that each state party (the signatory nation) undertakes "never in any circumstances" to develop, produce, stockpile or otherwise acquire or retain microbial or other biological agents or toxins of types and in quantities that have no justification for prophylactic, protective or other peaceful purposes. The signatory nations commit themselves not to produce weapons, equipment or means of delivery designed to use such infectious agents or toxins for hostile purposes or in armed conflicts. Moreover, each member of the convention has to destroy, or to divert to peaceful purposes, "as soon as possible all agents, toxins, weapons, equipment and means of delivery" which are in its possession or under its jurisdiction or control. Finally, a state party to this convention would not be allowed to transfer to any recipient whatsoever, directly or indirectly, and in any way to assist, encourage, or induce a state, group of states or international organizations to manufacture or otherwise acquire any of the agents, toxins, weapons, equipment or means of delivery.

Additionally, each state party has to take "any necessary measures to prohibit and prevent" the development, production, stockpiling, acquisition, or retention of the agents, toxins, weapons, equipment and means of delivery. They must be willing to consult one another and to cooperate in solving any problems, which may arise in relation to the convention. Hereby, consultation and cooperation can also be undertaken through appropriate international procedures within the framework of the United Nations and in accordance with its charter. If it is found that any other state party is acting in breach of obligations deriving from the provisions of the convention, a complaint can be lodged with the Security Council of the United Nations. Such a complaint should include all possible evidence confirming its validity, as well as a request for its consideration by the Security Council.

Remarkably, the BTWC convention not only refers to and reaffirms the Geneva Protocol of 1925 but also explicitly demands that each state party affirms the recognized objective of an effective prohibition of chemical weapons. To achieve this, the signatory nations are to undertake to continue serious negotiations with a view to reaching early agreement on effective measures for the prohibition of the development, production and stockpiling and for the destruction of such weapons, as well as on appropriate measures concerning equipment and means of delivery specifically designed for the production or use of chemical agents for weapons purposes. In order to monitor progress in pursuing the BTWC goals, it was concluded that five years after their entry into force, or earlier if it was requested by a majority a conference of states parties should be held at Geneva, Switzerland, to review the operation of the convention, including the provisions concerning negotiations on chemical weapons. Such review should also take into account any new scientific and technological developments relevant to this convention.

However, although very carefully prepared with regard to the future, the BTWC contains serious weak points. Research work on potential weapons is not controlled, and the possibility cannot be excluded that under the cover of civilian research, offensive, that is, forbidden BTW activities may be hidden. Likewise the procurement, production and storage of bio- and toxin weapons are permitted, if it serves preventive, protective or other peaceful purposes. This permission could also be abused for forbidden work, particularly since it may be supposed that effective preventive measures require knowledge of the offensive potentials of weapons. In addition, the development of vaccines may serve not only for the protection of opposing employments, but also the security of own offensive activities. So far, there are no regulations regarding how the adherence to the convention can be controlled.

In contrast, in 1992, after a decade of long and painstaking negotiations, the Conference on Disarmament agreed to the text of the Chemical Weapons Convention (CWC) which was then adopted by the General Assembly at its forty-seventh session, on November 30th 1992, in its resolution entitled "Convention on the Prohibition of the Development, Production, Stockpiling and Use of Chemical Weapons and on their Destruction" [6]. This Convention was the first disarmament agreement negotiated within a multilateral framework that provides for the elimination of an entire category of weapons of mass destruction. Its scope, the obligations assumed by states parties and the system of verification envisaged for its implementation are unprecedented. It prohibits all development, production, acquisition, stockpiling, transfer, and use of chemical weapons. It requires each state party to destroy any chemical weapons and chemical weapons production facilities it possesses, as well as any chemical weapons it may have abandoned on the territory of another state party. The verification provisions of the CWC not only affect the military sector but also the civilian chemical industry, world-wide, through certain restrictions and obligations regarding the production, processing and consumption of chemicals that are considered relevant to the objectives of the convention. They can be verified through a combination of reporting requirements, routine on-site inspections of declared sites and short-notice challenge inspections. The convention also contains provisions on assistance in case a state party is attacked or threatened with attack by chemical weapons and on promoting the trade in chemicals and related equipment among states parties.

The Secretary-General of the United Nations became the depositary of the CWC. The convention was opened for signature on January 13th 1993 in Paris by the Secretary-General of the United Nations with 130 state parties signing it (Figure 3.2). On October 31st 1996, Hungary became the 65th member to deposit its instrument of ratification, thus triggering the process of entry into force of the CWC 180 days later. Finally, the convention entered into force on April 29th 1997. In 2000, 135 states had joined the CWC, and a further 36 countries signed the contract. The Organization for the Prohibition of Chemical Weapons (OPCW) was established in The Hague. It is responsible for the implementation of the convention. The OPCW is mandated to ensure the implementation of its provisions, including those for international verification of compliance with it, and to provide a forum for consultation and cooperation among states parties.

Thus, the CWC became established, with a functional verification regime. This was not the case for the BTWC and, more critically, after the introduction of

Figure 3.2 Signature Ceremony of the Chemical Weapons Convention as the breakthrough in the ban of chemical weapons, 13–15 January, 1993, Paris, www.opcw.org/. With courtesy of OPCW.

genetic engineering in the late 1970s, its weak points became much more evident. Therefore, in 1986 the first attempt was made to strengthen this convention by agreements of a confidence-building information exchange. In 1991, this agreement was affirmed and extended. Nevertheless, it did not prove particularly effective because although it was politically relevant, it was not binding according to international law. Moreover, only a limited number of nations took part in the information exchange. The given information was often incomplete or in some cases obviously wrong, and there was no possibility of examining its correctness. Therefore, the partner countries of the convention decided to prepare an obligatory supplementary protocol. This document was mainly intended to contain regulations with which the convention could be strengthened and its observance could be monitored. In this way – analogous to the CWC regulations – violations of contract were supposed to be reliably deterred.

Since 1994, an international team of experts in, so far, more than 22 meetings lasting several weeks in Geneva has been working to prepare an appropriate document. The members of the group have been supported in their efforts by numerous bio-scientists, among them members of the Pughwash movement, the Federation of American Scientists and further non-government organizations. Also, scientists and other experts of the Federal Republic of Germany supported the work of the Geneva group. However, skeptics feared that an agreement over the contract text at short notice could not be counted on. Even after the group partially succeeded in overcoming the fundamental differences of opinion that still existed – for example, over national export controls, as well as release and execution of inspections – the supplementary protocol had to be accepted and ratified by a representative number of countries including the Depositary Governments, before it finally could come into force. In July 2001 the United States of America rejected the draft of the supplementary protocol because of possible endangering of their national security. So the question of how to effectively control the observance of BTWC remained open [7].

Within an important range the CWC and BTWC overlap: prominent toxin weapons fall into the area of application of both contracts. As the CWC with its appropriate control measures is much more stringent than the BTWC, the toxin agents listed in the CWC (botulinum toxin, ricin, and saxitoxin) are thus subjected to the robust prohibition regulations of chemical weapons.

3.3 Chemical/Biological Agent Characteristics

The core of chemical and biological weapons is the respective agent that may be weaponized in multiple forms. These (ultra-) toxic chemicals, (micro-) biological pathogenic organisms and toxins can be and have been employed with various equipments and means of delivery both in warfare and in terrorist attacks. The toxic effects and illnesses caused by such agents are well known. For example, nerve agents inhibit acetylcholinesterase and thus produce symptoms of increased cholinergic activity. Ammonia, chlorine, vinyl chloride, phosgene, sulfur dioxide, and nitrogen dioxide, tear gas, and zinc chloride primarily injure the upper respiratory tract and the lungs. Sulfur mustard and nitrogen mustard are vesicant (blister) and alkylating agents. Cyanide poisoning ranges from sudden-onset headache and drowsiness to severe hypoxemia, cardiovascular collapse, and death. Botulinum toxin produces severe paralysis. Staphylococcal enterotoxin B produces a syndrome of fever, nausea, and diarrhea and may produce a pulmonary syndrome if aerosolised. Ricin intoxication can manifest as gastrointestinal hemorrhage after ingestion, severe muscle necrosis after intramuscular injection, and acute pulmonary disease after inhalation (reviewed in [8, 9]).

Besides the most prominent chemical nerve agents like the G-agents sarin, soman cyclosarin and tabun, and post-World War II agent VX, as well as the blister agents sulfur and nitrogen mustard, a variety of biological agents have been developed as effective weapons for warfare. The United States, the former Soviet Union and other nations have, at various times in their histories, incorporated these weapons into their arsenals and developed plans to use them for both tactical and strategic goals. The most frequently cited biological agents are those that were the focus of the former United States and Soviet Union biological weapons programs. However, depending on the specific goals, a wider variety of agents could be used effectively. It is reasonable, nonetheless, to expect an enemy to choose those agents that are most reliable and whose preparation and delivery is best understood. This results in a relatively small list, the so-called "dirty dozen" [10, 11]. The resulting diseases in humans are listed in Table 3.1.

The nature of the diseases caused by biological agents and their modes of transmission vary considerably. However, it is useful to define several broad categories, which are likewise relevant for the chemical agents: (i) lethal agents will kill the victim in a relatively short period of time, where as (ii) incapacitants will render a victim disabled to one degree or another. The logistical support needed by the community to treat and sustain these individuals contributes to the utility of incapacitants on the battlefield as well as in the context of terrorism. (iii) Transmissible biological agents can be spread from one infected individual to another. The potentially uncontrolled nature of this type of disease can be a potent source of

Table 3.1 Diseases caused by biological agents.

Bacteria	***Viruses***	***Toxins***
Anthrax	Smallpox	Botulism
Plague	Viral Encephalitis (e.g. VEE)	Ricin intoxication
Tularemia	Viral Hemorrhagic Fevers (e.g. Marburg)	Staphylococcal Entero-toxin B (SEB) intoxication
Glanders and Meliodosis		
Brucellosis		
Q Fever		

fear as in, for example, small pox, but it can create problems for the user of the weapon, if they cannot protect themselves from the disseminating illness. (iv) Non-transmissible agents such as *Bacillus anthracis*, the causative agent of anthrax do not pose this risk.

The choice of an agent will be determined by several considerations. These comprise a number of non-technical factors including availability of the agent, experience of the operator, the nature of the target and the intended consequences. Additional technical issues involve ease of production of needed quantities of agent, and storage, stabilization and mode of delivery. Overcoming some of the challenges posed by these considerations requires substantial sophistication.

3.4 Asymmetric Warfare

With the end of the Cold War and the collapse of the Soviet Union, the focus of military planning has shifted from the possibility of a major war against a peer adversary to that of regional wars, ranging from major theatre wars against non-peers to smaller scale contingencies. At the same time, there has been growing concern about the proliferation of weapons of mass destruction following a burst of chemical weapons proliferation in the 1980s and the near-brush with Iraqi biological weapons in 1992. These factors caused rising concern about the utility of weapons of mass destruction in an asymmetric conflict. The notion that chemical and biological weapons might be also used by non-peer adversaries to attack or otherwise shape the dynamics of a regional war presented new and alarming possibilities [12]. Experts postulate that the future battlefield will be characterised by the use of asymmetric warfare, unconventional weapons and terrorist/guerrilla type tactics. In this scenario, chemical and biological warfare agents could become the choice of many. Given this, it is likely that the threat of chemical and biological weapons – although internationally banned – will become greater in the future rather than diminish [13]. This certainly includes increasingly horrifying assaults by non-state actors as recently demonstrated by the first sarin gas terrorism in the Japanese town Matsumoto in 1994 and the Tokyo subway attack in the following year which led to the deaths of 19 people, as well as to a large number of injuries [14].

Biological weapons, frequently called "the atom bomb of the poor man", seem to be an even higher threat in an asymmetric scenario. This is mainly to be attributed to the fact that as a means of mass destruction biological weapons have a high damage potential and are technically and financially more easily fabricated when compared with nuclear and chemical weapons. Nonetheless, the costs of biological weapons production are dependent on the type and quantity of necessary agents, and thus there are considerable differences in manufacturing expenses. Of importance when considering the asymmetric threat, is that countries that are not able to produce nuclear and chemical weapons can accomplish biological weapon programs. Also, terrorists and especially financially strong terrorist groups, may autonomously develop techniques and/or use proliferated governmental know-how and commercial hardware.

The most probable use of biological weapons in a military conflict is supposedly the spreading of bio-aerosols by means of explosives or other forms of aerosol generation, e.g. spraying from an aircraft. Principally, there seems to be no difference between a military or terrorist application technology. It can, however, be assumed that in a military application the quantities of agents are considerably higher than in a terror scenario. Large-scale biological warfare is technically complex and needs particular knowledge that used to be restricted to a limited number of specialists. Possibly, earlier attacks of terrorist groups like the Aum sect in 1995 failed for this reason. Nevertheless, the activities of the 1990s can be seen as the beginning of worrying violations of the taboo regarding the use of these agents and, as the deadly anthrax letters in the fall of 2001 in the United States demonstrate clearly, biological agents can be extremely dangerous in the hands of terrorists [11].

The military use of pathogens and other biological agents began in antiquity. Usage was documented in numerous records in the Middle Ages and such documentation continues in modern times [15]. The National Defense University in Washington DC, USA recently summarized the records of bio-terrorism and biocriminality since 1900 [16]. By the example of the activities of the Japanese Aum sect in the 1990s, it not only became evident how probable an incidence of terrorist attack was, but also with which technical obstacles and know-how gaps the terrorists were confronted at that time. The list of the Aum sect activities from 1993 to 1995 is remarkable: at the beginning of June 1993 botulinum toxin was sprayed from a vehicle; at the end of this month *Bacillus anthracis* spores were spread out from of the roof of a multi-storied building; in July 1993 the same spores were disseminated out of converted trucks in order to contaminate the Japanese parliament; and at the end of July 1993 they were again sprayed from trucks against the Emperor's Palace. Moreover, in 1993 a journey was made with a medical team of 40 to Zaire with the goal of coming into possession of *Ebola* virus; and so far, finally in March 1995, botulinum toxin was released in the Tokyo subway.

All of these attacks and projects failed. According to research results from the National Defense University, the Aum scientists obviously made mistakes either with the production or the dissemination of the agents [16]. It can be assumed that the sect used a strain of *Clostridium botulinum,* isolated by the group itself from the environment, which produced only little or no toxin. They possibly had similar problems with *Bacillus anthracis.* According to the statements of the Japanese authorities they–obviously because of unawareness – used a vaccination strain of the germ, which was comparatively harmless. There were also difficulties on the technical side e.g. with clogged nozzles of the used spraying devices. Likewise, theoretical and experimental knowledge was apparently not sufficient to handle the *Ebola* virus.

These failures, however, should not mislead about the fact that with the necessary know-how, with modern techniques and sufficient availability of financial means, biological weapons can be substantially more effectively developed, proliferated and also implemented. Renowned experts like Davis and Johnson-Winegar of the US Department of Defense warned in their publication on the anthrax terror, a year before September 11th 2001 and its well-known consequences, that the relative ease with which biological weapons can be procured together with other changes in the world could mark the direction towards a new type of war in the twenty-first century [17].

3.5
Threat Scenarios for Chemical/Biological Weapons

A prerequisite for the threat assessment of a weapon is the knowledge of the extent of its effect. In order to estimate the relative effectiveness of chemical and biological weapons, their overall damage potential compared with conventional and nuclear weapons has recently been assessed by numerical simulation [18]. In this study, a conventional warhead with a ton of explosive was computed to result in 6 dead and 13 wounded; a chemical weapon with 300 kg of sarin was computed to result in 200–3000 dead and a comparable number of wounded; and a nuclear warhead with an explosive yield of 20 kilotons would result in 40 000 dead and 40 000 wounded. A biological weapon with only 30 kg of anthrax exciters, however, was calculated to cause as many as 20 000–80 000 dead in the same computer model.

In reality, the most convincing example of the devastating consequences of chemical warfare was the horror of the killing in the trenches of World War I. Fortunately, this has not been proven on such a scale for biological weapons. Apart from the experiments by Japanese occupation forces there has been no clear-cut evidence of military employment of these weapons in practice in the last century [19]. However, the Sverdlovsk production plant accident of 1979, where at least 68 people died [20], and the terrorist anthrax attacks in the aftermath of the September 11th assaults in the United States gave a lasting impression on the horrifying killing potential of biological warfare agents.

A reasonable number of suitable chemical and biological agents can be delivered *via* air and a wide range of technical methods can be effectively used. The more sophisticated approaches can saturate very large areas, permitting the creation of strategic weapons. For tactical deployment, the variety of delivery methods is limited only by the operator's imagination. Overall, it may be assumed that in many cases an operator favours agents that can be delivered to the target most easily, even at the expense of some degree of effectiveness. This assumption will be less valid in the case of state-sponsored operations.

The two most likely chemical agents would be nerve gases such as sarin or VX, which short-circuit the nervous system, and mustard gas, which causes deadly internal and external blistering. The "ideal" biological weapon would be a modified natural disease with a high infection rate spread by the air that killed or disabled almost everyone but then died out after two days [21]. Thus, while some infection might be spread by airline travel anywhere in the world, the only massive outbreaks would be where the original infection was propagated. The attackers would have an effective vaccine to protect their leadership and other key personnel. Having some sort of targeting control over the infection, and having a reliable protection from the infection, are the two key parts of an effective biological weapon that have not existed until now. Moreover, considering the advanced knowledge in modern molecular biology and biotechnology the question comes up whether anyone would wish to misuse such information to create new biochemical weapons. As the genomics revolution proceeds, we may no longer maintain a differentiation between chemical and biological weapons and may have to view these as a continuous biochemical threat spectrum [22].

Besides the routine military assessments of chemical/biological threats, scenarios with different impact levels have ever more frequently been presented and discussed in civilian communities during the past years. One impressive example is that of Leonard Cole of Rutgers University in Newark, USA, whose books on biological and chemical warfare issues include *The Eleventh Plague: The Politics of Biological and Chemical Warfare* [23]: "The first sign of trouble might be rather mild–people showing up at doctors' offices or hospital emergency rooms with runny noses, teary eyes, headaches and fevers. Only the sheer number of these patients, not the severity of the initial symptoms, might suggest that something unusual is afoot. But the progression of these flu-like symptoms over a period of days into worse problems, such as bleeding, internal and external lesions, and labored breathing, might provide the first proof of the unspeakable: an attack on a civilian population with biological weapons."

Frank Cilluffo of the US Homeland Security Policy Institute called biological weapons "silent killers" because it could take days or weeks for symptoms to manifest themselves. An attack could remain unknown for some time, unless the perpetrators announced it. Because no one would know what had happened, many people who had been exposed, for example, to a transmissible disease such as smallpox, might unwittingly spread the virus to many more victims who had initially been spared (domino effect). Experts believe the two most likely biological agents would be the exciter of smallpox, *Variola major*, that killed millions of people throughout the centuries until it was declared eradicated worldwide two decades ago, and that of anthrax, *Bacillus anthracis*, spread by spores and generally confined to sheep, cattle, horses, goats and pigs. Smallpox kills only about 30 percent of those it infects, but is alarmingly infectious. Anthrax is not spread from person to person but kills about 90 percent of those it infects [8].

If eventually such a severe chemical/biological attack or assault occurred, appropriate post-exposure measures for mass casualties, contaminated sites, vehicles, (defence) material, and infrastructure would become critically urgent. The success of these measures would be strongly dependent on efficient decontamination methods for the agents. The respective technology existing to date, mainly in the military, is principally based on thermal and/or chemical treatments. These procedures usually need toxic components themselves, are mostly expensive, and difficult to handle. Therefore, new technologies are being sought, and notably the growing potential of biotechnology is recognized as a promising option to develop non-toxic detoxifying and disinfecting reactive components [24]. The results obtained so far, and reported in detail in the following chapters of this book, strongly suggest that biotechnological approaches may generally become more appropriate alternatives for existing protection measures against chemical and biological agents, not only because of the significant lowering of the logistical and operational burden for the military and civilian forces in charge, but also because of their being less hazardous, less corrosive, potentially applicable to skin, and environmentally compatible.

References

1 Prentiss, A.M. (1937) *Chemicals in War: A Treatise on Chemical Warfare*, McGraw-Hill, New York, NY, pp. 343–477, 533–566, 574, 685–689.

2 League of Nations (1925) Protocol for the Prohibition of the Use in War of Asphyxiating, Poisonous or Other Gases, and of Bacteriological Methods of Warfare. The Harvard Sussex Program on CBW Armament and Arms Limitation. Internet source http://fas-www.harvard.edu/~hsp/1925.html (Aug, 2006.).

3 United Nations (1972) Convention on the Prohibition of the Development, Production and Stockpiling of Bacteriological (Biological) and Toxin Weapons and on Their Destruction. United Nations General Assembly resolution 2826 (XXVI), New York. Full text available at http://www.brad.ac.uk/acad/sbtwc (Aug, 2006.).

4 United Nations (1986) Second Review Conference of the Parties to the Convention on the Prohibition of the Development, Production and Stockpiling of Bacteriological (Biological) and Toxin Weapons and on their Destruction. In: Final Documents, Part II, Final Declaration. BWC/CONF. II/13/II. Full text available at http://www.brad.ac.uk/acad/sbtwc (Aug, 2006.).

5 United Nations (2001) Protocol to the Convention of the Development, Production and Stockpiling of Bacteriological (Biological) and Toxin Weapons and on their Destruction. BWC/AD HOC GROUP/56-2, Annex B: 347–565. Full text available at http://www.brad.ac.uk/acad/sbtwc (Aug, 2006.).

6 United Nations (1993) Convention on the Prohibition of the Development, Production, Stockpiling and Use of Chemical Weapons. Full text available at http://www.opcw.nl (Aug, 2006.).

7 Geißler, E. (2001) *Schwarzer Tod und Amikäfer. Biologische Waffen und ihre Geschichte*. C.G. Roßberg, Frankenberg, Germany.

8 Kortepeter, M.G. and Parker, G.W. (1999) Potential Biological Weapons Threats. *Emerging Infectious Diseases*, **5**, 523–7.

9 Greenfield, R.A., Brown, B.R., Hutchins, J.B., Iandolo, J.J., Jackson, R., Slater, L.N. and Bronze, M.S. (2002) Microbiological, biological, and chemical weapons of warfare and terrorism. *The American Journal of the Medical Sciences*, **323**, 326–40.

10 Sohns, T. (1999) Defense against biological terrorism. *ASA Newsletter*, **74**, 23–8.

11 Dierstein, R. and Driks, A. (2006) The biological threat, in *Science and Culture Series* (eds A. Zichichi), World Scientific Publisher Co., Ltd, Singapore, pp. 432–6.

12 Roberts, B. (2000) *Asymmetric Conflict 2010*. Final Report of the Defense Threat Reduction Agency, Fort Belvoir, VA, 46 pages.

13 Roberts, B. (2004) *Nonproliferation Challenges Old and New*. The Counterproliferation Papers Future Warfare Series 24, USAF Counterproliferation Center, Air University, Maxwell Airforce Base, Alabama, USA, p. 40.

14 Seto, J. (2001) The sarin gas attack in Japan and the related forensic investigation, in *Treaty Enforcement and*

International Cooperation in Criminal Matters (eds R. Yepes-Enríquez and L. Tabassi), Cambridge University Press, pp. 301–7.

15 Geißler, E. and van Courtland-Moon, J. (eds) (1999) *Biological and Toxin Weapons: Research, Development and Use from the Middle Ages to 1945*, Oxford University Press, Oxford, UK.

16 Carus, W. (2001) *Working Paper: Bioterrorism and Biocrimes. The Illicit Use of Biological Agents Since 1900*, Center for the Counterproliferation Research, National Defense University, Washington, DC.

17 Davis, J. and Johnson-Winegar, A. (2000) The anthrax terror: DOD's number one biological threat. *Aerospace Power Journal*, **14** (4), 15–29.

18 Pearson, G. (2000) The prohibition of chemical and biological weapons, in *Verification of the Biological and Toxin Weapons Convention* (eds M. Dando, G. Pearson and T. Toth), Kluwer Academic Publishers, Dordrecht, NL, pp. 1–31.

19 Müller, H. (2001) Dealing with a headache: three scenarios and two dilemmas, in *The Role of Biotechnology in Countering BTW Agents* (eds A. Kelle, M. Dando and K. Nixdorff), Kluwer Academic Publishers, Dordrecht, Boston, London, pp. 1–7.

20 Wampler, R.A. and Blanton, T.S. (eds) (2001) Anthrax at Sverdlovsk, 1979. U.S. Intelligence on the Deadliest Modern Outbreak, November 15, 2001 edn. National Security Archive Electronic Briefing Book No. 61. Full text available at http://www.gwu.edu/ (Aug, 2006.).

21 Gray, C.H. (1998) *Postmodern War: The New Politics of Conflict*, Guilford Publications, Inc, New York.

22 Wheelis, M. and Dando, M. (2006) *Neurobiology: A Case Study of the Imminent Militarization of Biology*. International Review of the Red Cross.

23 Cole, L. (1998) *The Eleventh Plague: The Politics of Biological and Chemical Warfare*, W.H. Freeman, New York.

24 Dierstein, R., Gläser, H.-U. and Richardt, A. (2001) Search of biotechnology-based decontaminants or C/BW agents, in *The Role of Biotechnology in Countering BTW Agents* (eds A. Kelle, M. Dando and K. Nixdorff), Kluwer Academic Publishers, Dordrecht, Boston, London, pp. 257–65.

4
Biological Warfare Agents

Heiko Russmann and André Richardt

4.1
Introduction

When Ken Alibek, a Russian scientist who defected in 1992 from Russia to the United States, published his book in 1999 [1] about the extensive Soviet bio warfare program, even military experts and politicians did not fully recognize the looming threat of biological warfare agents. It was not until 2001, when the first "anthrax letters" emerged in the United States, that the public understood the risk emanating from this threat. Since then, many efforts to counter this threat have been made, but still significant deficiencies can be demonstrated.

For unscrupulous states and terrorist groups, biological warfare agents possess some priceless advantages. Development and production can easily be concealed, because no sophisticated equipment or internationally controlled products are imperatively necessary. The production of a biological warfare agent is cheap, compared to other non-conventional weapons; and as long as the means for an opponent to meet the biological threat are inadequate, the desire of some states or terrorist groups to access this technology will remain unchanged. However, the development of a weapon to include biological warfare agents requires detailed knowledge and is not easy to achieve.

This chapter divides between biological warfare agents and biological weapons and provides some basic information about different kinds of biological warfare agents.

4.2
Biological Warfare Agents and Biological Weapons: A Division

Confusion sometimes occurs when trying to differentiate between biological warfare agents and biological weapons. It is necessary to know the differences and to divide between biological warfare agents and biological weapons. A first definition defines biological warfare agents as a pathogen (virus, bacteria, fungus) or

Decontamination of Warefare Agents.
Edited by André Richardt and Marc-Michael Blum

ISBN: 978-3-527-31756-1

toxin that will cause harm to either humans, animals or crop and can thus be used in biological warfare or biological terrorism. In contrast, biological weapons are defined as a combination of a biological warfare agent and a disseminating device [2].

However, although many bacteria, viruses and toxins can cause disease or intoxication of humans, only a few would be effective if employed as a part in a biological weapon. There are some requirements for an ideal biological warfare agent to be used as a part of a biological weapon [3]:

- Availability or ease of production
- Stability after production in storage, weapons, and the environment
- Weaponized form of the agent
- Ease of dissemination
- Incapacitation and lethality
- Appropriate particle size in aerosol
- Susceptibility of intended victims and non-susceptibility of friendly forces [3].

The desired biological agent can be delivered in wet or dry form. The release can be done by the use of explosive devices, sprays and contamination of food and water [3–5]. The aerosolization of the agent is the most common form of delivery. There are studies about the infectiosity of aerosol clouds sprayed from devices [4]. A more sophisticated description is given in the next sub-chapter.

4.3
Types of Biological Warfare Agents

Biological agents, the effective component in a biological weapon, are virological, bacteriological, fungal or toxin preparations that are harmful to humans, animals or crops. In theory, any of thousands of known biological agents, capable of harming or infecting humans, animals or crops could be considered as a potential biological warfare agent. But not all pathogens or toxins are equally suitable for this purpose. Biological agents will necessarily need to have certain properties, beneficial for the potential attacker. An US military report [6] has summarized some of the most important properties (Table 4.1).

Besides the pathogenic effect, the stability of the agent is the most stringent selection criterion.

From the viewpoint of less developed countries, in terrorism or crime, additional issues become more important for the selection of an agent, like availability of the agent, simplicity of production and agent properties supporting easy dissemination.

Some agents, like for example, *Bacillus anthracis,* the pathogen causing anthrax, are endemic in most countries of the world. The frequent anthrax outbreaks among livestock could be used to collect samples potentially pathogenic to humans.

Table 4.1 Important properties of biological warfare agents.

Requirements	*Desirable characteristics*
• Consistently produce a given effect (death, disability or crop damage).	• Possible for the using forces to protect against.
• Be manufacturable on a large scale.	• Difficult for a potential enemy to detect or protect against.
• Be stable during production and storage in munitions and during transportation.	• A short and predictable incubation period.
• Be capable of efficient dissemination.	• A short and predictable persistency if the contaminated area is to be promptly occupied by friendly troops.
• Be stable after dissemination.	• Capable of infecting more than one kind of target (for example, man and animals) through more than one portal of entry, being disseminated by various means, producing desired psychological effects.

On the other hand, smallpox was eradicated the 1970s and official stocks of *Variola major*, the pathogen causing smallpox, exist in only two laboratories, in Russia and in the United States.

Comparing the efforts necessary for the production of biological warfare agents, in principle, virological agents require the highest skills and highest complexity in technical equipment, while toxins are mostly quite easy to produce.

A very important issue is the method of dissemination. There are different ways to deploy biological warfare agents [5]:

- Contamination of food and water supplies which are ingested by the victims
- Release of infectious vectors, such as fleas or mosquitoes which then transmit the pathogen
- Creation of an aerosol cloud that is then inhaled by the victim.

The most effective method of dissemination is as an aerosol because in that way the agent can travel for long distances and effectively infect contaminated individuals. The 1979 Sverdlovsk incident is a good example of the deadly efficiency of weaponized pathogen aerosols [7]. The difficulty is in generating an aerosol that is bio-available and infectious. To be bio-available the aerosol has to range in size between 1 and 10 μm particles in diameter, so that it is respirable (Figure 4.1).

For this purpose the agent can be milled in a dry form or sprayed as a slurry. Both methods will administer physical stress to the agent and might inactivate a significant fraction of the agent. Finally, some agents will not sufficiently withstand the environmental condition into which they are released and can thus be inactivated after a few feet of travel as an aerosol. For instance *Yersinia pestis*, the

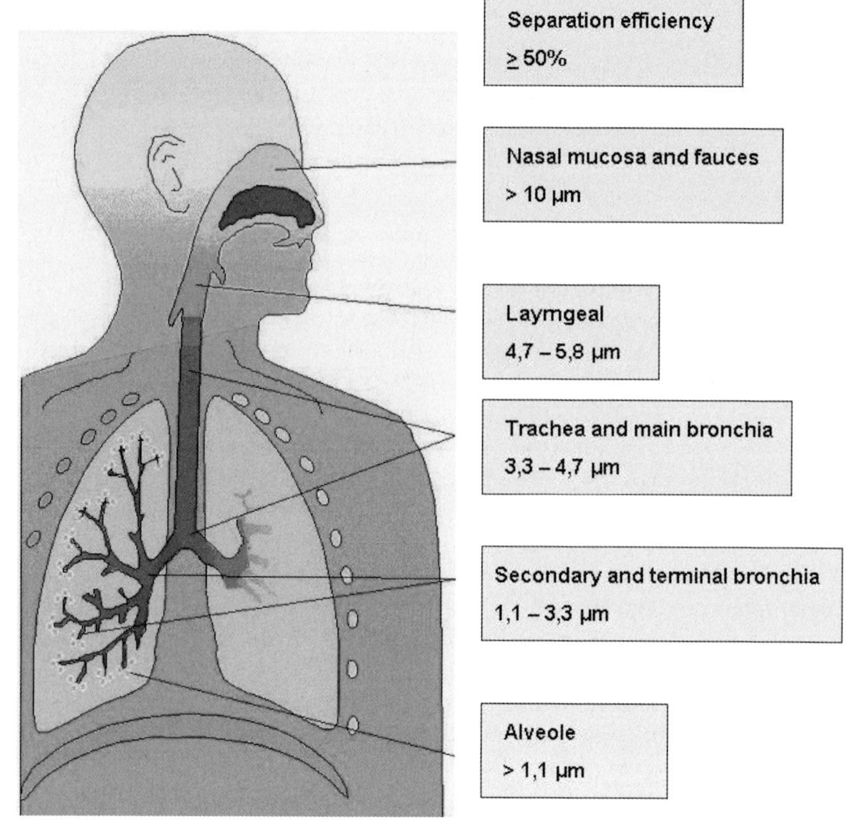

Figure 4.1 Lung separation efficiency of aerosols (reproduced from DIN ISO 7708 VDI 2463 Bl.1 Grasby & Anderson).

pathogen causing plague, is known to be easily inactivated by sunlight or drying. To use plague bacteria in an attack will necessarily require some expertise in stabilizing aerosolised bacteria. In the 1960s, the United States discarded plague as a warfare agent for their military arsenal due to this virtually insurmountable problem [8]. However, according to information from Ken Alibek, during the concealed Soviet offensive bioweapon program a highly pathogenic and stable strain of *Yersinia pestis* was developed and weaponized [1].

The concealed contamination of water and food is most likely to be used by terrorists or criminal individuals [5]. It is very difficult to contaminate enough food or water and to choose the proper location for the attack to gain a military advantage, since water treatment and cooking will significantly inactivate biological agents. The use of vectors, that is, organisms that do not cause disease themselves but spread an infection by conveying pathogens from one host to another, was tested and applied by Japan during World War II and further developed in the USSR. Since then, the technology for stabilizing biological agents has significantly advanced, so that for military purposes the release of biological agents as aerosol is more convenient.

Balancing all the criteria and evaluating for all possible pathogens, lists of highly probable biological warfare agents can be generated. Dependent on the weighting factor for each criterion, the ranking of individual pathogens and toxins and the length of the list might change, but in comparing these compilations certain pathogens and toxins appear in almost every list.

Anthrax is considered one of the most effective biological agents for several reasons. It forms spores that withstand environmental inactivation and dispersal stress. The pneumonic form of the infection is not transmittable and will not cause secondary infections in other people. Thus, the effect of the agent is confined to the target and will not spread any further, at worst only among the target's own troops. The onset of the infection is very fast, with a fatality rate of more than 80%, if not treated in time (Table 4.5). The most probable pathogens and toxins to be used as biological warfare agents and that are present in virtually every list, are called the "Dirty Dozen" (Table 4.2) [9]. The Dirty Dozen consist of six bacteria, three viruses and three toxins. All of the listed agents have once been tested for use as biological warfare agents. Interestingly, with the exception of smallpox, all of the listed microorganisms are originally animal pathogens, but will severely infect humans, too (Table 4.5).

After the anthrax letter attacks in 2001 it became evident that, in the case of an attack among the civilian population, challenges were similar to military scenarios, but options were different for health services and the first responder. The US Center for Disease Control (CDC) developed categories and generated a criteria list with a focus on civil demands:

Category A Diseases/Agents

The US public health system and primary healthcare providers must be prepared to address various biological agents, including pathogens that are rarely seen in the United States. High-priority agents include organisms that pose a risk to national security because they

Table 4.2 "Dirty Dozen" toxins and diseases [9].

Bacteria	***Viruses***	***Toxins***
Anthrax (*Bacillus anthracis*)	Smallpox (*Variola major*)	Botulinum toxin
Plague (*Yersinia pestis*)	Viral encephalitis (e.g. VEE)	Ricin
Tularemia (*Francisella tularensis*)	Viral hemorrhagic fevers (e.g. Ebola, Marburg disease)	Staphylococcal enterotoxin B
Glanders (*Burkholderia mallei*) and meliodosis (*Burkholderia pseudomallei*)		
Brucellosis (*Brucella sp.*)		
Q fever (*Coxiella burnetii*)		

- Can be easily disseminated or transmitted from person to person
- Result in high mortality rates and have the potential for major public health impact
- Might cause public panic and social disruption
- Require special action for public health preparedness.

Category B Diseases/Agents
Second highest priority agents include those that

- Are moderately easy to disseminate
- Result in moderate morbidity rates and low mortality rates
- Require specific enhancements of CDC's diagnostic capacity and enhanced disease surveillance.

Category C Diseases/Agents
Third highest priority agents include emerging pathogens that could be engineered for mass dissemination in the future because of

- Availability
- Ease of production and dissemination
- Potential for high morbidity and mortality rates and major health impact.

In the resulting list, all pathogens of the "Dirty Dozen" were listed in either Category A or B (Table 4.3).

A more comprehensive list resulted with the aim controlling proliferation and handling of ominous materials is the US Select Agent List. Select Agents are pathogens or biological toxins which have been declared by the US Department of Health and Human Services or by the US Department of Agriculture to have the "potential to pose a severe threat to public health and safety" (Table 4.4). A similar compilation of pathogens can be found in the German Military Arms List (Table 4.5).

4.4 The "Dirty Dozen"

As mentioned before, the most probable pathogens and toxins to be used as biological warfare agents are called the "Dirty Dozen" (Table 4.2). It consists of six bacteria, three viruses and three toxins and a short description with additional literature is given [9–11].

4.4.1 Bacteria

Anthrax, Plague, Tularemia, Glanders, Brucellosis and Q fever are described in the literature as the most likely diseases caused by biological warfare agents [9]. The pathogenesis for all six agents is well known [12].

Table 4.3 Diseases and pathogens/toxins according to CDC categories.

Category A	*Category B*	*Category C*
Anthrax (*Bacillus anthracis*)	Brucellosis (*Brucella species*)	Emerging infectious diseases such as Nipah virus and Hantavirus
Botulism (*Clostridium botulinum toxin*)	Epsilon toxin of *Clostridium perfringens*	
Plague (*Yersinia pestis*)	Food safety threats (e.g. *Salmonella sp., Escherichia coli O157:H7, Shigella*)	
Smallpox (*Variola major*)	Glanders (*Burkholderia mallei*)	
Tularemia (*Francisella tularensis*)	Melioidosis (*Burkholderia pseudomallei*)	
Viral hemorrhagic fevers (filoviruses [e.g. Ebola, Marburg[and arenaviruses [e.g. Lassa, Machupo[)	Psittacosis (*Chlamydia psittaci*)	
	Q fever (*Coxiella burnetii*)	
	Ricin toxin from Ricinus communiz (castor beans)	
	Staphylococcal enterotoxin B	
	Typhus fever (*Rickettsia prowazekii*)	
	Viral encephalitis (alphaviruses [e.g. Venezuelan equine encephalitis, eastern equine encephalitis, western equine encephalitis[)	
	Water safety threats (e.g. *Vibrio cholerae, Cryptosporidium parvum*)	

4.4.1.1 Anthrax

Bacillus anthracis is a Gram-positive and spore-forming bacillus and causes a zoonetic disease (Anthrax) in domesticated and wild animals. Contact with infected animals or inhalation of spores (Woolsorter's Disease) lead to anthrax in humans [13, 14]. For pathogenesis, three well-known virulence factors are important:

- Antiphagocytic capsule;
- Two protein exotoxins, the lethal and the edema toxin.

The importance of these factors was confirmed by different investigations [15–17]. The anthrax toxins possess two components: the A Domain with the toxic and the enzymatic activity, and the B Domain as a cell binding component (Figure 4.2). The roles of the toxins were investigated in detail [18–21].

Table 4.4 US Select Agents List according to US Health and Human Services (HHS), US Department of Agriculture (USDA).

HHS select agents and toxins	***USDA select agents and toxins***	***USDA plant protection and quarantine (PPQ) select agents and toxins***
Abrin	African horse sickness virus	*Candidatus* Liberobacter africanus
Cercopithecine herpesvirus 1 (Herpes B virus)	African swine fever virus	*Candidatus* Liberobacter asiaticus
Coccidioides posadasii	Akabane virus	*Peronosclerospora philippinensis*
Conotoxins	Avian influenza virus (highly pathogenic)	*Ralstonia solanacearum* race 3, biovar 2
Crimean-Congo haemorrhagic fever virus	Bluetongue virus (Exotic)	*Schlerophthora rayssiae* var *zeae*
Diacetoxyscirpenol	Bovine spongiform encephalopathy agent	*Synchytrium endobioticum*
Ebola virus	Camel pox virus	*Xanthomonas oryzae* pv. *oryzicola*
Lassa fever virus	Classical swine fever virus	*Xylella fastidiosa* (citrus variegated chlorosis strain)
Marburg virus	*Cowdria ruminantium* (Heartwater)	
Monkeypox virus	Foot-and-mouth disease virus	
Reconstructed replication competent forms of the 1918 pandemic influenza virus containing any portion of the coding regions of all eight gene segments (Reconstructed 1918 Influenza virus)	Goat pox virus	
Ricin	Japanese encephalitis virus	
Rickettsia prowazekii	Lumpy skin disease virus	
Rickettsia rickettsii	Malignant catarrhal fever virus (Alcelaphine herpesvirus type 1)	
Saxitoxin	Menangle virus	
Shiga-like ribosome inactivating proteins	*Mycoplasma capricolum*/M.F38/ *M. mycoides Capri* (contagious caprine pleuropneumonia)	
South American Haemorrhagic Fever viruses (Flexal, Guanarito, Junin, Machupo, Sabia)	*Mycoplasma mycoides mycoides* (contagious bovine pleuropneumonia)	
Tetrodotoxin	Newcastle disease virus (velogenic)	
Tick-borne encephalitis complex (flavi) viruses (Central European Tick-borne encephalitis, Far Eastern Tick-borne encephalitis, Kyasanur Forest disease, Omsk Hemorrhagic Fever, Russian Spring and Summer encephalitis)	Peste des petits ruminants virus	
Variola major virus (Smallpox virus) and Variola minor virus (Alastrim)	Rinderpest virus	
Yersinia pestis	Sheep pox virus	
	Swine vesicular disease virus	
	Vesicular stomatitis virus (Exotic)	

Table 4.5 B.W. Agent characteristics.

Disease	*Transmit Man to Man*	*Infective Dose (Aerosol)*	*Incubation Period*	*Duration of Illness*	*Lethality (approx. case fatality rates)*	*Persistence of Organism*	*Vaccine Efficacy (aerosol exposure)*
Inhalation anthrax	No	8000–50 000 spores	1–6 days	3–5 days (usually fatal if untreated)	High	Very stable – spores remain viable for >40 years in soil	2 dose efficacy against up to 1000 LD_{50} in monkeys
Brucellosis	No	10–100 organisms	5–60 days (usually 1–2 months)	Weeks to months	<5% untreated	Very stable	No vaccine
Cholera	Rare	10–500 organisms	4 hours – 5 days (usually 2–3 days)	≥1 week	Low with treatment, high without	Unstable in aerosols & fresh water; stable in salt water	No data on aerosol
Glanders	Low	Assumed low	10–14 days via aerosol	Death in 7–10 days in septicemic form	>50%	Very stable	No vaccine
Pneumonic Plague	High	100–500 organisms	2–3 days	1–6 days (usually fatal)	High unless treated within 12–24 hours	For up to 1 year in soil; 270 days in live tissue	3 doses not protective against 118 LD_{50} in monkeys
Tularemia	No	10–50 organisms	2–10 days (average 3–5)	≥2 weeks	Moderate if untreated	For months in moist soil or other media	80% protection against 1–10 LD_{50}
Q Fever	Rare	1–10 organisms	10–40 days	2–14 days	Very low	For months on wood and sand	94% protection against 3500 LD_{50} in guinea pigs

Table 4.5 continued

Disease	*Transmit Man to Man*	*Infective Dose (Aerosol)*	*Incubation Period*	*Duration of Illness*	*Lethality (approx. case fatality rates)*	*Persistence of Organism*	*Vaccine Efficacy (aerosol exposure)*
Smallpox	High	Assumed low (10–100 organisms)	7–17 days (average 12)	4 weeks	High to moderate	Very stable	Vaccine protects against large doses in primates
Venezuelan Equine Encephalitis	Low	10–100 organisms	2–6 days	Days to weeks	Low	Relatively unstable	TC 83 protects against 30–500 LD_{50} in hamsters
Viral Hemorrhagic Fevers	Moderate	1–10 organisms	4–21 days	Death between 7 and 16 days	High for Zaire strain, moderate with Sudan	Relatively unstable – depends on agent	No vaccine
Botulism	No	0.001 μg/kg is LD_{50} for type A	1–5 days	Death in 24–72 hours; lasts months if not lethal	High without respiratory support	For weeks in non-moving water and food	3 dose efficacy 100% against 25–250 LD_{50} in primates
Staph Enterotoxin B	No	0.03 μg/person incapacitation	3–12 hours after inhalation	Hours	<1%	Resistant to freezing	No vaccine
Ricin	No	3–5 μg/kg is LD_{50} in mice	18–24 hours	Days – death within 10–12 days for ingestion	High	Stable	No vaccine
T-2 Mycotoxins	No	Moderate	2–4 hours	Days to months	Moderate	For years at room temperature	No vaccine

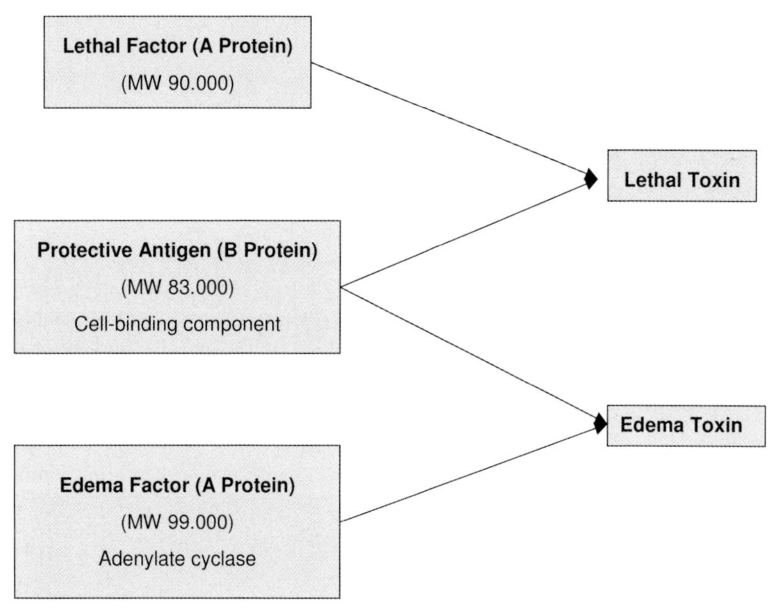

Figure 4.2 Composition of anthrax protein toxins.

Different forms of the disease are described:

- Inhalational Anthrax with a high mortality rate despite appropriate treatment
- Cutaneous Anthrax
- Oropharyngeal and Gastrointestinal Anthrax
- Meningitis [14, 22].

After a positive diagnosis, different antibiotics can be used for treatment and for prophylaxis [14, 15, 23]. Also, vaccines and other strategies have been intensively tested against anthrax [24–29].

4.4.1.2 **Plague**

Yersinia pestis is a Gram-negative, non-motile and non-sporulating coccobacillus and responsible for three great pandemics in the history of human disease:

- The First Pandemic in the sixth century [30, 31]
- The Second Pandemic (The Black Death) in the fourteenth century [32]
- The Third Pandemic in the twentieth century [33].

Plague is a zoonotic infection and can be transmitted from rodents. There are different virulence factors of *Y. pestis* encoded on the chromosome and the three plasmids [34–36]. It is essential to know that as few as 1 to 10 *Y. pestis* organisms are enough to infect rodents and humans. The clinical manifestations for *Y. pestis* as a biological weapon have been described in detail in medical handbooks and literature [36, 37], and can be divided into:

- Epidemic pneumonia by using aerolized bacteria as carriers
- Bubonic or septicemic plague by using flees as carriers.

After laboratory confirmation all patients with plague have to be isolated immediately and different antibiotics can be used for treatment and for prophylaxis [38]. Also, vaccines have been intensively tested against plague [39, 40].

4.4.1.3 Tularemia

As with Anthrax and Plague, Tularemia is a zoonosis caused by *Fransicella tularensis*. The infectious agent is Gram-negative, obligately aerobic and non-motile coccobacillus. The organism is highly infectious by the cutaneous and aerosol routes. Based on the clinical signs the disease tularaemia can be divided into two forms: (a) ulceroglandular form, as the most common form (75% of patients), and (b) typhodial form (25% of patients) [41]. Without antibiotic treatment patients have tularaemia over a period of months. However, with medical treatment the death toll can be calculated at under 2.5% [42]. For prophylaxis, vaccines were produced and one is effective against aerosol infection [43, 44].

4.4.1.4 Glanders and Meliodosis

Glanders is also a zoonosis and caused by *Burkholderia mallei*. The agent is Gram-negative and aerobic and possesses a high infection rate as an aerosol. Infection with glanders varies from asymptomatic acquisition to life-threatening pneumonia and bacteremia. Experience in treating glanders in humans is limited, and only a few antibiotics have been evaluated *in vivo*. Without any appropriate antibiotic treatment the mortality rate can be high. However, an effective oral antibiotic therapy is possible and can prevent relapse [14].

The agent *Pseudomonas (Burkholderia) pseudomallei* is aerobic and Gram-negative and is responsible for the disease meliodosis. The most likely clinical manifestations due to bioterrorism would be pulmonary infection due to aerosolized bacteria. Normally, without any treatment the disease causes a high death rate. However, therapy with different antibiotics over a minimum of eight weeks shows that an effective treatment of the disease is possible [14].

4.4.1.5 Brucellosis

Brucellosis, caused by different organisms of the genus *Brucella*, has a long history in the context of biological warfare [45]. Organisms of the genus *Brucella* are non-motile, Gram-negative coccobacilli. As with the other agents mentioned above, Brucellosis is a highly infectious zoonotic disease of wild and domesticated animals. Humans can be infected through the skin, respiratory tract, conjunctiva and gastrointestinal tract [46]. The clinical manifestations of brucellosis are diverse and the most important diagnostic tool for the identification of the disease is the history of potential exposition [45]. However, an appropriate treatment is possible with different antibiotics and with aminoglycosides [14]. For prophylaxis only proper protective clothing and the standard gas mask could protect personnel from brucellae. Vaccines for humans are not commercially available [45].

4.4.1.6 Q Fever

The last of the six bacteria is also responsible for a zoonotic disease. The agent *Coxiella burnetii* is classified in the family *Rickettsiaceae*, but is not a true member of the genus *Rickettsia* [47]. According to genetic analysis, its closest relative is *Legionella* [48]. *Coxiella burnetii* can infect humans by inhalation. The agent is extremely infectious and there is evidence that a single organism is enough for an outbreak of the disease [49]. After an incubation period of several days (10–40 days) the symptoms are similar as for a heavy flu. Infection with *C. burnetii* can lead to acute Q fever and to chronic Q fever. However, acute Q fever can be treated by several antibiotics and shortens the course of the disease [50]. Treatment of chronic Q fever is much more difficult. The mortality is higher than for acute Q fever and a therapy over 2 years is required [51]. However, vaccine prophylaxis for Q fever is possible and immunization confers protection for up to five years [47].

4.4.2 Viruses

The literature on viruses provides the interested reader with an overabundance of information on which of them might be used as a biological warfare agent [2]. With genetic engineering it could be possible to design even more infectious viruses [52]. Smallpox, viral encephalitides and viral hemorrhagic fever are diseases caused by such viruses, which have been considered as the most probable biological warfare viruses.

4.4.2.1 Smallpox

The last natural occurrence of smallpox was in Somalia in 1977 and the last laboratory-caused infection was reported in 1978 [53]. Two years later, the Certification of Smallpox Eradication was ratified by the World Health Organization (WHO). However, after the worldwide eradication of all known variola viruses two institutions still have variola strains with permission by the WHO: the Center for Disease Control and Prevention, Atlanta, USA; and Vector Laboratories, Koltsovo, Novosibirsk, Russia. In the literature, some reasons are given for retaining these smallpox stocks:

1. Cadavers in permafrost could release the virus [54].
2. Dry crypts could also release the virus [55].
3. Some unreported virus specimens still persist in nature [56].
4. Genetic engineering can be abused to merge the published sequence of variola with other *Orthopoxviruses* [52].

Therefore, these reasons prevent the destruction of the last variola strains.

Variola virus is stable and is effective for long periods outside the host. The incubation has been investigated in detail and also the molecular mechanism has been described in detail [57] (Figure 4.3).

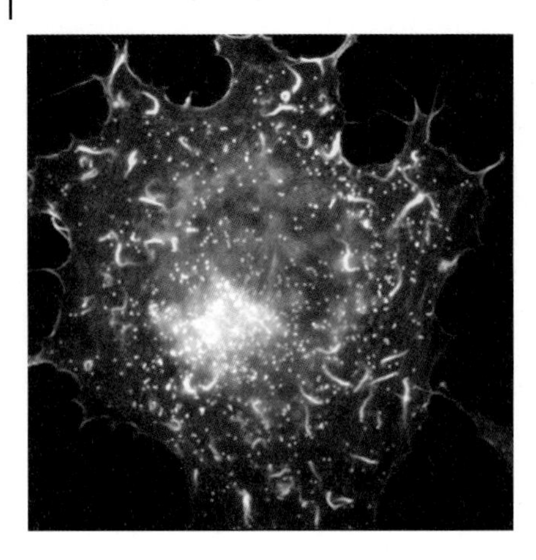

Figure 4.3 Interactions between host (cell) and pathogen (smallpox) on a cellular level during an infection [57].

The medical management must be aware that any confirmed smallpox case is an international emergency and must be reported immediately to the WHO. Active and passive immunoprophylaxis can be used to gain immunity against smallpox [56, 58]. For post-exposure prophylaxis, also vaccinia vaccination, vaccinia immune globulin and methisazone can be used with some efficacy [53, 58].

4.4.2.2 **Viral Encephalitides**

Venezuelan Equine Encephalitis (VEE), *Western Equine Encephalitis* (WEE) and *Eastern Equine Encephalitis* (EEE) are similar [59, 60]. These viruses are vectored in nature by mosquitoes. With four additional virus complexes they have been grouped into the *Alphavirus* genus. In contrast to smallpox, the pathogenesis of VEE, WEE and EEE infections is still unknown [61]. The infect dose is assumed by 10 to 100 viruses. It is important to know that no specific therapy and treatment exists for viral encephalitides. EEE with causes fatality rates up to 70% is the most severe. WEE is less virulent and encephalitis is rare by a VEE virus infection. However, although existing vaccines have some disadvantages, effective immunoprophylaxis would prevent outbreaks of VEE, WEE and EEE. Therefore, improved antibodies against alphaviruses are under development [62, 63].

4.4.2.3 **Viral Hemorrhagic Fevers (VHF Agents)**

Pathogens from four virus families contribute to the group of VHF agents: the *Flaviviridae, Arenaviridae, Bunyaviridae* and the *Filoviridae*. These are relatively simple RNA viruses with lipid envelopes. The exact nature of the viral hemorrhagic fever syndrome caused by these agents depends on routes of exposure, dose, host factors, strain characteristics and viral virulence. The medical management has to manage supervision and intensive care of patients with VHF syndromes. This management includes (a) supportive care, (b) treatment of shock and hypertension, (c) treatment of bleeding, (d) isolation, and (e) specific antiviral therapy [64].

Prophylaxis has been attempted by active and passive immunization without great success. New vaccines are under development. However, at the moment, there are no licensed vaccines to treat the diseases caused by VHF agents [65]. Antibiotics like ribavirin show some efficiency against hemorrhagic fever in the case of Crimean-Congo hemorrhagic fever (CCHF) [66].

4.4.3 Toxins

Some toxins have been identified as potential warfare agents. They have sometimes been claimed as biological and sometimes as chemical agents [67]. Even the Biological Weapons Convention circumscribes the agents as "Microbial or other biological agents, or *toxins*" [68]. However, in this book we follow the common road, and therefore toxins are biological warfare agents. Three of them are listed in the Dirty Dozen and a short description with additional literature is given.

4.4.3.1 Ricin Toxin

The recent discovery of ricin at different locations shows the ease with which the toxin can be obtained and used as a threat [69]. In the bean of the castor plant *Ricinis communiz*, the ricin toxin can be found and easily extracted. Ricin is a 60–65 kilodalton (kd) globular protein and consists as a heterodimer. A 32-kd A-chain is disulfide-linked to a 32-kd B-chain [70]. The toxin was characterized and crystallized [71]. It was found that the toxicity of ricin varies for animals [67]. The B-chain is a lectin and binds to galactosides of cell-surface carbohydrates, facilitating the entry of the toxin into the cytosol [72]. After entering the cytoplasma of a eukaryotic cell, the A-chain enzymatically attacks the 28S ribosomal subunit and the following inhibition of the protein synthesis results in the death of the cell [67, 69]. However, the pathological manifestations of ricin toxicity vary with the dose and the route of exposure [69]. The route of exposure can be (a) oral intoxication [73], (b) injection [74] and (c) inhalation [75]. For protection, prophylaxis or therapy, active immunization and the use of antispecific antibodies have shown the best results [69, 76].

4.4.3.2 Botulinum Toxins

Botulinum toxin, as a neurotoxin, poses a major bioweapon threat due to its (a) extreme lethality, (b) ease of production and transport, and (c) need for intensive care [77, 78]. Neurotoxins are the most toxic substances known today and humans are normally exposed to neurotoxins produced by *Clostridium botulinum*. Seven distinct antigenic types of botulinum toxin can be produced by bacteria, designated A through G [77]. The toxin is a simple dichain polypeptide with a 100-kd "heavy" chain and linked by a disulfide bond to a 50-kd "light" chain [79]. The light chain is an endopeptidase and attacks one of the fusion proteins (SNAP-25) at a neuromuscular junction. This attack prevents vesicles from anchoring to the membrane to release acetylcholine, resulting in flaccid paralysis of muscles and always beginning in bulbar musculature [77, 80]. Death results from pharyngeal and upper airway muscle paralysis and accessory respiratory muscle paralysis. Mortality has been decreased due to improved contemporary therapy consisting of supportive care, antitoxin administration and artificial ventilation [81]. If initiated in time, these treatments are quite effective. Depending on the rate and amount of toxin absorption, functional recovery can take several weeks to months. Therapy of botulinum poisoning can be done by timely administering of passive neutralizing botulinum antitoxins or antibodies [82].

4.4.3.3 **Staphylococcal Enterotoxin B (SEB)**

Staphylococcal enterotoxins (Ses) belong to a group of bacterial exotoxins that cause severe immunopathologies [83]. Especially, the use as an aerosol has severe implications [83, 84]. As a potential biological agent, SEB is one of seven enterotoxins produced by certain strains of the coagulase-positive S aureus bacteria [85]. This toxin is well known as the enterortoxin that most commonly causes classic food poisoning. The toxin consists of 239 amino acid residues and has a molecular weight of 28 kd. It is one of the six least antigenically distinct enterotoxin proteins that have been identified (A, B, C, D, E, G). Staphylococcal enterotoxin B (SEB) is classified as an exotoxin, since it is excreted by an organism [85]. SEB normally exerts its effect on the intestinal tract when ingested and therefore is characterized as an enterotoxin. Similar to the streptococcal pyrogenic exotoxins and toxic syndrome toxin 1 (TSST-1) SEB is classified as a "superantigen" because of its profound effects on the immune system. SEB has been studied as a potential biological agent of war, because (a) it easily can be aerosolized, (b) is very stable, and (c) can cause widespread systemic damage, multiorgan system failure, and even shock and death when inhaled at very high dosages [85]. It was shown that prophylactic treatment by passive vaccination is able to protect animals from the agent [85–87].

References

1 Alibek, K. and Handelman, S. (1999) *Biohazard: The Chilling True Story of the Largest Covert Biological Weapons Program in the World–Told from Inside by the Man Who Ran It.* Delta, ISBN 0-385-33496-6.

2 Hawley, R.J. and Eitzen, E.M. (2001) *Biological Warfare–A Primer for Microbiologists,* Annual Reviews of Microbiology, **55**, 235–253.

3 Eitzen, E. (1997) Use of biological weapons, in *Textbook of Military Medicine: Medical Aspects of Chemical und Biological Warfare* (eds R. Zajtchuk and R.F. Bellamy), Office of the Surgeon General, US Dept of the Army, Washington, DC., Chapter 20: pp. 437–50.

4 McNally, R.E., Morrison, M.B., Stark, M. *et al.* (1993) *Effectiveness of Medical Defense Interventions against Predicted Battlefield Levels of Bacillus Anthracis.* Science Applications International Corp, Joppa, MD.

5 Alibek, K. (2001) *Mighty Microbe,* Defense Review, F33612-91-D-0652, No. 0002, p. 44.

6 US Departments of Army and Air Force Military Biology and Biological Agents (1964) Departments of Army and Air Force Technical Manual 3–216/Air Force Manual 355–6, 12 March.

7 Wilkening, D.A. (2006) Sverdlovsk revisited: modeling human inhalation anthrax, *Proceedings of the National Academy of Sciences of the United States of America*, **103**, 7589–94.

8 Henderson, D.A. (1999) The looming threat of bioterrorism. *Science*, **283**, 1279–82.

9 Sohns, T. (1999) Defense against biological terrorism. *ASA Newsletter*, **5**, 99.

10 Russmann, H. (2003) Toxine. *Bundesgesundheitsblatt Gesundheitsforschung Gesundheitsschutz*, **46**, 989–96.

11 Richardt, A. and Russmann, H. (2006) Biologische Kampfstoffe: Nachweis und Bekämpfung – Strategien gegen Milzbrand, Pocken & Co. *Biologie in Unserer Zeit*, **5**, 322–9.

12 Hepburn, M.J., Purcell, B.K. and Paragas, J. (2007) Pathogenesis and sepsis caused by organisms potentially utilized as biologic weapons: opportunities for targeted intervention. *Current Drug Targets*, **8**, 519–32.

13 Mock, M. and Fouet, A. (2001) Anthrax. *Annual Review of Microbiology*, **55**, 647–71.

14 Pustelnik, T. (2004) Militärbiologie – eine militärische und militärmedizinische Herausforderung. *Vielfalt in Uniform Band*, **3**, 112–38.

15 Friedlander, A. (1997) Anthrax, in *Textbook of Military Medicine: Medical Aspects of Chemical und Biological Warfare* (eds R. Zajtchuk and R.F. Bellamy) Office of the Surgeon General, US Department of the Army, Washington, DC, pp. 467–78.

16 Ivins, B.I., Ezzell, J.W., Jemski, J., Hedlund, K.W., Ristroph, J.D. and Leppla, S.H. (1986) Immunization studies with attenuated strains of *Bacillus anthracis*. *Infection and Immunity*, **52**, 454–8.

17 Leppla, S.H. (1991) The anthrax toxin complex, in *Sourcebook of Bacterial Protein Toxins* (eds J.E. Alouf and J.H. Freer), Academic Press, London, England, pp. 277–302.

18 Stanley, J.L. and Smith, H. (1961) Purification of factor I and recognition of a third factor of the anthrax toxin. *Journal of General Microbiology*, **26**, 49–66.

19 Klimpel, K.R., Arora, N. and Seppla, S.H. (1994) Anthrax toxin lethal factor contains a zinc metalloprotease consensus sequence which is required for lethal toxin activity. *Molecular Microbiology*, **13**, 1093–100.

20 Pezard, C., Berche, P. and Mock, M. (1991) Contribution of individual toxin components to virulence of *Bacillus anthracis*. *Infection and Immunity*, **59**, 3472–7.

21 Cataldi, A., Labruyere, E. and Mock, M. (1990) Construction and characterization of a protective antigen-defizient *Bacillus anthracis* strain. *Molecular Microbiology*, **4**, 1111–7.

22 Akula, A., Gomedhikam, J.P., Kota, M.K., Bodhanapu, V.B., Kalagara, M. and Kota, B.P. (2005) Anthrax: an overview. *International Journal of Risk Assessment and Management*, **2005** (5), 76–94.

23 Doganay, M. and Aydin, N. (1991) Antimicrobial susceptibility of *Bacillus anthracis*. *Scandinavian Journal of Infectious Diseases* **23**, 333–5.

24 Shlyakhov, E.N. and Rubinstein, E. (1994) Human live anthrax vaccine in the former USSR. *Vaccine*, **12**, 727–30.

25 Cohen, H.W., Sidel, V.W. and Gould, R.M. (2000) Prescriptions on bioterrorism have it backwards. *BMJ (Clinical Research Ed)*, **320**, 1211.

26 Coulson, N.M., Fulop, M. and Titbal, R.W. (1994) *Bacillus anthracis* protective antigen expressed in Salmonella typhimurium SL 3261, afford protection against spore challenge. *Vaccine*, **12**, 1395–401.

27 Ivins, B., Fellows, P. and Pitt, L. (1995) Experimental anthrax vaccines: efficacy of adjuvants combined with protective antigen against an aerosol *Bacillus anthracis* spore challenge in guinea pigs. *Vaccine*, **13**, 1779–94.

28 Mourez, M., Kane, R., Modgridge, J., Metallo, S., Deschatelets, P., Sellmann, B.R., Whitesides, G.M.R.J. (2001) Designing a polyvalent inhibitor of anthrax toxin. *Nature Biotechnology*, **19**, 958–61.

29 Sellman, B.R., Mourez, M. and Collier, R.J. (2001) Dominant-negative mutants of a toxin subunit: an approach to therapy of anthrax. *Science*, **292**, 695–7.

30 Bayliss, J.H. (1980) The extinction of bubonic plague in Britain. *Endeavour*, **4**, 58–66.

31 Mee, C. (1990) How a mysterious disease laid low Europe's masses. *Smithsonian*, **20**, 66–79.

32 McEvedy, C. (1988) The bubonic plague. *Scientific American*, **Feb**, 118–23.

33 Ampel, N.M. (1991) Plagues–What's past is present: thoughts on the origin and history of new infectious diseases. *Reviews of Infectious Diseases*, **13**, 658–65.

34 Brubaker, R.R. (1991) Factors promoting acute and chronic diseases caused by Yersiniae. *Clinical Microbiology Reviews*, **4**, 309–24.

35 Straley, S.C., Skrzypek, E., Plano, G.V. and Bliska, J.B. (1993) Yops of *Yersinia* ssp pathogenic for humans. *Infection and Immunity*, **61**, 3105–10.

36 McGovern, T.W. and Friedlander, A.M. (1997) Plague, in *Textbook of Military Medicine: Medical Aspects of Chemical und Biological Warfare* (eds R. Zajtchuk and R.F. Bellamy), Office of the Surgeon General, US Department of the Army, Washington, DC, pp. 479–502.

37 Gage, K.L., Lance, S.E., Dennis, D.T. and Montenieri, J.A. (1992) Human plague in the United States: a review of cases from 1988–1992 with comments on the likelihood of increased plague activity. Border. *Epidemiological Bulletin*, **19**, 1–10.

38 Bonacorsi, S.P., Scavizzi, M.R., Guiyoule, A., Amouroux, J.H. and Carniel, E. (1994) Assessment of a fluoroquinolone, three betalactams, two aminoglycosides, and a tetracycline in treatment of murine *Yersinia pestis* infection. *Antimicrobial Agents and Chemotherapy*, **38**, 481–6.

39 Centers for Disease Control (1982) Immunization practices advisory committee. Plague vaccine. *Morbidity and Mortality Weekly Report*, **31**, 301–4.

40 Oyston, P.C.F., Williamson, E.D., Leary, S.E., Eley, S.M., Griffin, K.F. and Titball, R.W. (1995) Immunization with live recombinant Salmonella typhimurium aroA producing F1 antigen protects against plague. *Infection and Immunity*, **63**, 563–8.

41 Evans, M.E. and Friedlander, A.M. (1997) Tularemia, in *Textbook of Military Medicine: Medical Aspects of Chemical und Biological Warfare* (eds R. Zajtchuk and R.F. Bellamy), Office of the Surgeon General, US Department of the Army, Washington, DC, pp. 503–12.

42 Rohrbach, B.W. and Westerman, E. (1991) Istre GR. Epidemiology and clinical characteristics of tularemia in Oklahoma 1979 to 1985. *The Southern Medical Journal*, **84**, 1091–6.

43 Tigertt, W.D. (1962) Soviet viable Pasteurella tularensis vaccines: a review of selected articles. *Bacteriological Reviews*, **26**, 354–73.

44 Hornick, R.B. and Eigelsbach, H.T. (1966) Aerogenic immunization of man with live tularemia vaccine. *Bacteriological Reviews*, **30**, 532–8.

45 Hoover, D.L. and Friedlander, A.M. (1997) Brucellosis, in *Textbook of Military Medicine: Medical Aspects of Chemical und Biological Warfare* (eds R. Zajtchuk and R. F. Bellamy), Office of the Surgeon General, US Department of the Army, Washington, DC, pp. 513–22.

46 Buchanan, T.M., Hendricks, S.L., Patton, C.M. and Feldman, R.A. (1974) Brucellosis in the United States, 1960–1972: an abattoir-associated disease, III: epidemiology and evidence for acquired immunity. *Medicine*, **53**, 427–39.

47 Bryne, W.R. (1997) Q fever, in *Textbook of Military Medicine: Medical Aspects of Chemical und Biological Warfare* (eds R. Zajtchuk and R.F. Bellamy), Office of the Surgeon General, US Department of the Army, Washington, DC, pp. 523–37.

48 Weisburg, W.G., Dobson, M.E., Samuel, J.E. *et al.* (1981) Phylogenetic diversity of the rickettsiae. *Journal of Clinical Microbiology*, **13**, 603–5.

49 Tigertt, W.D. and Benenson, A.S. (1956) Studies on Q fever in man. *Transactions of the Association of American Physicians*, **69**, 98–104.

50 Sobradillo, V., Zalacain, R., Capelastegui, A., Uresandi, F. and Corral, J. (1992) Antibiotic treatment in pneumonia due to Q fever. *Thorax*, **47**, 276–8.

51 Raoult, D. (1993) Treatment of Q fever. *Antimicrobial Agents and Chemotherapy*, **37**, 1733–6.

52 Finkel, E. (2001) Engineered mouse virus spurs bioweapon fears. *Science*, **291**, 585.

53 Her Majesty'S Stationery Office (1980) *Report of the Investigations Into the Cause of* *the 1978 Birmingham Smallpox Occurrence*, HMSO, London, England.

54 Zuckerman, A.J. (1984) Paleontology of smallpox. *Lancet*, **2**, 1454.

55 Baxter, P.J., Brazier, A.M. and Young, S.E.J. (1988) Is smallpox a hazard in church crypts? *British Journal of Industrial Medicine*, **45**, 359–60.

56 McClain, D.J. (1997) Smallpox, in *Textbook of Military Medicine: Medical Aspects of Chemical und Biological Warfare* (eds R. Zajtchuk and R.F. Bellamy), Office of the Surgeon General, US Department of the Army, Washington, DC, pp. 539–60.

57 Frischknecht, F. (2003) The history of biological warfare. *EMBO Reports*, **4**, 47–52.

58 Baxby, D. (2002) Studies in smallpox and vaccination, *Reviews in Medical Virology*, **12** (4), 201–9.

59 Donald, J. Netolitzky, D.J., Schmaltz, F.L., Parker, M.D. *et al.* (2000) Complete genomic RNA sequence of western equine encephalitis virus and expression of the structural genes. *The Journal of General Virology*, **81**, 151–9.

60 Nagata, L.P., Hu, W.G., Parker, M., Chau, D. *et al.* (2006) Infectivity variation and genetic diversity among strains of Western equine encephalitis virus. *The Journal of General Virology*, **87**, 2353–61.

61 Schlesinger, S. and Schlesinger, M.S. (eds) (1986) *The Togaviridae and Flaviviridae*, Washington University School of Medicine, St. Louis, MO; Plenum Press, New York and London, p. 210 ff.

62 Griffin, D.E., Ubol, S., Despres, P., Kimura, T. and Byrnes, A. (2001) Role of antibodies in controlling alphavirus infection of neurons. *Current Topics in Microbiology and Immunology*, **260**, 191–200.

63 Dong, M., Zhang, P.F., Grieder, F., Lee, J. *et al.* (2003) Induction of primary virus-cross-reactive human immunodeficiency virus type 1-neutralizing antibodies in small animals by using an alphavirus-derived in vivo expression system. *Journal of Virology*, **77**, 3119–30.

64 Jahrling, P.B. (1997) Viral hemorrhagic fevers, in *Textbook of Military Medicine: Medical Aspects of Chemical und Biological Warfare* (eds R. Zajtchuk and R.F. Bellamy), Office of the Surgeon General, US Department of the Army, Washington, DC, pp. 591–602.

65 Borio, L., Inglesby, T., Peters, C.J., Schmaljohn A.L. *et al.* (2002) Hemorrhagic fever viruses as biological weapons. *The Journal of the American Medical Association*, **287**, 2391–405.

66 Mardani, M., Keshtkar Jahromi, M., Holakouie Naieni, K. and Zeinali, M. (2003) The efficacy of oral ribavirin in the treatment of Crimean-Congo hemorrhagic fever in Iran. *Clinical Infectious Diseases: An Official Publication of the Infectious Diseases Society of America*, **36**, 1613–8.

67 Franz, D.R. (1997) Defense against toxin weapons, in *Textbook of Military Medicine: Medical Aspects of Chemical und Biological Warfare* (eds R. Zajtchuk and R.F. Bellamy), Office of the Surgeon General, US Department of the Army, Washington, DC, pp. 603–20.

68 United Nations (1972) Convention on the Prohibition of the Development, Production and Stockpiling of Bacteriological (Biological) and Toxin Weapons and on their Destruction. United Nations General Assembly resolution 2826 (XXVI), New York, Full text available at http://www.brad.ac.uk/acad/sbtwc (Oct, 10, 2007).

69 Audi, J., Belson, M., Patel, M., Schier, J. and Osterloh, J. (2005) Ricin poisoning a comprehensive review. *The Journal of the American Medical Association*, **294**, 2342–51.

70 Balint, G.A. (1974) Ricin: the toxic protein of castor oil seeds. *Toxicology*, **2**, 77–102.

71 Rutenberger, E., Katzin, B., Collins, E. *et al.* (1991) The crystallographic refinement of ricin at 2.5 Å resolution. *Proteins*, **10**, 240–50.

72 Sandvig, K. and van Deurs, B. (2000) Entry of ricin and Shiga toxin into cells: molecular mechanisms and medical perspectives. *The EMBO Journal*, **19**, 5943–50.

73 Rauber, A. and Heard, J. (1985) Castor bean toxicity re-examined: a new perspective. *Veterinary and Human Toxicology*, **27**, 498–502.

74 Ramsden, C.S., Drayson, M.T. and Bell, E.B. (1989) The toxicity, distribution and execretion of ricin holotoxin in rats. *Toxicology*, **55**, 161–71.

75 Griffiths, G.D., Rice, P., Allenby, A.C. *et al.* (1995) Inhalation toxicology and histophathology of ricin and abrin toxins. *Inhalation Toxicology*, **19**, 247–56.

76 Smallshaw, J.E., Richardson, J.A., Pincus, S. *et al.* (2005) Preclinical toxicity and efficacy testing of RiVax, a recombinant protein vaccine against ricin. *Vaccine*, **23**, 4775–84.

77 Arnon, S.S., Schechter, R.S., Inglesby, T.V. *et al.* (2001) Botulinum toxin as a biological weapon: medical and public health management. *The Journal of the American Medical Association*, **8**, 1059–70.

78 Caya, J.G., Rashmi A. and Miller, J.E. (2004) *Clostridium botulinum* and the clinical laboratorian: a detailed review of botulism, including biological warfare ramifications of botulinum toxin *Archives of Pathology and Laboratory Medicine*, **128**, 653–62.

79 Lacy, D.B., Tepp, W., Cohen, A.C., DasCupta, B.R. and Stevens, R.C. (1998) Crystal structure of botulinum neurotoxin type A and implications for toxicity. *Nature Structural Biology*, **5**, 898–902.

80 Montecucco, C. (ed.) (1995) Clostridial neurotoxins: the molecular pathogenis of tetanus and botulism. *Current Topics in Microbiology and Immunology*, **195**, 1–278.

81 Maselli, R.A. and Bakshi, N. (2000) American Association of Electrodiagnostic Medicine case report 16: botulism. *Muscle and Nerve*, **23**, 1137–44.

82 Arnon, S.S. (1998) Infant botulism, in *Textbook of Pediatric Infectious Diseases*, 4th edn (eds R.D. Feigin and J.D. Cherry), WB Saunders Co, Philadelphia, PA, pp. 1570–7.

83 Roy, C.J., Warfield, K.L. and Welcer, B.C. *et al.* (2005) Human leukocyte antigen-DQ8 transgebic mice: a model to examine the toxicity of aerosolized staphylococcal enterotoxin B. *Infection and Immunity*, **73**, 2452–60.

84 Mattix, M.E.R., Hunt, R.E., Wilhelmsen, C.L., Johnson, A.J. and Baze, W.B. (1995) Aerosolized staphylococcal enterotoxin B-induced pulmonary lesions in rhesus monkeys (*Macaca mulatta*). *Toxicologic Pathology*, **23**, 262–8.

85 Ulrich, R.G., Sidell, S., Taylor, T.J., Wilhelmsen, C.L. and Franz, D.R. (1997) Staphylococcal enterotoxin B and related pyrogenic toxins, in *Textbook of Military Medicine: Medical Aspects of Chemical und Biological Warfare* (eds R. Zajtchuk and R. F. Bellamy), Office of the Surgeon General, US Department of the Army, Washington, DC, pp. 621–30.

86 Bavari, S., Ulrich, R.G. and LeClaire, R.D. (1999) Cross-reactive antibodies prevent the lethal effects of *Staphylococcus aureus* superantigens. *The Journal of Infectious Diseases*, **180**, 1365–9.

87 LeClaire, R.D., Hunt, R.E. and Bavari, S. (2002) Protection against bacterial superantigen staphylococcal enterotoxin B by passive vaccination. *Infection and Immunity*, **70** (5), 2278–81.

5
Chemical Warfare Agents

Alexander Grabowski, André Richardt and Marc-Michael Blum

5.1
Introduction

A lot of confusion exists regarding terms like chemical weapons and chemical warfare agents (CWAs). Many chemical compounds are toxic, but not all chemicals are warfare agents. Also, a chemical warfare agent is only one component of a chemical weapon. Quoting the famous alchemist, physician and astrologer Paracelsus (1493–1541) only adds to this confusion: "*All substances are poisons; there is none which is not a poison. The right dose differentiates a poison. . . .* " In this chapter we will try to clarify and eliminate these discrepancies and we want to describe the most important chemical warfare agents in more detail.

5.2
Chemical Warfare Agents and Chemical Weapons: A Differentiation

Some help comes in form of the treaty text of the Chemical Weapons Convention (CWC) which includes a definition that defines a chemical weapon as a toxic chemical which is further defined as "any chemical which through its chemical action on life processes can cause death, temporary incapacitation or permanent harm to humans or animals." The Convention text contains annexes that list the chemicals covered by the CWC. This helps us to discriminate chemical agents from other chemicals which are not considered warfare agents although used in military conflict, namely explosives that consist of chemicals but act due to shock-wave and heat generation, and incendiary agents like napalm or white phosphorus that act due to heat generation and fire. But the definition of the CWC is not without problems, as the toxic chemical (the agent) is identical with the chemical weapon. A definition that is more helpful in practical terms will differentiate between the agents and the chemical weapon which consists of the toxic chemical and any technical device needed to disperse the agent. This can be a detonator in

Decontamination of Warefare Agents.
Edited by André Richardt and Marc-Michael Blum

ISBN: 978-3-527-31756-1

an artillery shell or an aircraft mountable spray tank for aerosol generation, to name just two examples.

The remaining obstacle is to clearly define what makes a toxic chemical a chemical warfare agent. Unfortunately, this is not easily done, although in order to be effectively used as warfare agents toxic chemicals must posses several distinct properties. One is easy penetration into the human body, which is most easily achieved by inhalation as vapor or aerosol or penetration through the skin. Stability of the chemical is important as well. If used in artillery shells that disperse the agent using an explosive detonator, the agent must be thermostable. It should also be resistant against light and moisture for some time. The third property is, of course, a high toxicity, and protection against these agents should be as difficult as possible. Finally, the agents should be easy and cheap to produce and handling should be possible without serious risk for personnel deploying the weapon. Most of the early agents that were used in World War I, the first major chemical war in history, were chemicals with an industrial application for which production facilities existed. This is true for agents like chlorine, phosgene or cyanide. The use of agents with no civilian application only occurred a few years later with the introduction of mustard. Nerve agents like tabun, sarin or VX, although side products of research programs for new pesticides, did continue this trend.

Since the Iran/Iraq war, which ended in 1988, there has been no use of chemical weapons by official military forces. However, there is ample evidence that chemical weapons could be produced or captured, stockpiled and used by terrorist or guerilla movements. The use of the nerve agent sarin in the attack on the Tokyo subway by the Aum Shinrikyo cult in 1995 is an example of this. There is also evidence that the results of research in the biomedical field as well as research into new chemicals could be abused for military or terrorist intentions [18].

Defenses against chemical warfare agents include protective measures, effective medical therapy in case of intoxication and the ability to detoxify and decontaminate the agent in order to remove the threat. To achieve this, broad knowledge about the chemical constitution and the properties of the different warfare agents currently known is essential, as well as intensive monitoring of the scientific developments in the chemical and life sciences. This chapter presents the most important warfare agents currently known, giving a short description of their relevant properties, their mechanism of action and estimates toxicity.

5.3 Classification

To classify CWAs systematically, several different schemes exist. These include grouping by physical properties, chemical substance class, physiological effect or toxicological effects, to name just the more important [14]. The most common and generally accepted scheme orders agents by their target organism in the human body and physiological effects. (Figure 5.1). It should be noted, however, that the military importance of several listed agents has vanished over time, since effective

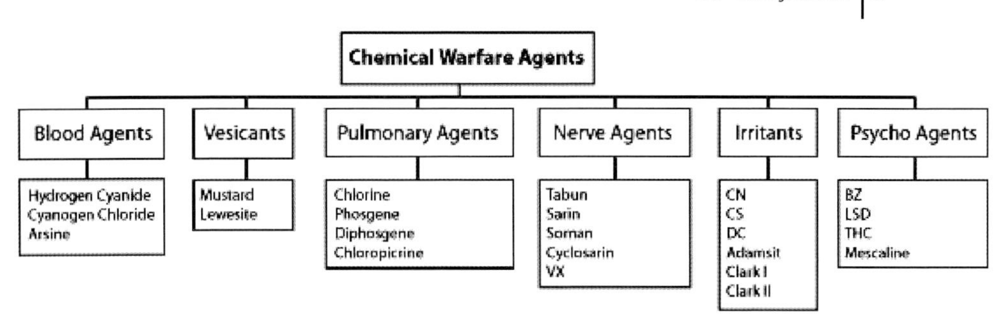

Figure 5.1 Classification of chemical warfare agents and typical representatives.

protection is achievable or because other agents are more effective. While these agents may be considered obsolete in military conflict, their use against an unprotected civilian population is possible and their effects might still be devastating.

5.3.1
Nerve Agents

Nerve agents form the most deadly group among the Chemical Warfare Agents. Their development dates back to the time between the World Wars. The German chemist Gerhard Schrader, working at the I.G. Farben laboratory in Wuppertal-Elberfeld (now Bayer AG), synthesized several organophosphorus compounds in 1936 in search for new insecticides. Among these compounds was Ethyl N,N-dimethylphosphoramidocyanidate that was a very efficient insecticide but also showed high toxicity in mammals. Known under name tabun or as "GA" according to the American naming scheme, this compound was tested at Elberfeld and reported to the German Army Procurement Office (Heereswaffenamt) according to existing law at the time that required companies to report findings of potential military importance. Tabun was later produced in bulk amounts at a special manufacturing plant at Dyhernfurt near the city of Breslau (today Wrocłav), but was never used during World War II. Schrader continued his work and discovered the more toxic compound 2-(fluoro-methyl-phosphoryl)oxypropane also known as sarin or "GB." Sarin obtained its name from the scientists involved in its synthesis and development: *S*chrader, *A*mbors, *R*üdiger and von der L*in*de [1]. Towards the end of the war an even more toxic agent was found, but its development did not progress beyond the laboratory level. This time it was the group of Richard Kuhn at the Kaiser-Wilhelm Institute for medical research that made the discovery of 3-(fluoro-methyl-phosphoryl)oxy-2,2-dimethyl-butane also known as soman or "GD." The first letter "G" in the American naming scheme stands for German or Germany as the country of origin for these compounds. A nerve agent "GC" does not exist, "GE" is the code for Ethylsarin and "GF" for Cyclohexylsarin which was used in the Iran/Iraq war. The group is known as "G-type" nerve agents [2].

Even more toxic than the G-agents is the second class of nerve agents known as the V-agents where "V" is thought to stand for "victory" or "venomous." As with tabun, the development of these agents started with research work on new pesticide compounds. The chemist Ranajit Ghosh, working for the ICI Plant Protection Laboratories in the UK, did discover the compound O,O-Diethyl-S-(2-(diethylamino)ethyl) phosphorothioate also known under its trademark name Amiton or the code "VG." ICI did indeed sell Amiton for some time but the compound was withdrawn being too toxic for agricultural use. This research resulted in the discovery of the nerve agent O-ethyl S-(2-diisopropylaminoethyl) methylphosphonothiolate better known as VX. V-agents are less volatile than G-agents and pose a severe contact risk as they are readily absorbed through the skin (Table 5.1). The toxicity of VX on skin is about 1000 times greater than that of sarin [1]. VX was stockpiled by the USA while the Soviet Union produced a slightly different V-agent known as Russian VX or R-VX. A very good history of the development of the nerve agents can be found in a recent book by Jonathan Tucker [3].

Intoxication by these agents may occur percutaneously, by ingestion or by inhalation, and symptoms of poisoning can be detected rapidly. These can range from immediate effects in case of severe poisoning to 15–20 minutes in case of very light intoxication. Symptoms include miosis, headache, difficulty in breathing, salivation, coughing, production of foamy sputum, severe spasms, loss of consciousness, paralysis of the respiratory center and finally, death [4, 5].

The primary target for the action of nerve agents is the enzyme Acetylcholinesterase (AChE). The enzyme catalyses the hydrolysis of acetylcholine to choline in the synaptic cleft of the cholinergic nerve system which is released after a nerve signal. If acetylcholine is not removed by AChE the signal remains triggered and

Table 5.1 Chemical and physical properties of nerve agents adapted from [4, 5].

Properties	*Tabun (GA)*	*Sarin (GB)*	*Soman (GD)*	*VX*
Boiling point	230–246 °C	147–158 °C	190–198 °C	298 °C$_{decomp}$
Vapor pressure	4.9×10^{-2} hPa at 20 °C	2.8 hPa at 20 °C	0.53 hPa at 25 °C	9×10^{-4} hPa at 20 °C
Density vapor (compared to air)	5.6	4.9	6.3	9.2
Density liquid	1.08 g/ml at 25 °C	1.10 g/ml at 25 °C	1.02 g/ml at 25 °C	1.01 g/ml at 20 °C
Density volatility	576–610 mg/m^3 at 25 °C	16–22 mg/m^3 at 25 °C	3.0–3.9 mg/m^3 at 25 °C	10.5 mg/m^3 at 25 °C
Appearance	Clear to brown liquid	Clear liquid	Clear liquid	Clear to yellow liquid
Odor	Fruity (if pure)	Odorless	Fruity (camphor)	Odorless
Solubility in water	9.8 g/100 g at 25 °C	Miscible	2.1 g/100 g at 20 °C	Miscible <9.5 °C; Slight at 25 °C
Solubility in other solvents	Soluble in most organic solvents	Soluble in all solvents	Soluble in some solvents	Soluble in all solvents

the nerve system being jammed cannot transmit further signals. Nerve agents inhibit AChE by phosphorylating an essential serine residue in the AChE active site. The inhibited form of AChE completely loses its enzymatic activity against acetylcholine. A detailed discussion of the chemistry of AChE inhibition can be found in all major textbooks of toxicology (e.g. [6]).

Gerhard Schrader, who continued work on organophosphorus pesticides after the war, came up with a general description of an organophosphorus compound that can inhibit AChE. This description is known as the Schrader formula that is depicted in the figure below (Figure 5.2).

The following essential structural features can be identified: (1) Oxygen or sulfur directly bonded to the phosphorus atom forming a phosphoryl- or thiophosphoryl-group. (2) R_1 ad R_2 being either an alkoxy, alkyl, N-alkylamino or N,N-dialkylamino. (3) The leaving group X is typically a halogen or pseudohalogen, aryloxy-, thioxy- or phosphate-group [7]. The choice of the right leaving group is a delicate task. The compound must be able to phosphorylate the serine residue in AChE but must be stable enough in aqueous media to reach the target without being hydrolyzed.

The close structural proximity of the nerve agents to industrial pesticides also poses problems for verification measures under the Chemical Weapons Convention. The production facilities for nerve agents and pesticides are very similar and clandestine weapons programs can be camouflaged as legal pesticide production. Only intrusive inspections, including the analysis of samples taken on site, can distinguish between a harmless factory for agricultural products and a production plant for nerve agents.

Finally, it should be noted that the handling of unitary chemical ammunition was problematic and even dangerous due to container corrosion over time, and also due to the agent leaking from the ammunition. This led to research efforts to produce the agent inside of the artillery shell, bomb or missile warhead from less toxic precursors during flight. Efforts to stockpile this binary chemical ammunition ended only with the end of the Cold War and were subject of controversial debate.

5.3.2 Vesicants

Vesicants, or blister agents, target the human skin and mucous membranes. The most prominent member of this family is Bis(2-chloroethyl) sulfide, better known as sulfur mustard (Figure 5.3). Several letter codes exist for mustard depending

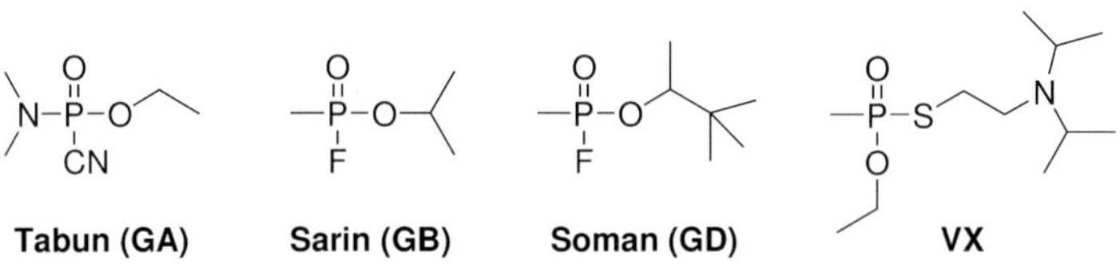

Tabun (GA) **Sarin (GB)** **Soman (GD)** **VX**

Figure 5.2 Chemical structures of the most common nerve agents.

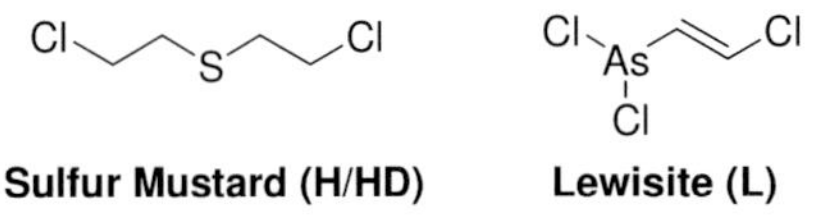

Figure 5.3 Chemical structures of Sulfur Mustard and Lewisite.

Table 5.2 Important chemical and physical properties of vesicants.

Properties	*Sulfur Mustard (HD)*	*Phosgene Oxime (CX)*	*Lewisite (L)*
Boiling Point	~230 °C	~130 °C	~190 °C
Vapor Pressure	~9.6 × 10^{-2} hPa at 20 °C	~17 hPa at 40 °C	~0.53 hPa at 20 °C
Density Vapor	5.4	<4.0	7.1
Density liquid	~1.3 g/ml at 20 °C	Not determined	~1.9 g/ml at 20 °C
Volatility	~610 mg/m^3 at 25 °C	~1.8 mg/m^3 at 20 °C	~4.4 mg/m^3 at 20 °C
Appearance	Yellow to brown liquid	Clear (liquid or solid)	Clear, oily liquid
Odor	Mustard, garlic	Intense	Geranium
Solubility in water	~9.2 × 10^{-2} g/100 g at 22 °C	~70%	Slight
Solubility in other solvents	Complete in most organic solvents like CCl_4	Very soluble in most organic solvents	Soluble in most organic solvents
Persistence in soil	Persistent	2 hours	Days
Persistence on materiel	Hours to days	Nonpersistent	Hours to days
Vapor (mg x min/m^3)	LC_{t50}: 1500 inhaled	LC_{t50}: 3000 (estimated)	LC_{t50}: 1200–1500 (estimated)
Liquid	LD_{50}: 100 mg/kg	No estimate	LD_{50}: 40–50 mg/kg

on its purity or whether it is used in mixtures with other agents. Typically, the code "H" or "HD" (D for distilled) is used. Closely related to sulfur mustard are compounds where sulfur is replaced by nitrogen or oxygen (nitrogen and oxygen mustard). Another important vesicant is the arsenic compound Lewisite (Figure 5.3), and to a lesser extent, some halogenated oximes. Mustard was first used during the later phase of World War I by German forces in 1917 and about a year later in 1918 by allied forces. The first use by Germany led to the use of the letter H for mustard which was termed the "Hun Stuff."

Intoxication by these agents may occur through all three classic ways: percutaneously, by ingestion and inhalation. Compared to nerve agents the number of fatalities is rather low and these mainly occur if the lung is directly damaged by inhaling the agent as an aerosol (Table 5.2). As indicated by the name, vesicants cause blisters on the skin. This effect might be delayed in case of sulfur mustard where the earliest burns show about 4 hours after exposure and up to 2 hours might pass before symptoms are visible. These symptoms are reddening of the skin, followed by the formation of small blisters ("strings of pearls"). These small blisters then grow to form large blisters filled with lymph [8]. Large-scale necrosis is the next step [9]. If the victim survives, recuperation might take months. If ingested, the effects show within less than 30 minutes and are mostly fatal. Small erythema accompany very painful vomiting and nausea [10]. The effects of Lewisite take a shorter time, both to visualize and to heal. The effect of Lewisite is almost immediate. Pain upon beginning of the effects of all agents is often excruciatingly strong [11].

The effects of mustards are mostly due to the fact that they form strongly alkylating intermediates. Lewisite reacts with liponic acid contained in body tissue, thus disabling the coenzyme function of this compound [12]. Due to the low amounts of agent needed to cause burns of the skin and the long recovery time for wounded personnel, sulfur mustard especially was widely used in the late phase of World War I. In World War II and during the Cold War, large amounts of mustard were stockpiled, but the importance of mustard compared to nerve agents declined over the decades mainly due to better protective gear. As mustard will be highly effective against unprotected civilian population, is relatively easy to manufacture and would yield a formidable terror effect due to the terrible wounds caused, this agent is far from obsolete and remains a severe risk in today's asymmetric conflicts. Abandoned mustard remains a risk as well [19]. After the World Wars, large amounts of chemical ammunition were dumped in the shallow waters of the North Sea and Baltic Sea [17]. Mustard leaking from corroded shells or containers only hydrolyses at the direct interface to the seawater, forming a skin of hydrolysis products protecting the remaining mustard from destruction. These lumps of partly hydrolyzed agent are sometimes washed ashore or retrieved from the seabed by Baltic fisherman, so that mustard casualties from picking up these amber-like looking lumps are reported every now and then.

5.3.3
Pulmonary Agents

This group contains derivatives of halogenated carbonic acid, halogenated nitroalkanes and interhalogen compounds. The main agents of this group are chlorine, chloropicrin (PS), phosgene (CG) and diphosgene (DP) (Figure 5.4).

Intoxication with phosgene and diphosgene occurs exclusively via inhalation. The onset of effects is often delayed, sometimes for several hours. The most serious effect of intoxication is that of a pulmonary odema. Secondary infections, occurring while already recuperating from the main intoxication, are mostly fatal.

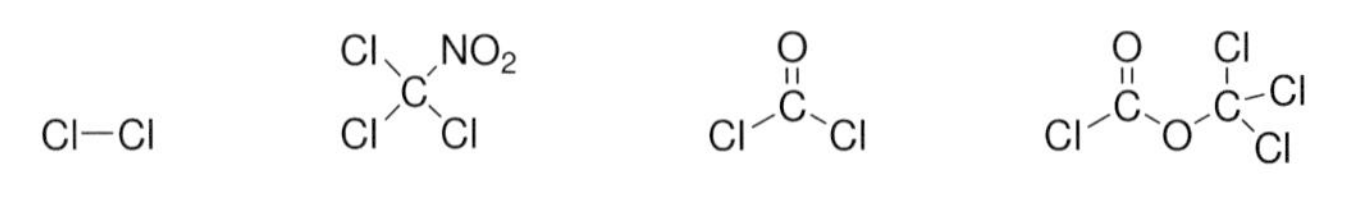

Chlorine **Chloropicrin (PS)** **Phosgene (CG)** **Diphosgene (DP)**

Figure 5.4 Chemical structures of pulmonary agents.

Chlorine was the first chemical warfare agent used in World War I. Chlorine, that was used in blow attacks releasing the gas from many buried cylinders while the wind would carry it towards the enemies trenches, quickly lost its military importance after more effective agents were introduced that could be delivered by artillery and mortar shells. Choking agents are of little importance in modern military scenarios, although the great amounts of chlorine produced by the chemical industry make it an attractive target for terrorist use. This was already demonstrated by insurgents' attacks on chlorine transports in Iraq, although at present little is known regarding the number of casualties due to the release of chlorine compared to the casualties inflicted by the explosives used in the attacks.

5.3.4 Blood Agents

The term blood agent is a rather misleading term for this group of agents as only arsine, AsH_3 (AS) affects the blood directly. The other members of the group are hydrogen cyanide, HCN (AC) and cyanogen chloride, CNCl (CK); these agents affect cellular respiration. Cyanogen chloride does release cyanide in the body and toxicological effects are comparable to AC, but CK also produces local irritant effects.

Cyanide combines with ferric iron in cytochrome a3 (the last component of the cytochrome oxidase electron transport chain in mitochondria) and inhibits this enzyme. This prevents intracellular oxygen use and the generation of cellular ATP, ceasing aerobic energy production. Anaerobic production of ATP continues as pyruvate is converted to lactate. Metabolic acidosis ensues rapidly, not due to lactic acidosis, but rather as a result of cells continuing to produce hydrogen ions through use of ATP without the balance of consumption by oxidative phosphorylation. While all organ systems are impacted, the most oxygen-dependent ones are the most affected (brain, heart). Inhaled arsine is distributed rapidly and causes massive red blood cell hemolysis that can potentially lead to global cellular hypoxia.

Blood agents were used in World War I and cyanide was used in gas chambers by the Nazi regime as a method for mass murder. The military importance of this group of agents vanished rapidly as their use in open terrain requires the delivery of large amounts of agent to the target to achieve the required saturation of agent. Rapid dispersal and dilution by wind and ascending air currents contribute to the fast reduction of initially high concentrations. As incorporation by ingestion and via the skin are only minor routes, the main way of entry into the body being via inhalation, modern protective masks are effective means of protection against these agents.

Hydrogen cyanide remains a threat as it is a major industrial chemical and could possibly be released from storage tanks or transport vessels during conflict either accidentally or by purpose. Unprotected civilian population again is the group that would be affected most severely.

5.3.5 Riot Control Agents

Riot control agents, including lachrymators and irritants, produce temporary incapacitation by irritation of the eyes and the upper respiratory tract. The use of these agents in war is prohibited by the Chemical Weapons Convention, although their use as a riot control agent by law enforcement agencies is legal in many countries [16]. The lethal dose for these agents is high but their incapacitating properties take effect even at low concentration, which implies a high safety ratio (Table 5.3). The group can be subdivided into lachrymatory agents, causing impaired vision up to incapacitation (e.g. Bromoacetone, Chloroacetophenone, CS, CN), (Figure 5.5) and agents irritating the upper respiratory system (e.g. Clark I and Clark II).

Effects of intoxication appear almost immediately when no protection is available. These effects include swelling of the eyes, mild to serious lachrymation, painful irritation of the upper respiratory tract and coughing. Effects vanish rapidly after removal from the contaminated area and recovery is usually rapid.

There is a significant diversity regarding the chemical constitution of these agents, although halogens and pseudohalogens are frequently encountered. The incapacitating effect intensifies in the order F<Cl<Br<I and is strongest with the pseudohalogen CN. While lachrymators seem to obtain their specific properties from the presence of carbonyl groups and (pseudo)halogens, most members of

Table 5.3 Chemical and physical properties of CS and CN.

Properties	*o-Chlorobenzylidene Malononitrile (CS)*	*1-Chloroacetophenone (CN)*
Boiling point	~310 °C	~250 °C
Vapor pressure	~4.7 × 10^{-4} hPa at 20 °C	~5.5 × 10^{-3} hPa at 20 °C
Volatility	~0.7 mg/m^3 at 25 °C	~34 mg/m^3 at 20 °C
Appearance and odor	White powder with pungent odor (pepper)	Fragrant
Solubility in water	Insoluble	Insoluble
Solubility in organic solvents	Soluble	Soluble
Persistency in soil	Varies	Short
Persistency on materiel	Varies	Short
Biologically effective amount: Aerosol (mg x min/m^3)	LC_{t50}: 600 000 IC_{t50}: 3–5	LC_{t50}: 7000–14 000 IC_{t50}: 20–40

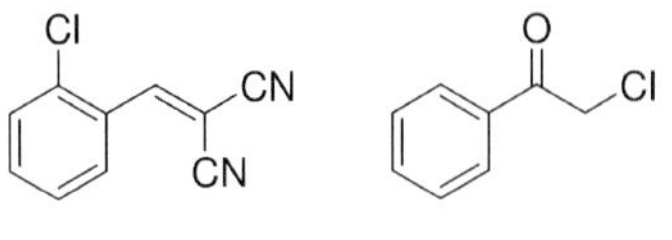

CS CN

Figure 5.5 Chemical structures of irritants CS and CN.

the other group are organic arsenic compounds. Some of these irritating agents like Clark I and Clark II were used in World War I in combination with choking agents, using the irritant as a mask breaker causing the soldier to remove his protective mask and expose himself to the deadly agent. This use became obsolete when masks were equipped with effective particle filters. Today, only a few of these agents are still in use, most of them as riot control agents. Examples include o-Chlorobenzylidenmalodinitrile (CS) and Chloroacetophenone (CN). These compounds are more and more replaced by Oleum Capiscorum (OC, "Pepper Spray"). The active ingredient is the natural product capsicain, which occurs in several plants including chili peppers and is an inflammatory, not an irritant. The substance is more predictable than CS or CN and only short time exposure is necessary as it causes immediate closure of the eye.

5.3.6 Psycho Agents

In a military context, incapacitation is understood to mean inability to perform one's military mission. A suitable nonlethal weapon fitting the criteria – military, medical, and budgetary- should be selected by the following criteria: persistence, effectiveness, relative low toxicity, and predictability, to name the most important. Psychochemical agents fit in this scheme [15]. Psychogenic substances can be divided into four categories: stimulants, depressants, psychedelics, and deliriants. Among others, they contain atropin-mimetics such as Chinuclidin-3-yl-benzilat (BZ), Lysergic Acid Diethylamide (LSD), and derivatives of Tetrahydrocannabinol, THC (Figure 5.6).

The effects are widespread and vary greatly from person to person, both in kind and intensity. They comprise general clouding of consciousness. Optical and acoustic hallucinations often occur simultaneously; after intoxication with benzilates, acoustic hallucinations outweigh the other effects. Also, intense tremor, states of fear, combined with refusal of advice and support are typical symptoms. The effects vanish after a relatively short period of time. However, "flash-backs," meaning the recurring of symptoms hours after recuperation, have been reported, for example, after intoxication with LSD. Therefore, intoxicated persons need continuous supervision. Psychogenics carry with them an aura of non-lethality and

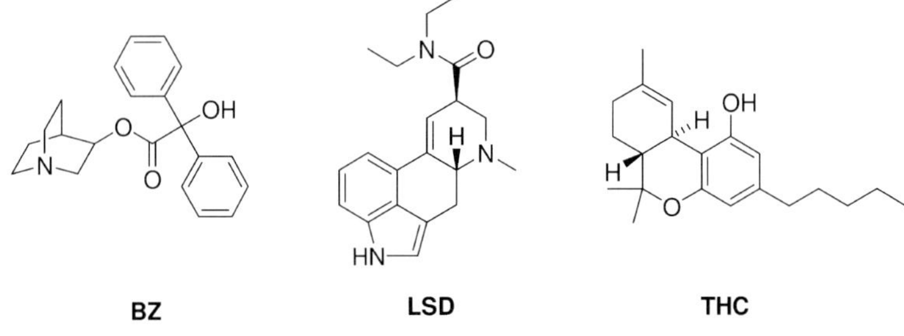

Figure 5.6 Chemical structures of BZ, LSD and THC.

therefore tend to be more easily accepted by the public than deadly chemical agents. Except for BZ, however, no militarization of such substances has been reported, although laboratory research has been performed on several candidates. Psychogenics would be used as sabotage agents rather than offensive/defensive combat weapons. Several negative properties prevent their widespread use. It is rather difficult to evenly disperse solid, crystalline substances over a large area and the problem of having to deal with a large number of adversaries in various states of psychosis obviously outweighs the advantages of using a "non-lethal" agent.

5.4 Summary

The unique properties of chemical warfare agents make them a permanent threat for military personnel as well as the civilian population. A number of books, reviews, and articles have dealt with these properties and their abuse in conflicts. Although chemical warfare agents have not been used after the Iraq/Iran conflict and are banned by the Chemical Weapons Convention, there is still evidence that terrorist groups have an interest in developing and using chemicals for their purposes. Therefore, exact knowledge about their toxicology, persistence and other properties is of essential importance for military forces and civilian first responders to react with appropriate counteractive measures for minimizing the damage to the population.

References

1 Harris, R. and Paxman, J. (1982) *A Higher Form of Killing*, Hill and Wang, New York, NY; p. 53.

2 Robinson, J.P. (1971) The rise of CB weapons, in *The Problem of Chemical and Biological Warfare*, Humanities Press, New York, Vol. 1: p. 71.

3 Tucker, J.B. (2006) *War of Nerves–Chemical Warfare from World War I to Al-Qaeda*, Pantheon Books, New York.

4 Sidell, F.R. (1997) Nerve agents, in *Mediacl Aspects of Chemical and Biological Warfare*, TMM Publications, Washington, pp. 129–79.

5 Wiener, M.D. and Hoffmann, R.S. (2004) Nerve agents: a comprehensive review. *Journal of Intensive Care Medicine*, **19**, 22–37.

6 Koelle, G.B. (1970) Anticholinesterase agents, in *The Pharmacological Bais of Therapeutics*, 4th edn (eds L.S. Goodman and A. Gilman), Macmillan, New York, NY, p. 446.

7 Clark, V.M., Hutchinson, D.W., Kirby, A.I. and Warren, S.G. (1964) Phosphorylierungsmittel–Bauprinzip und Reaktionsweise. *Angewandte Chemie*, **76**, 704–12.

8 Renshaw, B. (1946) Mechanisms in production of cutaneous injuries by sulfur and nitrogen mustards, in *Chemical Warfare Agents, and Related Chemical Problems*, Parts 3–6. Office of Scientific Research and Development, National Defense Research Committee, Washington, DC, Div 9, pp. 478–520.

9 Papirmeister, B., Feister, A.J., Robinson, S.I. and Ford, R.D. (1991) *Medical Defense Against Mustard Gas: Toxic Mechanisms and Pharmacological Implications*, CRC Press, Boca Raton, FL.

10 Anslow, W.P. and Houck, C.R. (1946) Systematic pharmacology and pathology of sulfur and nitrogen mustards, in *Chemical Warfare Agents, and Related Chemical Problems*, Parts 3–6. Office of Scientific

Research and Development, National Defense Research Committee, Washington, DC, Div 9, pp. 440–78.

11 Goldman, M. and Dacre, J.C. (1989) Lewisite: its chemistry, toxicology, and biological effects. *Reviews of Environmental Contamination and Toxicology*, **110**, 75–115.

12 Trammel, G.L. (1992) Toxicodynamics of organoarsenic chemical warfare agents, in *Chemical Warfare Agents* (ed. S.M. Somani), Academic Press, San Diego, CA, pp. 255–70.

13 Bestwick, F.W. (1983) Chemical agents used in riot control and warfare. *Human Toxicology*, **2**, 247–56.

14 Franke, S. et al. (1967) Franke, Militärchemie, Bd. 1, Berlin, Deutscher Militärverlag.

15 Furmanski, F. and Dando, M.R. (2006) Midspectrum incapacitant programs, in *Deadly Cultures: Biological Weapons from 1945 to the Present* (eds M. Wheelis, L. Roza and M. Dando), Harvard University Press, Cambridge, pp. 236–51.

16 Hu, H., Fine, J., Epastein, P. and Kelsey, K. (1989) Tear gas–harassing agent or toxic chemical weapon? *The Journal of the American Medical Association*, **262**, 660–3.

17 Kaffka, A.V. (1996) *Sea-Dumped Chemical Weapons: Aspects, Problems, Solutions*, Kluver Academica Publishers, Dordrecht.

18 Wheelis, M. and Dando, M. (2006) Neurobiology: a case study of the imminent militarization of biology. *International Review of the Red Cross*.

19 Willems, J.L. (1989) Clinical management of mustard gas casualties. *Annales Medicinae Militaris. Belgicae*, **3S**, 1–61.

Additional Web Links

US Army Medical Research Institute of Chemical Defense (USAMRICED).

http://www.brooksidepress.org/Products/OperationalMedicine/DATA/operationalmed/Manuals/RedHandbook/001TitlePage.htm

http://www.rand.org/pubs/monograph_reports/MR1018.5/index.html

The Stockholm International Peace Research Institute, Biological and Chemical Weapon Research:

http://www.sipri.org/contents/cbwarfare

http://www.brad.ac.uk/acad/ntw/research_reports/docs/BDRC_ST_Report_No_8.pdf (Last accessed Oct, 18, 2007).

6
Decontamination of Biological Warfare Agents

Bärbel Niederwöhrmeier and André Richardt

6.1
Introduction

Decontamination of biological agents is different to decontamination of chemical agents. Normally, if spores like anthrax endospores can be killed by a method, it is likely that a sufficient killing rate can also be achieved for the other biological agents. Therefore, this chapter will focus on decontamination of endospores. However, the key questions especially for biological agents are, "is clean clean enough?" or "how clean is clean enough?" Is it possible to determine what level of decontamination is necessary to meet regulatory needs in the case of a biological incident in a civilian building, or is it sufficient enough to give an estimation? Therefore, in this chapter a short overview of the special requirements for an effective decontamination or inactivation of endospores and building decontamination methods, especially for endospores as the main threat, is given.

6.2
Resistance of Spores Against Stress and Basics of Inactivation

In the case of a biological incident and the need for the best decontamination procedure, the decontamination strategy can be by chosen considering three factors (a) surface decontamination, (b) area (space) decontamination and (c) the potential resistance of the agent against the chosen decontaminant [1–3]. From this point of view, it is not only necessary to have mature decontamination technology, which can be used for decontamination, but also a comprehensive understanding of the biological agent itself. An understanding of the different points of attack for the decontaminant is also necessary, especially for endospores, as the most resistant biological agents [2]. If endospores can be destroyed effectively, the other biological agents can also be inactivated easily, with the exception of biological toxins. For this reason endospores have been the main target in the research for effective decontamination methods for biological agents in the last decades.

Decontamination of Warefare Agents.
Edited by André Richardt and Marc-Michael Blum

ISBN: 978-3-527-31756-1

The ability to form endospores (further written as spores) is widely distributed among bacteria, but the largest group of spore forming bacteria is the family of Bacillaceae [4]. Spores are resistant stages in the life cycle of several gram-positive bacteria. In electron microscopy they show a thin outer spore coat, a thick spore coat, a thick spore cortex and an inner spore membrane surrounding the spore contents (Figure 6.1). They are formed inside the vegetative cell in response to nutritional debrivation and enable the organism to survive without metabolizing. The spores are more resistant than the parental, vegetative cells to the lethal effects of heat, drying, freezing, radiation or toxic chemicals.

Bacterial spores can survive in the environment for a very long time and they resist physical and chemical stress. Several methods have already been tested for the inactivation of *Bacillus* spores: for example, heat, acid pHs, ultrasonic waves, irradiation, high pressure, formaldehyde, hypochlorite solutions [6]. Although the methods regarding spore production, inoculum size, concentration and other variables are not consistent between the experiments, each experiment provides some specific information of value. So variations in time, temperature, concentration, pH and relative humidity may affect the sporicidal activity of various agents [6].

Earlier experiments show that an association of heat with *pressure* is very efficient for the inactivation of *Bacillus* species spores [7]. For *Bacillus anthracis* Sterne spores this effect was confirmed by Cléry-Barraud *et al.* [8]. The combination of heat and pressure resulted in complete destruction of *Bacillus anthracis* spores with a D-value of approximately 4 minutes after pressurization at 500 MPa and 74 °C.

The thermal resistance of *Bacillus anthracis* and potential surrogates was determined by Montville *et al.* [9]. This study demonstrated, that *Bacillus anthracis* spores are usually heat resistant and that spores from validated *Bacillus species* are appropriate surrogates for thermal resistance studies. Spores of various *Bacillus species* are extremely resistant to heat [10]. Therefore, the heat resistance of native and demineralized spores of *Bacillus subtilis* sporulated was investigated at different temperatures [11]. The following data about the heat resistance of *Bacillus anthracis* were extracted from the National Food Processors Association bulletin "Anthrax, what it is and addressing inquiries." The results indicated that sporulation temperature increases heat resistance by increasing the mineralization of the spores.

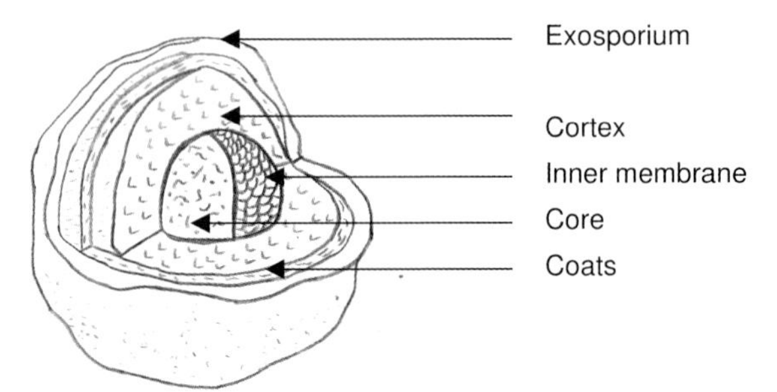

Figure 6.1 Schematic representation of the internal structure of a bacterial spore–reproduced from [5].

The vegetative cells of *Bacillus anthracis* are less resistant than the spores:

D-value is the needed time to achieve 90% mortality of a standard spore indicator and is a measure of the killing time.

Moist heat:

- vegetative cells survived 5 but not 5.5 minutes at 65 °C and 50 but not 60 seconds at 80 °C.
- with spores the D-value of 2.5–7.5 minutes at 90 °C was determined.

Dry heat:

- vegetative cells are inactivated by 2 hours at 92–100 °C
- spore suspension on glass: death time 60 minutes at 120 °C.

The resistance of *Bacillus subtilis* spores to dry heat depends of the water content in the spores. The resistance increases with increasing water content [12].

Blatchley *et al.* [13] conducted experiments to examine the effectiveness of ultraviolet radiation (UV_{254}) and gamma-radiation for the inactivation of *Bacillus* spores. Most of the experiments were done using *Bacillus cereus* spores, a limited number using *Bacillus anthracis* Sterne spores. They observed that spores dried on surfaces were more resistant to UV_{254} than the same spores in aqueous suspension. Gamma-radiation was shown to accomplish similar rates of inactivation for spores in aqueous suspension and for spores dried on surfaces. With the conditions used in these experiments, the spores of *Bacillus anthracis* and *Bacillus cereus* were more sensitive to UV_{254} radiation in aqueous suspension than the spores of *Bacillus subtilis*. Similar results were described by Nicholson and Galeano [14]. They show that standard UV treatments that are effective against *Bacillus subtilis* spores are likely also sufficient to inactivate *Bacillus anthracis* spores. A further study was performed to determine the effectiveness of gamma-radiation on *Bacillus anthracis* spores suspended in water using a Cobalt 60 (Co. 60) source and single and fractionary irradiation doses. Using *Bacillus anthracis* Sterne in a concentration of 1.5×10^9 spores in 1 ml, the sporicidal doses gamma-radiation amount to 25.0 kGy (single dose) or 41.5 kGy (fractionary dose) [15]. Todoriki *et al.* [16] demonstrated the modification of radiation resistance of *Bacillus* spores by water. The resistance of spores to radiation increased with increasing moisture content.

The kinetics or the inactivation of *Bacillus subtilis* spores by ozone alone and ozone with UV irradiation were investigated by Urakami *et al.* [17]. The spore inactivation by ozone combined with UV irradiation was more effective than by ozone alone in the presence of organic compounds.

These examples show clearly that an effective inactivation of spores is possible and a sufficient cleaning of a building should be possible.

6.3 Technical Methods for Decontamination of Spore Infested Buildings

After the events of September and October 2001 with anthrax letters, there was an increased concern due to the possibility that terrorists could deliver toxic chemicals or biological agents into a building. A fast reaction would be required to protect building occupants and to decontaminate the building for re-opening. Some of the methods described for C-decontamination could also be used for B-decontamination. However, the focus in this section is to provide an overview of the decontamination of buildings by gases. An overview of the efficiency of chemicals, gases and radiation on the inactivation of spores was listed by Spotts-Whitney *et al.* [6]. This overview provides a basis for further research regarding improved methods for remediation of environments contaminated with *Bacillus anthracis*. Most of the existing data concerns laboratory or food industry settings and is mostly based on other *Bacillus* species.

6.3.1 Ethylene Oxide

Ethylene oxide is an odourless gas at room temperature and can be used for several applications. It is known as a useful decontaminant and is widely used in hospital and biomedical sterilization applications due to the fact that it is highly penetratative of different surfaces. During anthrax remediation efforts on Capitol Hill and in other federal mail facilities, as well as at the National Broadcasting System (NBC) offices in New York City, ethylene oxide was used as an effective disinfectant. For re-use of critical items off-site ethylene oxide chambers were used for successful sterilization. However, ethylene oxide is a highly reactive molecule with flammable and explosive vapours. Vapours with little as 3% ethylene oxide in air can be flammable and explosive [18]. Further, due to the fact that ethylene oxide is highly reactive, with unprotected exposure there is evidence that (a) acute exposure to ethylene oxide can cause nausea, vomiting and death and (b) chronic exposure is responsible for irritation of skin, eyes, mucous membranes, cataracts, and problems in brain function [19]. Also, there is increasing evidence that exposure to ethylene oxide could lead to lung, liver, and kidney damage [20]. At last, ethylene oxide is rated as a Group B1 (probable) human carcinogen. Therefore, to reduce the release of ethylene oxide from the critical items treated during the remediation efforts, an additional heating step can be used to decrease the risk. However, the risk for human health and the flammability of ethylene oxide and therefore, the limitation of its use to carefully controlled chambers, leads to the conclusion that this technology, in comparison to other technologies, has no applicability for the fumigation of buildings. The only possible use could be the sterilization of critical items under proven conditions [19].

6.3.2 Chlorine Dioxide Gas

Chlorine dioxide (ClO_2) reacts as an oxidizing agent. There is evidence that the main target of its oxidizing action is the protein of a bacteria or virus. Although the exact mechanism of this process is not yet fully characterized, the oxidation process of the proteins should lead to functional disruption. A lot of data show clearly the effect of ClO_2 as a biological sterilizer.

6.3.2.1 Description of the Technology

It is important to know that chlorine dioxide gas is generated at the decontamination side and is injected into the sealed building areas. After the injection the gas remains in place for the required period of time, normally 12–24 hours. After treatment it is necessary to neutralize the highly toxic chlorine dioxide. One method for the neutralization is the circulation of the building air through a neutralization solution like sodium sulfite/bisulfite. To generate chlorine dioxide gas, different methods can be used (Table 6.1).

To produce ClO_2 gas, one of the two parent compounds, chlorite (ClO_2^-) or chlorate (ClO_3^-), have to be used. However, the manufacturer has to keep in mind two basic concerns when he wants to choose and to determine a synthesis method for chlorine dioxide:

1. The amount of gas needed
2. The safety concerns of the precursors and by-products [19].

For these reasons, the electrochemical generation of chlorine dioxide from a ClO_2^- salt seems the safest method for the on-site production of chloride dioxide gas and shows the fewest safety concerns. However, for decontamination purposes a reactive gas is required and is therefore dangerous not only to the target like spores, but also to human beings. Therefore, it is necessary to have not only the

Table 6.1 Methods for generation of chlorine dioxide gas, adapted from [19].

ClO_2 from Sodium Chlorite	***ClO_2 from Sodium Chlorate***
Electrochemical oxidation of chlorite $ClO_2^- \rightarrow ClO_2 + e^-$	Reduction of chlorates by sulfur dioxide (Mathieson Process) $2\ NaClO_3 + H_2SO_4 + SO_2 \rightarrow 2\ ClO_2 + 2\ NaHSO_4$
Reaction of sodium hypochlorite and sodium chlorite (Sabre Technologies) $NaOCl + 2\ NaClO_2 + 2\ HCl \rightarrow 2\ ClO_2 + 3\ NaCl + H_2O$	Reduction of chlorates by acidification in the presence of oxalic acid $2\ HClO_3 + H_2C_2O_4 \rightarrow 2\ ClO_2 + 2\ CO_2 + 2\ H_2O$
Acidification of chlorite $5\ ClO_2^- + 4\ H^+ \rightarrow 4\ ClO_2 + 2\ H_2O + Cl^-$	Action of sulfuric acid on chlorates $NaClO_3 + NaCl + H_2SO_4 \rightarrow ClO_2 + \frac{1}{2}\ Cl_2 + Na_2SO_4 + H_2O$
Oxidation of chlorite by persulfate $2\ NaClO_2 + Na_2S_2O_8 \rightarrow 2\ ClO_2 + 2\ Na_2SO_4$	$NaClO_3 + 2HCl \rightarrow ClO_2 + \frac{1}{2}\ Cl_2 + NaCl + H_2O$
Oxidation of chlorite by chlorine $2\ NaClO_2 + Cl_2 \rightarrow 2\ NaCl + 2\ ClO_2$	$3\ NaClO_3 + 2\ H_2SO_4 + 0.85\ CH_3OH \rightarrow 3\ ClO_2 + Na_3H(SO_4)_2 + H_2O + 0.05\ CH_3OH + 0.6\ CHOOH + 0.2\ CO_2$
Action of acetic anhydride on chlorite $4\ NaClO_2 + (CH_3CO)_2O \rightarrow 2\ ClO_2 + NaClO_3 + NaCl + 2\ CH_3CO_2Na$	$NaClO_3 + \frac{1}{2}\ H_2O_2 + H_2SO_4 \rightarrow ClO_2 + NaHSO_4 + H_2O + \frac{1}{2}\ O_2$

Table 6.2 Advantages and disadvantages for the use of chloride dioxide gas, adapted from [19].

Advantages	*Disadvantages*
A well documented disinfectant for spores, vegetative bacteria and viruses.	The gas is unstable. It must be constantly produced to have the target concentration for the required time.
Rapid natural breakdown of ClO_2 eases its removal after application.	The gas must be generated on-site. The equipment for the generation is significant.
The gas is very soluble and stable in water.	The killing efficiency decreases significantly at relative humidity levels below 70%. Therefore, maintenance of humidity is critical.
The gas is very effective on porous and non-porous surfaces. It reaches all regions within an enclosure except for the hardest to reach, isolated areas (e.g. closed employee lockers).	A large volume of liquid waste materials is generated.
The gas is commercially generated by several methods.	Resulting from condensation collateral damage could happen to the surfaces of machinery and electrical systems.
The gas odor can be detected by humans at very low concentrations (0.1 ppm).	

advantages, but also the disadvantages of gaseous bioremediation in mind (Table 6.2).

6.3.2.2 Applications for Chlorine Dioxide

In the early 1800s Sir Humphrey Davy was the first to report the reaction of sulfuric acid with potassium chlorate [21]. In a short period of time, different potential applications for chlorine dioxide was discovered (Table 6.3). The vast majority of applications of chlorine dioxide utilize the gas dissolved in water.

The gaseous form of chlorine dioxide is the form most frequently employed as a fumigant. Chlorine dioxide gas was registered by the EPA as an antimicrobial sterilant in the 1980s. The application of gaseous chlorine dioxide for building contamination was demonstrated in the remediation of the Hart Senate Office Building (HSOB), the Brentwood Mail Processing and Distribution Center (P&DC), and the Trenton P&DC following the 2001 release of anthrax spores through the US mail.

6.3.2.3 Fumigation of Spore Infested Buildings

Before the 2001 anthrax events, there was no experience for the use of chlorine dioxide gas for the decontamination of spore infested buildings. Therefore, exemptions were given by the United States Environmental Protection Agency (EPA) for

Table 6.3 Applications for chlorine dioxide gas.

Application	*References*
Water odor	[22]
Water purification	[23–27]
Paper processing industry	[28]
Fruit and vegetable processing industry	[29–32]
Dairy, poultry and beef industries	[33–37]
Industrial waste processing facilities	[38, 39]
Antimicrobial sterilant	[19, 31, 32, 40]

different sites that were disinfected with ClO_2 gas [19]. The conditions and procedure for ClO_2 fumigation of a building and the treatment for anthrax are specified on EPA's web site [19]. The following initial activities have to taken into action to avoid further spread of the spores from the infested building:

1. Sealing of the HSOB to prevent further spread of the spores.
2. Environmental sampling to define the extent of the contamination.

After a further spread of the spores is prevented, the gaseous sterilant (chlorine dioxide gas) chosen as the central component of the remediation strategy is released into the sealed space. The effectiveness of the fumigation can be determined in two ways:

1. Spore strips containing surrogates for the anthrax (*Bacillus anthracis*) spores to verify a sufficient decontamination concentration of ClO_2
2. Environmental sampling to verify whether any surviving *B. anthracis* spores remain at the sampling location. Environmental sampling methods included:
   a. surface sampling, using wet wipe and vacuum techniques
   b. aggressive air sampling, that is, high-volume sampling of the site air after room surfaces had been agitated (blown) in an effort to resuspend any spores that might have been present.

This solution was chosen to decontaminate Senator Daschle's suite in the Hart Senate Office Building (HSOB). The result shows clearly that decontamination or fumigation of spore infested buildings needs not only manpower, but also that it is a time consuming process and that a significant amount of equipment and a high standard level of the required technology is needed.

6.3.3 Formaldehyde Solution or Gas

Paraformaldehyde is a polymerized form of formaldehyde $(CH_2O)_n$ and is a stable white crystalline powder. Because pure formaldehyde is unstable at ambient conditions, paraformaldehyde is used as a readily-usable form of formaldehyde at use sites. Then, formaldehyde gas is generated during heating.

Formaldehyde solution or gas has been used as a disinfectant and chemical sterilant for a long period and has been used for the disinfection of safety cabinets and laboratories [41, 42]. The high influence of temperature and relative humidity on the formaldehyde concentration to obtain a high reduction of spores count was demonstrated by several investigations [43]. But the growing evidence for carcinogenic properties of formaldehyde connected with a lot of regulations limited its use, although it can be neutralized with ammonia after fumigation. The resistance of spores to formaldehyde was also investigated by Spicher and Peters (1981) [44]. They determined the dependence of the effect of formaldehyde upon the duration of exposure. Compared with *Bacillus subtilis, Bacillus stearothermophilus* showed a higher resistance. The antimicrobial properties of formaldehyde are believed to result from its reactivity in the alkylation of proteins, nucleic acids, and DNA and RNA [45]. Both paraformaldehyde and formaldehyde have been used in decontamination for more than 30 years.

6.3.3.1 Description of the Technology

As mentioned before, paraformaldehyde is used to generate formaldehyde gas for decontamination. A typical procedure for the use of paraformaldeyde is

1. Isolation of the material or area being sterilized.
2. The use of hot plates to sublimate the paraformaldehyde.
3. Use of fans to distribute the vapor within the space for a specified time period.
4. Introduction of a compound (e.g. ammonia) that will neutralize the formaldehyde vapor once treatment is complete [42].

One technology is the use of formaldehyde generator, where the heating of the paraformaldehyde can take place inside a closed system and is generated at the decontamination side. Finally, the resulting formaldehyde is injected into the sealed room to be treated [46, 47].

6.3.3.2 Technical Maturity

Vaporization of paraformaldehyde is fully mature and commercialized. This technology can be routinely used for sanitizing and disinfecting rooms and equipment in the health services industry and in biological laboratories [48, 49]. Also, paraformaldehyde technology has been applied to rooms and small spaces (such as laboratories and safety cabinets), and to individual equipment items enclosed in chambers or tents.

Table 6.4 Application of formaldehyde by different users, adapted from [19].

User	*Application*
Resin manufacturers	Adhesives and binders in consumer and industrial applications
Household and domestic dwellings	Disinfectant, sanitizer, fungicide, and microbicide
Barber and beauty shop equipment	Sanitizer and fungicide
Dentistry	Fixative
in ships and ship holds	Disinfectant, sanitizer, fungicide, and microbicide
non-food/non-feed-transporting trucks	Disinfectant, sanitizer, fungicide, and microbicide
on bedding and clothing	Disinfectant, sanitizer, fungicide, and microbicide

6.3.3.3 **Applications of the Technology**

Paraformaldehyde is used for different applications as a source of either gaseous formaldehyde or solution formaldehyde (Table 6.4).

It is unclear which of these applications use aqueous formaldehyde and which use gaseous formaldehyde.

6.3.3.4 **Concerns for the User**

From a practical view, concerns for paraformaldehyde are similar to those for formaldehyde. This is because paraformaldehyde will generate gaseous formaldehyde during storage or use. Formaldehyde has been identified as a probable human carcinogen. However, there is growing evidence of carcinogenicity in humans through inhalation exposure, and sufficient evidence for carcinogenicity in experimental animals [2, 19]. One great disadvantage is the out-gassing of formaldehyde from porous surfaces over a long period of time and therefore can act as a possible carcinogen especially in closed rooms and buildings. The accepted exposure limits are several orders of magnitude below the concentrations used for decontamination as discussed. Therefore, after decontamination, complete neutralization of formaldehyde in the air as well as the removal of any chemical which may have condensed onto surfaces is necessary. However, the out-gassing over weeks and months from different porous surfaces is still a problem and is difficult to avoid. Furthermore, formaldehyde is a flammable, colourless gas (lower explosive limit of 7%) with a pungent odor, all of which are additional concerns during use [19].

6.3.4
Vapourized Hydrogen-Peroxide (VHP)

The use of aqueous hydrogen peroxide as a decontaminant has a long history [50, 51]. Vapourized Hydrogen-peroxide *(VHP)* is a safe broad spectrum and material compatible disinfectant which is used for fumigation and sterilization processes [52]. It is a dry process and is used at a much lower concentration than alternative oxidizing agents based on liquids. VHP has been widely used for the disinfection of enclosed areas and spaces including isolators, laminar flow cabinets, clean rooms and biological safety cabinets. The antimicrobial efficiency of VHP against a wide range of pathogens and environmental micro-organisms has been demonstrated [53–57].

The results show clearly that *Bacillus* strains, small viruses without envelope and mycobacteria are more resistant against VHP and therefore, longer decontamination time is needed as compared with, for example, viruses with lipid-envelope (Figure 6.2).

6.3.4.1 Description of the Technology

A number of hydrogen peroxide vapour generation systems are commercial available for small-scale chamber sterilization. Several of these have been adapted for potential use in the fumigation of larger volumes, applicable to buildings. In all cases, the hydrogen peroxide vapour is generated from a concentrated aqueous solution of hydrogen peroxide (35% H_2O_2) (Figure 6.3).

The vapour can be generated by controlled heating of the liquid, in a manner that reduces decomposition of the H_2O_2 and other methods as heated aerosolizers. The peroxide decays with time and it is thus necessary to continuously supply fresh peroxide into the space at a rate sufficient to maintain the desired concentration. Typical H_2O_2 vapour concentrations are about 0.3 mg/L and require a minimum of 2 to 6 hours of contact time to destroy anthrax spores [61]. After the fumigation process, the H_2O_2 generator is turned off, and hydrogen peroxide vapour is withdrawn from the space and passed over a catalyst to convert it into water and oxygen, thus leaving no toxic residue [62] (Figure 6.3).

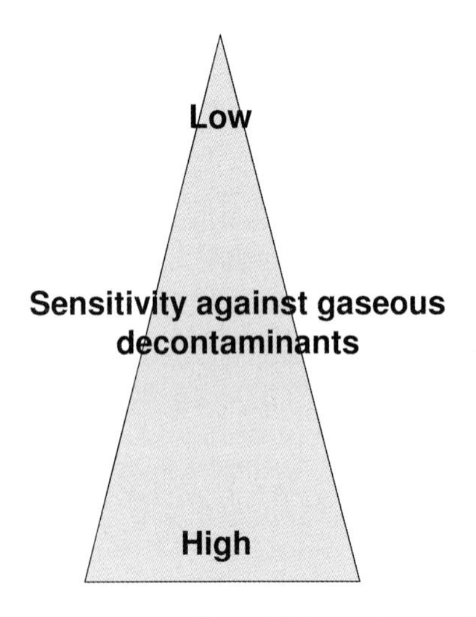

bacterial spores	e.g.: B. anthracis, B. subtilis
small viruses	e.g: Parvoviridae, Picornaviridae
mycobacteria	e.g: Mycobacterium tuberculosis
mould fungus	e.g: Aspergillus niger
gramnegative bacteria	e.g: Burkholderia cepacia, Proteus vulgaris
Large viruses	e.g: Adenovirus, Poxviridae
grampositive bacteria	e.g: Legionella pneumophila Lactobacillus casei
Viruses with envelope	e.g: Herpesviridae, Orthomyxoviridae

Figure 6.2 Increasing sensitivity against gaseous decontaminants [1, 3, 58–60].

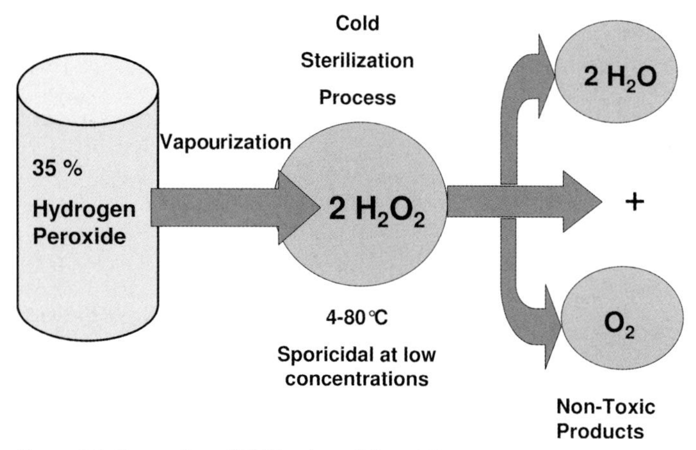

Figure 6.3 Generation of VHP, adapted from [60].

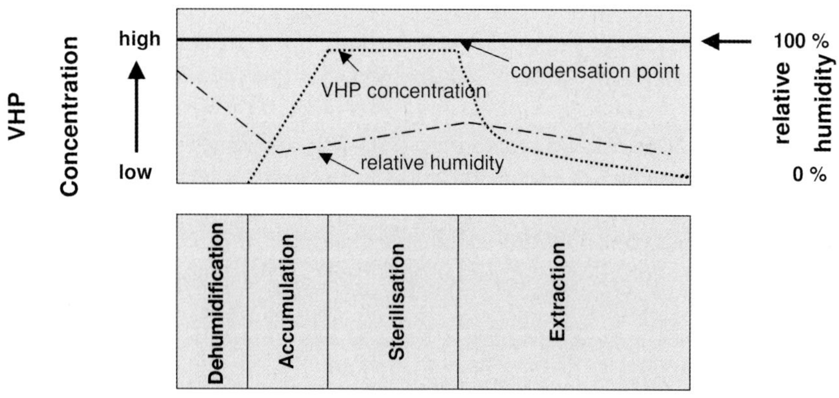

Figure 6.4 STERIS sterilization process for decontamination of spaces with VHP [59].

Relative humidity (RH) is an important parameter in determining the performance of hydrogen peroxide vapour, although the optimal RH level varies with the specific H_2O_2 process. The STERIS process, discussed below, maintains a low humidity in the space (below 40% RH at the start of fumigation), in an effort to keep the peroxide in the vapour phase for improved penetration of substrate surfaces. By comparison, the BIOQUELL process permits higher RH values, attempting to achieve "micro-condensation" of a thin film of peroxide over the surface to be decontaminated (Figure 6.4).

There are three main advantages to this technology application:

1. Effective against viruses, bacteria and spores.
2. Technology is available for implementation and additional research.

Table 6.5 Hydrogen peroxide application, adapted from [19].

User	*Application*
Home use	Disinfecting minor cuts and scrapes (minor concentrations) hair bleaching treatments (more concentrated)
Indoor use on surfaces in food establishments, medical facilities, and home bathrooms	Disinfecting of different surfaces and destruction of anthrax
Enclosed areas in commercial, institutional, and industrial settings	Hydrogen peroxide in the vapour form is registered as a pesticide
Rooms and biologic safety cabinets	Space decontamination with vapour
Industrial settings	Sterilization of sealed enclosures including scientific workstations, isolators, pass-through rooms, and medical and diagnostic devices
Pharmaceutical industry	Plasma sterilizers with VHP

3. The end products, after catalytic breakdown, are harmless (water and oxygen).

However, the user must keep in mind one great disadvantage of this technology. The high reactivity vapour can break down upon contact with certain materials with a high degree of surface area. This could be galvanized steel or a porous surface such as paper, and could catalyze the conversion of the active gas to water and oxygen [19]. For hydrogen peroxide as a strong oxidizing agent a wide variety of applications exists (Table 6.5).

As a summary, regarding different decontamination technologies, especially for spores, a sufficient cleanup of infrastructure is achievable from the point of view of decontamination itself. However, the key question "How Clean is Clean enough?" has also to be answered, and it is essential to take the right measurements.

6.4
Is Clean Clean Enough or How Clean Is Clean Enough?

To determine the required decontamination level, or cleanup, it is essential to address (a) public perception of risk to health, (b) political support, (c) public acceptance of recommendations based on scientific results and criteria and (d) economic concerns [63, 64]. As described and defined in the chapter "Biological Warfare Agents," incubation period, infectious dose and fatality rate for biological warfare agents differ [65, 66]. Therefore, from the scientific point of view also different threshold limit values for biological warfare agents should also exist. However, the public and politicians normally insist on zero living organisms after decontamination of an incident with biological warfare agents, especially in public buildings. Even, when they know that for example *Bacillus anthracis* is indigenous in different locations without any incident [63, 64, 67].

According to the fact that, for example, natural inhalation of anthrax is extremely rare, zero concentration of biological agents is in many cases not a necessity and sometimes also not possible. Another point that is important to understand is the need for an overarching concept to answer the questions "Is Clean Clean enough?" or "How Clean is Clean Enough?" The key methods to verify an effective and successful decontamination are sampling, identification and verification. If poor methods for validation are used, even decontamination methods with inadequate destroying rate of the biological agent could present the impression of a safe cleanup [63].

However, at this point, it is enough for the reader to understand that sufficient methods for decontamination of biological agents with advantages and disadvantages are available. But he should keep in mind, that the decontamination process is only one part in an efficient protection strategy against biological and chemical warfare agents. This issue will be more discussed in the Chapter 14 "Road Ahead."

References

1 Favero, M.S. and Bond, W.W. (1991) Sterilization and antisepsis in the hospital, in *Manual of Clinical Microbiology*, 5th edn (eds A. Balows, W.J. Hausler, K.L. Herrmann, H.D. Isenberg and H.J. Shadomy), American Society for Microbiology, Washington, pp. 183–200.

2 Hawley, R.J. and Eitzen, E.M. (2001) Biological weapons – a primer for microbiologists. *Annual Review of Microbiology*, **55**, 235–53.

3 Maillard, J.Y. (2002) Bacterial sides for biocide action. *Journal of Applied Microbiology Symposium Supplement*, **92**, 16–27.

4 Buchanan, R.E. and Gibbons, N.E. (1974) *Bergey's Manual of Determinative Bacteriology*, Williams and Wilkins, Baltimore.

5 Foster, S.J. and Johnstone, K. (1990) Pulling the trigger: the mechanism of bacterial spore germination. *Molecular Microbiology*, **4** (*1*), 137–41.

6 Spotts-Whitney, E.A., Beatty, M.E., Taylor, T.H., Weyant, R., Sobel, J., Arduino, M.J. and Ashford, D.A. (2003) Inactivation of *Bacillus anthracis* spores. *Emerging Infectious Diseases*, **9**, 623–7.

7 Raso, J., Palop, A., Pagan, R. and Condon, S. (1998) Inactivation of *Bacillus subtilis* spores by combining ultrasonic waves under pressure and mild heat treatment. *Journal of Applied Microbiology*, **85**, 849–54.

8 Cléry-Barraud, C., Gaubert, A., Masson, P. and Vidal, D. (2004) Combined effects of high pressure and temperature for inactivation of *Bacillus anthracis* spores. *Applied and Environmental Microbiology*, **70**, 635–7.

9 Montville, T.J., Dengrove, R., De Swiano, T., Bonnet, M. and Schaffner, D.W. (2005) Thermal resistance of spores from virulent strains of *Bacillus anthracis* and potential surrogates. *Journal of Food Protection*, **68**, 2362–6.

10 Gerhardt, P. and Marquis, R.E. (1989) Spore thermoresistance mechanisms, in *Regulation of Prokaryotic Development* (eds I. Smith, R. Slepecky and P. Setlow), American Society for Microbiology, Washington, DC, pp. 17–63.

11 Palop, A., Sala, F.J. and Condón, S. (1999) Heat resistance of native and deminerlized

spores of *Bacillus subtilis* sporulated at different temperatures. *Applied and Environmental Microbiology*, **65**, 1316–9.

12 Drummond, D.W. and Pflug, I.J. (1970) Dry-Heat Destruction of Bacillus subtilis spores on surfaces: effect of humidity in an open system. *Applied and Environmental Microbiology*, **20**, 805–9.

13 Blatchley, E.R., Meeusen, A., Aronson, A.I. and Brewster, L. (2005) Inactivation of Bacillus spores by Ultraviolet or Gamma Radiation. *Journal of Environmental Engineering*, **131**, 1245–52.

14 Nicholson, W.L. and Galeano, B. (2003) UV resistance of Bacillus anthracis spores recisited: validation of *Bacillus subtilis* spores as UV surrogates for spores of *B. anthracis* Sterne. *Applied and Environmental Microbiology*, **69**, 1327–30.

15 Mizak, L. and Mierzejewski, J. (1993) Gamma radiation of *Bacillus anthracis* spores. *Medycyna Doswiadczalna I Mikrobiologia*, **55**, 315–23.

16 Todoriki, S., Furuta, M., Nagai, T. and Hayashi, T. (2000) Modification of radiation resistance of *Bacillus* spores by water. *Radiation Physics and Chemistry*, **57**, 531–4.

17 Urakami, I., Mochizuki, H., Inaba, T., Hayashi, T., Ishizaki, K. and Shinriki, N. (1997) Effective inactivation of *Bacillus subtilis* spores by a combination treatment of ozone and UV irradiation in the presence of organic compounds. *Biocontrol Science*, **2**, 99–103.

18 US Department of Health and Human Services (2005) NIOSH Pocket Guide to Chemical Hazards, Washington, DC, DHHS (NIOSK) publication No. 2005-147, http://www.cdc.gov/niosh/npg/pdfs/2005-149.pdf (September 2005).

19 Office of Research and development, National Homeland Security Research Center (2005) EPA, Building Decontamination Alternatives, EPA/600/R-05/036, http://www.epa.gov/nhsrc/pubs/reportBuildDecon052705.pdf (March 2005).

20 US Department of Health and Human Services (1993) *Hazardous Substances Data Base, National Toxicology Program,* National Library of Medicine, Bethesda, MD.

21 Davy, H. (1811) On a Combination of Oxymuratic Gas and Oxygen Gas, *Philosophical Transactions*, **101**, 155.

22 McCarthy, J.A. (1945) Chlorine dioxide for the treatment of water supplies. *Journal of New England Water Works Association*, **59**, 252.

23 Synan, J.F., MacMahon, J.D. and Vincent, G.P. (1975) Chlorine dioxide, a development in the treatment of potable water. *Journal of American Water Works Association*, **91**, 566.

24 Bernard, M.A., Snow, W.B. and Olivieri, V.P. (1976) Chlorine dioxide disinfection temperature effects, *The Journal of Applied Bacteriology*, **30**, 159

25 Bernard, M.A., Snow, W.B., Olivieri, V.P. and Davidson, B. (1976) Kinetics and mechanism of bacterial disinfection by chlorine dioxide. *Applied Microbiology*, **15**, 257.

26 Ridenour, G.M. and Armbruster, E.H. (1949) Bactericidal effects of chlorine dioxide. *Journal – American Water Works Association*, **41**, 537.

27 Aieta, E.M. and Berg, J.D. (1996) A review of chlorine dioxide in drinking water treatment. *Journal–American Water Works Association*, **78**, 62.

28 Balcer, W.L., Dawson, L.E. and Lechowich, R.V. (1973) Influence of chlorine dioxide water treatment on numbers of bacteria associated with processed Turkey. *Poultry Science*, **52**, 1053.

29 Anon (1977) Chlorine dioxide gains favor as effective sanitizer. *Journal of Food Engineering*, March, 143.

30 Costilow, R.N., Uebersax, M.A. and Ward, P.J. (1984) Use of chlorine dioxide for controlling microorganisms during the handling and storage of fresh cucumbers. *Journal of Food Science*, **49**, 296.

31 Du, J., Han, Y. and Linton, R.H. (2002) Inactivation by chlorine dioxide gas (ClO_2) of *Listeria monocytogenes* spotted onto different apple surfaces. *Food Microbiology*, **19**, 481–90(10) ((AA)).

32 Du, J., Han, Y. and Linton, R.H. (2002) Chlorine Dioxide Gas Kills Dangerous Biological Decontaminants, Department of Food Schience, Purdue University, http://news.uns.purdue.edu/UNS/html4ever/020912.Linton.chlorinediox.html (Last accessed Nov, 1, 2007.).

33 Oliver, S.P., King, S.H., Torre, P.M. and Sordillo, L.M. (1989) Prevention of bovine mastitis by a postmilking teat disinfectant containing chlorous acid and chlorine dioxide in a soluble polymer gel. *Journal of Dairy Science*, **72**, 3091.

34 Baran, W.L. (1973) Dawson LE and lechowich RV, influence of CHLORINE dioxide water treatment on numbers of bacteria associated with processed Turkey. *Poultry Science*, **52**, 1053.

35 Lillard, H.S. (1979) Levels of chlorine and chlorine dioxide of equivalent bactericidal effects in poultry processing water. *Journal of Food Science*, **44**, 1594.

36 Thiessen, G.P., Usborne, W.R. and Orr, H.L. (1984) The efficacy of chlorine dioxide in controlling salmonella contamination and its effect on product quality of chicken broiler carcasses. *Poultry Science*, **63**, 647.

37 Emsweiler, B.S., Kotula, A.W. and Rough, D.K. (1976) Baterial effectiveness of three chlorine soxurces used in beef carcass washing. *Journal of Animal Science*, **42**, 1445.

38 Rauh, J.S. (1979) Disinfection and oxidation of wastes by chlorine dioxide. *Journal of Environmental Sciences*, **22**, 42.

39 Freymark, S.G. and Rauh, J.S. (1978) Selective oxidation of industrial wastewater contamination by chlorine dioxide. *Proceedings of the Mid-Atlantic Industrial Waste Conference*, **10**, 120.

40 Applegate, B., Han, Y., Linton, R.H. and Nelson, P.E. (2003) Decontamination of bacillus thuringiensis spores on selected surfaces by chlorine dioxide gas. *Journal of Environmental Health*, **66**, 16–20.

41 Everall, P.H. (1982) Problems in the disinfection of class I microbiology safety cabinets. *Journal of Clinical Pathology*, **35**, 245–63.

42 Munro, K. (1999) A comparative study of methods to validate formaldehyde decontamination of biological safety cabinets. *Applied and Environmental Microbiology*, **65**, 873–6.

43 Cross, G.L.C. and Lach, V. (1990) The effects of controlled exposure to formaldehyde vapour on spores of Bacillus globigii NCTC 10073. *The Journal of Applied Bacteriology*, **68**, 461–9.

44 Spicher, G. and Peters, J. (1981) Resistenz mikrobieller Keime gegenüber Formaldehyd. I. Abhängigkeit des mikrobiziden Effektes von der Konzentration und der Einwirkungsdauer des Formaldehyds. *Zentralblatt der Bakteriellen Hygiene, I. Abteilung Originale B*, **174**, 133–50.

45 Wickramanayake, G.B. (1990) Decontamination technologies for release from bioprocessing facilities: Part VI. Verification of wastewater decontamination. *CRC Critical Reviews in Environmental Control CCECAU*, **19** (6), 539–55.

46 Grantham, J.I. (1980) Apparatus for biological decontamination and subsequent neutralization of a space, US Patent, Patent number: 4241020.

47 Rogers, J.V., Choi, Y.W., Richter, W.R., Rudnicki, D.C., *et al.* (2007) Formaldehyde gas inactivation of *Bacillus anthracis, Bacillus subtilis*, and *Geobacillus stearothermophilus* spores on indoor surface materials. *Journal of Applied Microbiology*, **103**, 1104–1112. (Online Early Articles). doi: 10.1111/j.1365-2672.2007.03332.x.

48 Coldiron, V.R. and Janssen, H.E. (1984) Safe decontamination of hospital autopsy rooms and ventilation system by formaldehyde generation. *American Industrial Hygiene Association Journal*, **45**, 136–7.

49 Fink, R., Liberman, D.F., Murphy, K., lupo, D. and Israeli, E. (1988) Biological safety cabinets, decontamination or sterilization with paraformaldehyde. *American Industrial Hygiene Association Journal*, **49**, 277–9.

50 Schrodt, M. (1883) Ein neues Konservierungsmittel für Milch und Butter. *Milch-Zig*, **13**, 785.

51 von Bockelmann, B. and von Bockelmann, I. (1986) Aseptic packaging of liquid food products: a literature review. *Journal of Agricultural and Food Chemistry*, **34**, 384–92.

52 Rogers, J.V., Sabourin, C.L.K., Choi, Y.W., Richter, W.R., Rudnicki, D.C., Riggs, K.B., Taylor, M.L. and Chang, J. (2005) Decontamination assessment of Bacillus anthracis, Bacillus subtilis, and Geobacillus stearothermophilus spores on indoor surfaces using a hydrogen peroxide gas generator. *Journal of Applied Microbiology*, **99** (4), 739–48.

53 Heckert, R.A., Best, M., Jordan, L.T., Dulac, G.C., Eddington, D.L., Sterritt, W.G. (1997) Efficacy of vaporized hydrogen peroxide against exotic animal viruses. *Applied and Environmental Microbiology*, **63**, 3916–8.

54 Kokubo, M., Inoue, T. and Akers, J. (1998) Resistance of common environmental spores of the genus Bacillus to vapor hydrogen peroxide vapor. *PDA Journal of Pharmaceutical Science and Technology*, **52**, 228–31.

55 Hall, L., Otter, J.A., Chewins, J. and Wengenack, N.L. (2007) Use of Hydrogen Peroxide Vapor for Deactivation of *Mycobacterium tuberculosis* on a Biological Safety Cabinet and a Room, *Journal of Clinical Microbiology*, **45**, 810–815.

56 Kahnert, A., Seiler, P., Stein, M., Aze, B., McDonnell, G. and Kaufmann, S.H.E. (2005) Decontamination with vaporized hydrogen peroxide is effective against Mycobacterium tuberculosis. *Letters in Applied Microbiology*, **40**, 448–452.

57 Meszaros, J.E., Antloga, K., Justi, C., Plesnicher, C. and McDonnell, G. (2005) Area fumigation with hydrogen peroxide Vapor. *Applied Biosafety*, **10**, 91–100.

58 Russell, A.D., Furr, J.R. and Maillard, J.Y. (1997) Microbial susceptibility and resistance to biocides: an understanding. *ASM News*, **63**, 481–7.

59 STERIS (2003) Bio-Dekontamination mit VHP Monographie "Technische Daten" (Europa), 1–17.

60 Richardt, A. and Russmann, H. (2006) Biologische Kampfstoffe: Nachweis und Bekämpfung – Strategien gegen Milzbrand, Pocken & Co. *Biologie in Unserer Zeit*, **5**, 322–9.

61 Klapes, N.A. and Vesley, D. (1990) Vapor-phase hydrogen-peroxide as a surface decontaminant and sterilant. *Applied and Environmental Microbiology*, **56**, 503–6.

62 Lauderback, J, Fraser, J, Gustin, E, McDonnell, G and Williams, K. (2002) Point-of-Manufacture Sterilization Pharmaceutical and Medical Packaging News, http://www.devicelink.com/pmpn/archive/02/10/005.html (Last accessed Nov, 1, 2007.).

63 Raber, E., Jin, A., Noonan, K., McGuire, R. and Kirvel, R.D. (2001) Decontamination issues for chemical biological warfare agents: how clean is clean enough? *International Journal of Environmental Health Research*, **11**, 128–48.

64 Raber, E., Carlsen, T., Folks, K., Kirvel, R., Daniels, J. and Bogen, K. (2004) How clean is clean enough? Recent developments in response to threats posed by chemical and biological warfare agents. *International Journal of Environmental Health Research*, **14**, 31–41.

65 Benenson, A.S. (ed.) (1990) *Control of Communicable Disease in Man*, 15th edn. American Public Health Association.

66 Ingelsby, T., Henderson, D.A., Bartlett, J.G., Ascher, M.S., Eitzen, E., *et al.* (1999) Anthrax as a biological weapon: medical and public health management. *The Journal of the American Medical Association*, **281**, 1735–45.

67 Cox, C.S. and Wathes, C.M. (eds) (1995) *Bioaerosols Handbook*, Lewis Publishers, New York.

7 Decontamination of Chemical Warfare Agents

Hans-Jürgen Altmann and André Richardt

7.1 Introduction

To understand the ongoing need for research in the field of new decontaminants and to get a feeling for the urgent need for new mild decontamination solutions for special purposes, in this chapter a guide is given into the world of decontamination and some principles of decontamination are shown. Although the focus is the decontamination of chemical warfare agents, an introduction into basic decontamination solutions for biological warfare agents, especially for infrastructure contaminated with spores as the most difficult biological agent to destroy, is also given.

Should personnel or equipment be exposed to chemical or biological agents, proper decontamination will allow contaminated persons or property to return into service. Proper decontamination minimizes costs and makes reactions to a chemical or biological attack possible. Also, the probability of attacks with chemical and biological warfare agents changed after the end of the Cold War. Now, a new scenario is on the horizon. The possibility of terrorists attacks with biological and chemical warfare agents and a high death toll for civilians can no longer be denied. Therefore, decontamination is not old fashioned and obsolete and research in the area of decontamination is necessary to develop new generations of decontaminants and decontamination procedures which safely destroy contaminants on personnel, equipment, vehicles, sensitive equipment and protect the environment.

7.2 Definition of Decontamination

At the beginning of this chapter it is further necessary to give a definition of decontamination, to understand the total process of decontamination.

Decontamination of Warefare Agents.
Edited by André Richardt and Marc-Michael Blum

ISBN: 978-3-527-31756-1

Decontamination is a process of cleansing to remove contamination, or the possibility of contamination. It means the removal of NBC-agents by physical means or by chemical neutralization or detoxification, but does not necessarily mean the complete destruction of the contaminants. However, to achieve complete and safe detoxification in a short period of time is the ultimate aim of this process.

Decontamination can be roughly divided into two major parts:

- Decontamination of personnel
- Decontamination of material, infrastructure and equipment.

Persons and/or material suspected of being contaminated is usually separated from non-contaminated personnel or material and led into a decontamination site, where the contamination is removed by special treatment.

Personal decontamination is decontamination of personnel by employing their own decontamination capacities in form of personal decontamination kits, for example.

Casualty decontamination refers to the decontamination of casualties, and personnel decontamination usually refers to decontamination of non-casualties.

Material decontamination refers to

- The decontamination of equipment such as vehicles
- The decontamination of personal equipment for example protective masks suits and gloves
- The decontamination of infrastructure such as roads, buildings and harbors.

Decontamination as a general term can be broken down into three different parts:

- Decontamination of CW-agent
- Decontamination of BW-agents
- Decontamination of nuclear contamination.

In the following paragraphs, the only subject will be decontamination of CW-agents. However, some of the methods might also be useful for the decontamination of biological agents.

7.3 CW-Agent Decontamination

The procedures and substances presented in this paragraph are not an exhaustive evaluation of all decontamination chemicals, technologies and decontaminant formulations available, but will lead the reader to additional literature in order to find an optimized decontamination technology for a specific problem. The procedures and decontaminants available can be broken down into four different groups:

1. Water and water based decontaminants
2. Non aqueous decontaminants
3. Emulsions and foams
4. Vapor based decontaminants.

All these groups have advantages and disadvantages in decontamination efficiency, costs, material compatibility and in their influence on the environment. Therefore, knowledge of the different methods is essential when, in case of a contamination, decisions have to be made regarding which decontamination procedures can be used most effectively.

7.3.1
Water and Water Based Decontaminants

The most common and, in general, the most environmentally friendly decontaminant would be water and water based decontaminants with a high content of water.

7.3.1.1 Water

Water would be the ideal solvent to remove contaminants from a surface. It is inexpensive, not harmful to humans nor the environment and usually available in large amounts. The main disadvantage of water is its poor reactivity with most of the CW-agents and its very poor ability to dissolve these organic substances. CW-agents thickened with polymers make the problems even worse.

The ability of water to dissolve CW-agents and decompose them by hydrolysis can be improved by adding surface-active substances to the water. These "surfactants" improve the ability of water to remove CW-agents from surfaces and to dissolve them more rapidly, increasing the rate of hydrolysis.

But, in most of the cases, even with the improved ability of water-surfactant-mixtures to hydrolyze the hazardous material, the reaction velocity is not high enough to clean up a surface within acceptable time limits [1].

7.3.1.2 Water-Soluble Decontamination Chemicals

Due to the problems with water as decontaminant, so-called active chemistry is needed to improve the effectiveness of water-based systems and to speed up the decontamination reaction. Table 7.1 shows some common water soluble chemical products, which are able to speed up the reaction velocity of aqueous solutions and improve the ability of water to decontaminate surfaces.

The biggest disadvantage of water based decontaminants is their inability to penetrate paint and plastic layers and therefore they are not able to extract CW-agents dissolved in the surface layer or to dissolve thickened agents.

Using the decontaminants listed, one must be aware that in most of the applications residual agent will stay within the surface to be decontaminated. Dependent on surface material and temperature, the agent will desorb more or less rapid, so that the vapors coming out of the surface will pose a vapor hazard. To the user,

Table 7.1 Water solulable decont chemicals, a selection.

Decontaminant	Active ingredient	Application	material compatibility
Bleach	Na-hypochlorite	Terrain, equipment	Corrosive
HTH(High Test Hypochlorite) [2]	Ca-hypochlorite	Terrain, equipment	Highly corrosive
Chlorinated lime	Ca-hypochlorite	Infrastructure, equipment	Corrosive
Super tropical beach (STB) [2]	Ca-hypochlorite	Infrastructure, equipment	Corrosive
Peroxodisulfate/ Sodiumchlorite [1]	ClO_2	Water, B-decont	Corrosive, less corrosive than hypochlorite
Isocyanuric acid, Sodium and Potassium salts [1, 3]	Hypochlorite; in combination with organic solvents and as aqueous solution	Equipment, water, ointments	Slightly corrosive,
Sodium Hydroxide	Alkaline pH	Equipment	Non corrosive, may be harmful to synthetic material
Lithium Hydroxide	Alkaline pH	Equipment	Destroys alkyd paints, difficult to handle
Hydrogen Peroxide	Oxygen in stat. nasc.	Equipment	Destroys dyes
COTS peroxide cont. household detergents	Oxygen in stat. nasc, alkaline pH	Equipment	Destroys dyes, harmful to skin
COTS surfactants	improvement of agent solubility	Equipment, clothing, human skin	May be harmful to skin
COTS house hold detergents	Improved solubility of agent, slightly alkaline pH	Clothing, equipment as hasty decont, human skin	May be harmful to skin
BX 24 [4]	Hypochlorite (Fichlor based)	Equipment	Corrosive
Decon Green [5–8]	Peroxide based	Equipment	Low corrosivity

COTS = commercial off the shelf.

this means that in many cases MOPP-levels (Mission Oriented Protective Posture (MOPP) procedures) can only be reduced, but not canceled. It has to be kept in mind, that using the substances listed as aqueous solution, the decontamination efficiency of these systems can only be increased by

- Adding mechanical force, such as rubbing or scrubbing
- Combining the use of mechanical force with surface active substances to improve solubilization.

Therefore, a lot of labor and logistical burden have to be invested to make the decontamination system work. More labor in chemical protection suits, especially scrubbing the surface with mechanical force, automatically means more decontamination personnel resources or more exhausted personnel after the job is done.

7.3.2 Non-Aqueous Decontaminants

One advantage of non-aqueous decontaminants is the smaller logistical burden for the user. Non-aqueous decontaminants can be prefabricated and used immediately after a chemical or biological attack. The average surface concentration necessary to decontaminate is normally lower than with aqueous systems and therefore the costs/m^2 surface area should be lower. Due to the fairly high prices of the COTS (commercially of the shelf) products the advantage of decontaminant savings does not really decrease the m^2–costs.

Non-aqueous decontaminants and formulations such as DS2 have the ability to deal with grease contaminated surfaces and have an improved ability to dissolve and decontaminate thickened agents [2, 5, 9]. Mechanical support in the way of scrubbing the surfaces helps to make the formulations more effective and to decontaminate even thickened agent contaminated surfaces. The use of non-aqueous decontaminants is directly linked to problems resulting from the use of large amounts of organics. Issues include:

- Only usable in combination with organic solvents [1]
- Flammability
- For some formulations upper and lower explosion limits are established [10]
- Use of toxic or harmful organic solvents
- Environmental problems after use
- Toxic waste after decontamination
- Storage and handling problems
- COTS products are harmful to paints, plastics and elastomers [2, 11];
- Fairly high costs of the available COTS products in comparison to water soluble decontamination chemicals.

The biggest advantage is the "water free" use of these chemicals and formulations, which means that the logistical burden of the transport organization is decreased due to not needing large amounts of water for mixing the decontaminant solutions. For the removal of decontaminants used after the decontamination reaction has taken place, a certain amount of water is still necessary.

Non-aqueous decontaminants (Table 7.2) are normally ready to use or premixed, so that only small handling operations are necessary immediately before use.

Table 7.2 Non aqueous decont–chemicals and decont formulations.

Deconta-minant	Active ingredient	Application	Material compatibility
Chloramine T [1]	Hypochlorite	Equipment, clothing, human skin in ointments	Slightly corrosive
Monochloramine T [1]	Hypochlorite; not applicable for GB,VX,GD; solulable in water and alcohol	Clothing, human skin in ointments	Less corrosive than STB and relatives
Chloramine B [1]	Chlorine; only in combination with organic solvents,	Equipment	Corrosive
Hexachloromelamine [1]	Chlorine; only in combination with organic solvents,	Equipment	Corrosive
DANC (chlorinated Hydantoine in $C_2H_2Cl_4$) [2]	Chlorine; Not applicable for GB decont.	Sensitive equipment	Less corrosive than hypochlorite
DS2 [11, 12]	2-Methoxyethanol and NaOH	Equipment	Harmful to paints, plastics and elastomers
GDS 2000 [10]	Diethylenetriamine, several amino alcohols and Na -alcoholates	Equipment	Harmful to alkyd paints, plastics and elastomers
GD5 [13]	2-Amino-ethanol, KOH	Equipment	Harmful to alkyd paints, plastics and elastomers
GD 6 [13]	2-Amino-ethanol, Potassium-amino-ethoxid	Equipment	Harmful to alkyd paints, plastics and elastomers
RSDL[14–17]	Sodium –(2,3-Butanedione-monooximate) combined with Polyethyleneglycol and Tetraglyme	Human Skin, wounds; sensitive equipment	Slow reaction with HD
Phase transfer catalysts [18]	for example N.N-hexadecan-Methyl-diethanolammoniumbromide	Sensitive equipment, human skin	Very slow reactions with HD

7.3.3 Macro- and Microemulsions

To combine the advantages of the aqueous phase and the organic phase a lot of research was done in the field of emulsions. The properties of emulsions and microemulsions to be used as CB-decontaminants were investigated and new formulations were developed [19, 25, 26].

7.3.3.1 Macroemulsions

Macroemulsions or "Emulsions" combine the advantages of the aqueous and non-aqueous decontaminants. Two forms are possible:

- Water in oil emulsions
- Oil in water emulsions.

The preferable form is the water in oil form, due to the fact that the continuous phase is oil, which means solvent. The solvent type used has to match the solubility parameters or Hildebrand Solubility Parameters of the CW-agents as close as possible. The Hildebrand Solubility Parameter [20] is a dimensionless number which is available for many solvents and other organic compounds and can be calculated from the Hildebrand expression [20]

$$\delta = \left(\frac{\Delta E}{V}\right)^{1/2} = \left(\frac{D(\Delta Hv - RT)}{M}\right)^{1/2}$$

where δ is the Hildebrand parameter, ΔH_v the heat of vaporization, the temperature in Kelvin, M the molecular weight and D the density and R the gas constant.

The values drawn from this expression can be used to predict the solubility of most solutes in most solvents and even of CW-agents in solvents. However, the chapter microemulsions will give a more detailed explanation of the whole working principle. To understand the schematic decontamination mechanism of water in oil macroemulsions (Figure 7.1) is essential to tailor the decontaminant for an efficient and fast neutralization of the toxic compound.

During the decontamination process the oily phase is in direct contact with the surface to be decontaminated and has the ability to dissolve CW-agents and thickeners, remove them from the surface, penetrate the surface and extract CW agent out of paint layers or plastic material. The dissolved agents are transported to the boundary layer between oil and water phase and surfactants with phase transfer capacity transport the agent into the water droplet, where decomposition takes place [21, 22].

The oil in water emulsions are quite effective in cleaning up surface areas, but not able to penetrate into paint layers and porous material. Figure 7.2 shows the schematic mechanism of the decontamination process.

The extraction effectiveness is limited due to water being the surface contact media and the surfactants used as emulsifiers must have phase transfer properties.

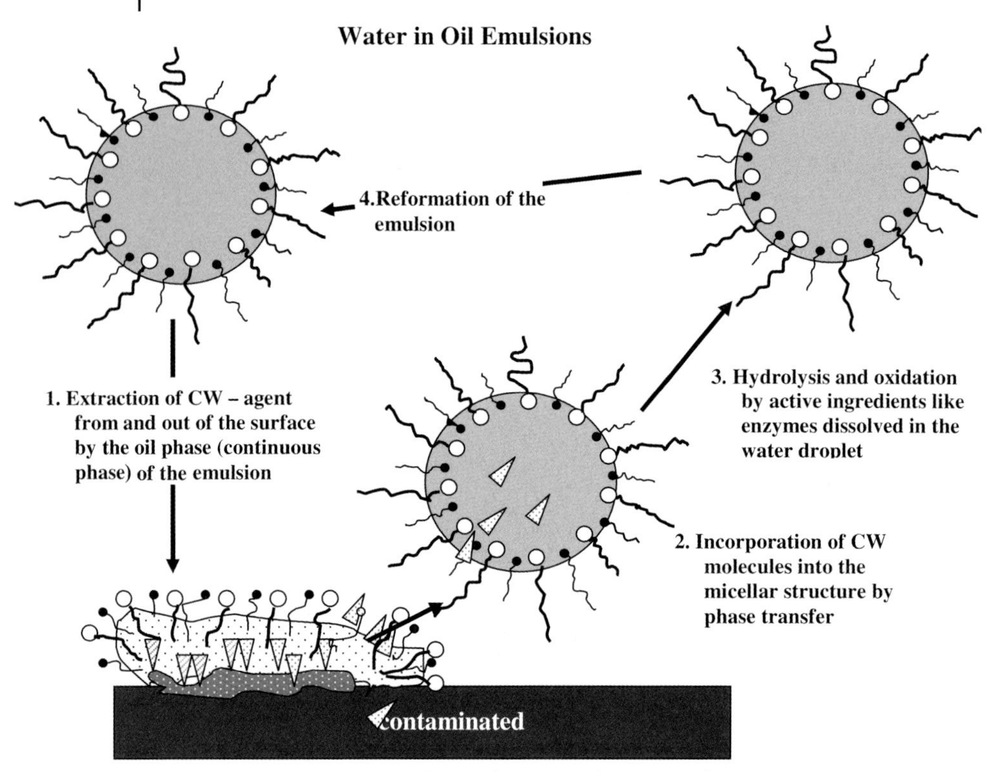

Figure 7.1 Decontamination scheme of water in oil macro emulsions.

If they don't have these properties, the reaction velocities are fairly low and the overall effectiveness is not better than using aqueous systems such as bleach- or HTH-solutions.

There are a few decontamination emulsions that have been introduced into the military inventory (Table 7.3). One example is the German Emulsion, introduced into the German Army, and a relative of the German Emulsion, the Xylene Emulsion of the Austrian Army. Both decontamination formulations are water in oil emulsions. They use Tetrachloroethylene (resp. Xylene) as solvent and HTH or Fichlor as active substance [3, 10, 13]. The main differences between them are the different solvents and the different active chlorine carriers. Both of the solvents have advantages and disadvantages. Tetrachloroethylene is the better choice as solvent for CW agents, as it is not flammable, but it has environmental problems due to the fact that its half-life in soil is up to 6 years and some of the metabolites are toxic [23, 24].

Xylene is a flammable liquid with a fairly low flash point of 21 °C and has a half-life in soil of two to four weeks. The effectiveness of both emulsion types against CW agents is well tested. The formulations are used in the German and Austrian Armies and the formulations are suitable for equipment decontamina-

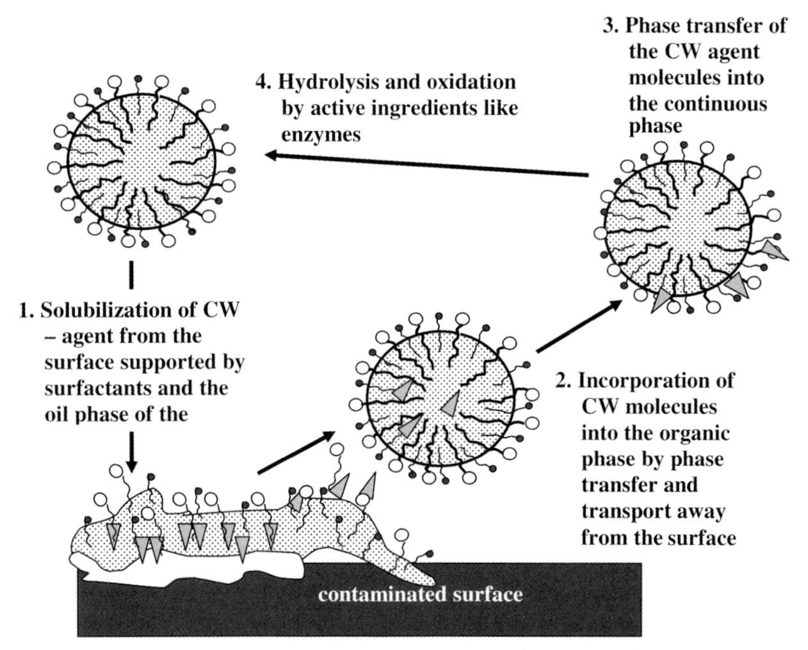

Figure 7.2 Decontamination scheme of oil in water emulsions.

Table 7.3 Selected formulations of macro emulsions introduced into military inventories [25, 26].

Emulsion type	Active substance and concentration	Type surfactant and concentration	Type of solvent and concentration	Amount of water
C8 Emulsion	HTH 7.5%	Anionic/nonionic 50/50, 1%	15%	76.5%
Xylene Emulsion	HTH 1% or Fichlor 1. 2%	Anionic/nonionic 50/50, 1%	10%	77%

tion, even under difficult conditions, including extremely dirty vehicles, thickened agent contamination and so on.

7.3.3.2 **Microemulsions**

It is well established that two immiscible liquids can be formed into a macroemulsion by using surfactants or surfactant mixtures. Microemulsions are relatives or improvements of the macroemulsion with a macroscopically homogeneous look and microscopically heterogeneous structure. This unique class of optically clear or opaque, thermodynamically stable and normally low viscous solutions have been subject to intensive research due to their properties and technological importance [27]. Microemulsions have the advantage over macroemulsions to have up to 100 times larger interfacial surface areas, which leads to characteristic properties such as

- Low viscosity
- Ultralow interfacial tension
- High capacity to solubilize both oil- and water soluble compounds.

This makes them more effective and more flexible in solubilization and decomposition of CW agents because, due to the larger inner surface area chemical reactions are faster and more effective, and it is possible to introduce less corrosive and environmentally harmful chemicals into the decontaminant formulation. The properties of microemulsion systems and their increased effectivity should lead to a minimization of reactive compounds in decontaminant formulations and a decreasing amount of decontaminant necessary to clean up a given surface area. This could lower the costs and the logistical burden of decontamination procedures in comparison to systems based on aqueous solutions or even macroemulsions. In the area of decontamination, several developments of microemulsions for decontamination purposes have taken place [28, 29]. One interesting development is a microemulsion formulation to which different reactive compounds, even enzymes, can be added [30–32].

This makes a decontamination system based on microemulsions very flexible and offers the possibility of developing tailored decontamination solutions for different applications.

Laboratory tests and published reports show that the implementation of enzymes active against G-agents as reactive compounds is possible and effective [29, 30]. A detailed description of theory and possible applications of microemulsions is given in Chapter 12.

7.3.4 Foams and Gels

Foams and gels are other forms of application which can be used for decontamination purposes.

7.3.4.1 Foams

Foam systems do have some advantages over "conventional" decontamination solutions.

A foam system can be easily tailored to the needs of the user and fire fighting foam systems, for example, show a broad band of foam properties tailored to the needs. Foam has some properties which are very important for the formulation of decontaminants, such as:

- Sticking on nearly any kind of surface
- Slowly decomposing under "bleeding" fresh decontamination solutions

- Less use of consumables due to the low density of foam and the "adjustable" residence time on a surface.

These properties are preferable for decontamination systems, but difficult to realize with decontamination solutions on the market. A foam system can produce "wet" or "dry" foams dependent on the type and concentration of surfactants or surfactant mixtures. The preferable form is wet foam, due to the slowly bleeding of this foam type. This bleeding makes certain, that the surface to be decontaminated is permanently supplied with fresh decontaminant or active chemistry, while the foam layer decomposes slowly. Wet foam sticks on every kind of surface independent of its orientation in space, which means it sticks on horizontal or vertical surfaces and even on overhead surfaces. Because of the decomposition of the foam bubbles, fresh decontamination solution wets the surfaces and decontamination can take place. An ideal foam system should combine the positive properties of emulsions or microemulsions, such as higher viscosity than aqueous solutions, an integrated solvent to improve extraction capacity from non-resistant surfaces, with the properties of foam. Combining these properties leads to a three phase foam, where an emulsion or microemulsion builds the foam cells and bleeding immediately takes place when this decontaminant is sprayed on a surface. The first commercially available system is the foamed C8 emulsion [33, 34], which brings the decontaminant consumption from a former $3 l/m^2$ pure emulsion down to $0.5–1 l/m^2$ foamed emulsion. Another well-tested system is the Canadian CASCAD foam system. Originally designed for decontamination of military equipment it is one of the eight decontaminants identified in a study of the Joint Fixed Site Decontamination program of the US Government [5]. An evaluation of different foam based decontamination systems shows that different types are on the market (Table 7.4).

Table 7.4 Selected and commercial available foam decont systems.

Decontaminant	Active ingredient	Application	Material compatibility
CASCAD [35–37]	Hypochlorite (Fichlor) and surfactants	Equipment decontamination, blast reduction,	Less corrosive than HTH-solutions
Sandia Foam [38–40]	Peroxide based	Equipment, buildings	Low corrosivity
Easy Deon [38–40]	Peroxide based, formulation equals SandiaFoam	Equipment, buildings	Low corrosivity
MDF 200 [38–40]	Peroxide based, formulation equals SandiaFoam	Equipment, buildings	Low corrosivity
Decon Schaum [33, 34]	Tensides plus active ingredients	Equipment, buildings	Low corrosivity
Foamed C8 Emulsion [33, 34]	Hypochlorite macro emulsion foamed	Equipment	Corrosive

7.3.4.2 Gels

A *gel* is a colloidal system in which a porous network of interconnected nanoparticles spans the volume of a liquid medium. In general, gels are apparently solid, grease-like materials. Both by weight and volume, gels are mostly liquid in composition and thus exhibit densities similar to liquids, however gels have the structural coherence of a solid. An example of a common gel is edible gelatin. Many gels display thixotropy, which means they become fluid when agitated, but resolidify when resting [41]. The fact, that gels have an internal liquid phase makes them interesting for the development of decontaminants. There are a few developments found in the open literature (Table 7.5), but commercially available decontamination gels are not known at present. This might be an area of research for the future, much as the microemulsion area is in the present.

Table 7.5 Decontamination gels.

Decontaminant	Active ingredient	Application	Material compatibility
L-Gel [19, 41, 42]	Oxone ($KHSO_3$*$KHSO_4$*K_2SO_4)	Equipment, paintes surfaces	Low corrosivity

7.3.5 Selected CW-Agents and Decontamination Reaction Mechanisms

For the detoxification of CW-agents, oxidizing and hydrolysis mechanisms can be used. Understanding these mechanisms is essential to the optimization of, and to tailoring the decontamination reaction for the technical application.

7.3.5.1 Sulfur Mustard (HD)

Mustard, also known as H, Yperite or Lost is a color- and odorless liquid with the chemical name 1,1 dichloro-diethyl-sulfide:

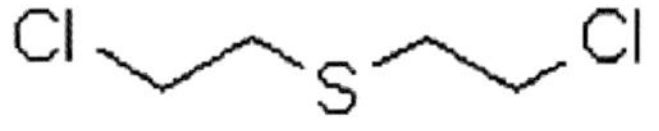

Cl S Cl

It was invented in the late 1800s [43]. The first mass production took place in Germany in 1917 and mustard was used first by Germany during WW I at the Ypern battlefield in Belgium [44]. It is a strong vesicant [45–47].

Hydrolysis

The first step of the hydrolysis reaction is the attack of the sulfide to form a sulfonium ion:

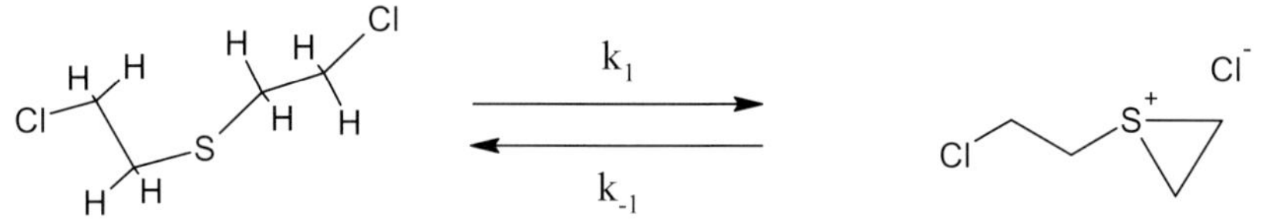

H H H H Cl Cl S H H H H $\underset{k_{-1}}{\overset{k_1}{\rightleftharpoons}}$ Cl S^+ Cl^-

Water is now able to attack the ring carbon, opening the ring and building a mixture of hemi mustard and hydrogen chloride:

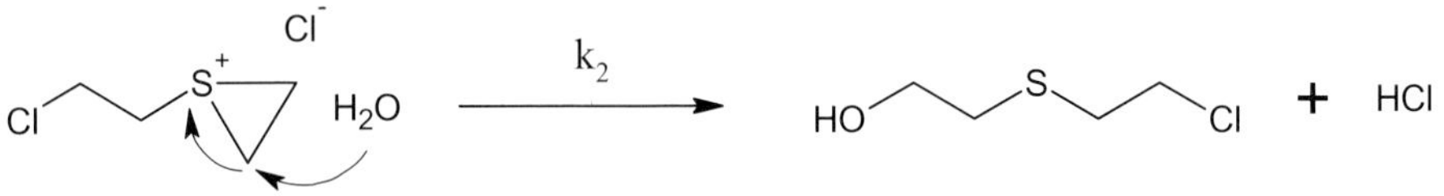

Hemi mustard is also a vesicant which reacts with water similar to the original mustard to form thiodiglycol and hydrogenchloride:

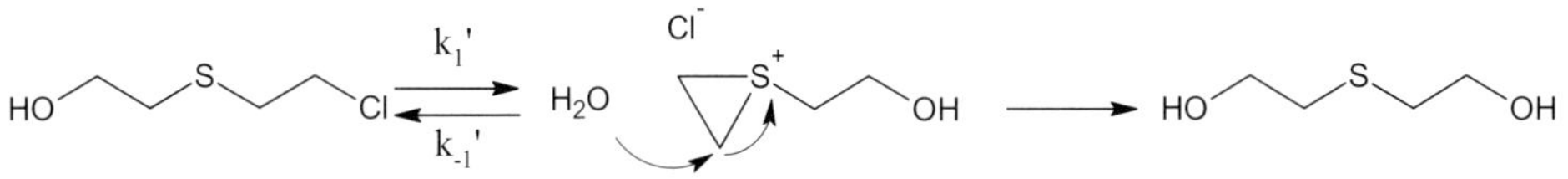

The ring forming intermediate product reacts in a parallel reaction line by internal displacement and forms 1,4 thioxane and hydrogenchloride:

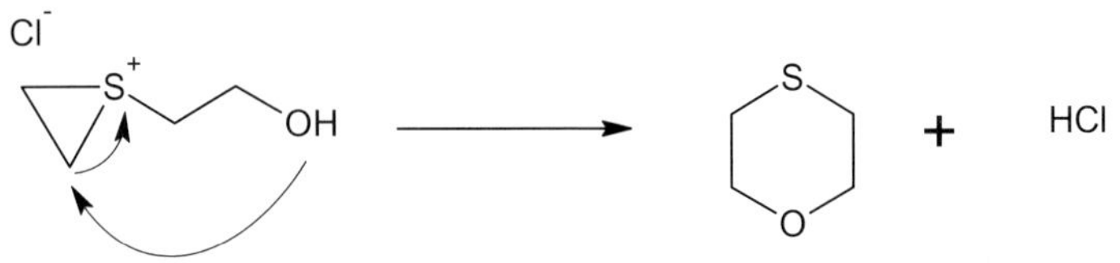

The ratio between thiodiglycol and thioxane produced by the hydrolysis reaction is described 4 : 1 in literature [48, 49]. The rates of hydrolysis reaction for mustard are fairly high; the limiting factor is clearly the amount of mustard dissolved in water per time unit. Therefore, during the development of an overall decontaminant for all chemicals, the physical removal and the solvation process for mustard are critical issues. Mustard is very persistent in an aqueous environment and the dissolution velocity of mustard is low. This has consequences for decontamination processes using plain water in a way, that thorough decontamination of mustard contaminated surfaces with plain water and even with water plus detergents is not very effective [1, 46, 47, 49, 50].

7.3.5.2 **Sarin (GB)**

Sarin and relatives were invented in 1938 by G. Schrader *et al.* The development as a CW-agent including a family of munitions took place in Germany in WWII. Large-scale production plants were ready to use in 1944 and about 500 tons were produced and loaded into munitions. After WWII large-scale production took place in the US and the former Soviet Union. Sarin is a color- and odorless very toxic liquid. Its chemical name is isopropyl methylphosphonefluoridate:

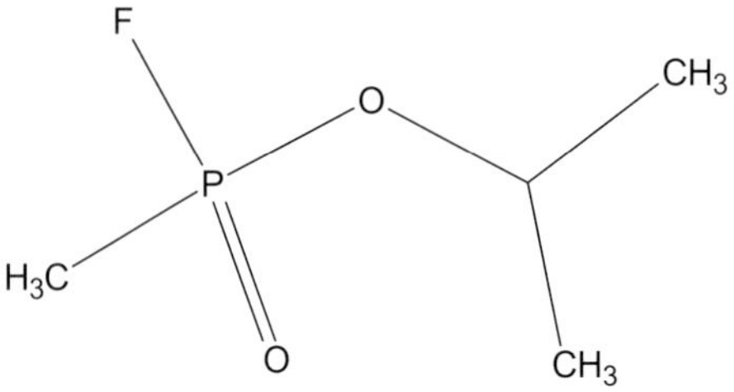

It acts as a strong cholinesterase inhibitor like other nerve agents. Decontamination is relatively easy due to the fact that hydrolysis can be accelerated by using basic aqueous solutions such as Na_2CO_{3-}, NaOH or KOH–solutions [1, 8, 19, 46, 47, 49].

Hydrolysis

The hydrolysis of sarin is strongly dependent on pH. Half-life at pH 1 is 15 minutes, for pH 5.

165 hours and half-life at pH 13 is 0.3 second. Sarin is miscible with water in any ratio and at neutral pH the rate of hydrolysis is 50 hours. The decomposition reaction with water at neutral pH follows [1]

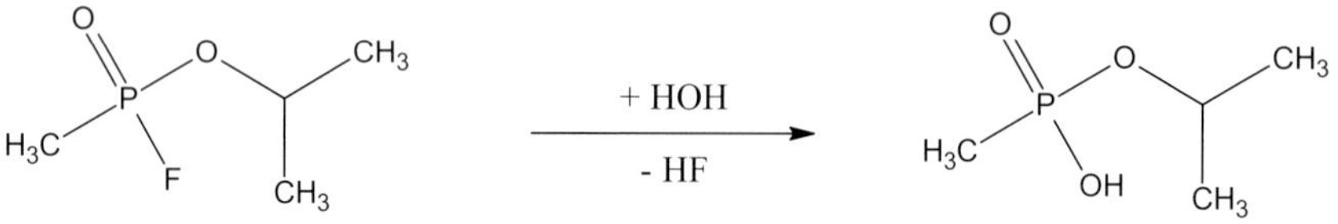

Hydrolysis can be accelerated by increasing pH, adding anionic-, nucleophilic catalysts or metalchelate complexes [1, 8, 19, 49].

7.3.5.3 Soman (GD)

Soman (GD) was developed in 1944 in Germany by R. Kuhn, laboratory testing was in progress at the end of WWII.

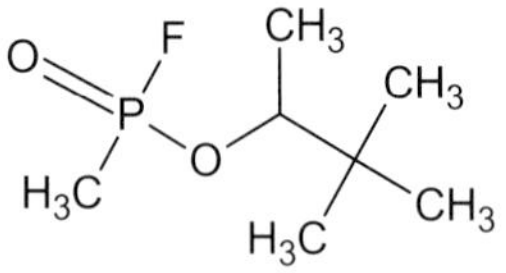

Soman is one of the most toxic compounds used as a CW-agent. It is a colorless liquid with a fruity odor, the industrial product is yellow–brown with a camphor-like odor. It was mass produced and stockpiled by the former Soviet Union [1, 18, 46, 47].

Hydrolysis

Soman is attacked by hydrolysis in acidic, neutral and basic media. The end product is fluoride and pinacolyl–methylphosphonate as the initial products [1, 18, 46, 51]:

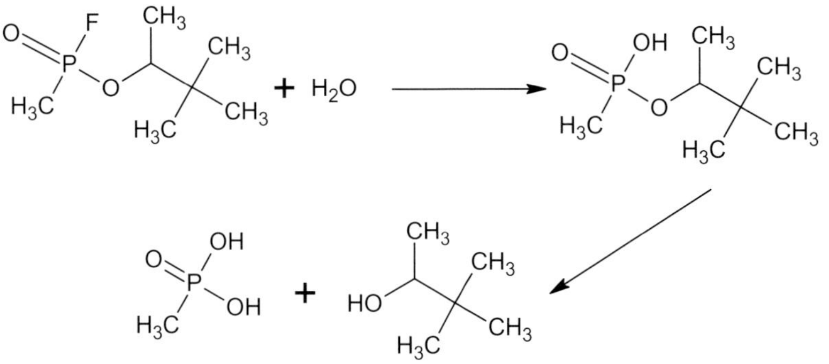

Due to the chemical similarity to sarin (GB) the hydrolysis of soman will be catalyzed by the hypochlorite anion. Heavy metal complexes, such as copper (II) accelerate the soman hydrolysis to for decontamination purposes useful rates [9, 49].

Na_2CO_3-, KOH- and NaOH-solutions are used in decontamination procedures for the fast decomposition of soman with half-life times of approximately less than 1 minute [1, 19, 51, 52].

7.3.5.4 **VX**

The phosphorylthiocholine class was discovered in 1952–1953 by three researchers, Tammlin of Sweden, Schader of Bayer and Ghosh of ICI. Intensive research of the US Army on this class of compounds led to the development of VX at Edgewood Arsenal [1, 19, 46]. It is a very persistent, odorless, amber-colored liquid, similar in appearance to motor oil. Although VX is many times more persistent than the G-agents, it is very similar to GB in mechanism of action and effects. The molecular weight of VX is 211.2. The substance with the structure formula

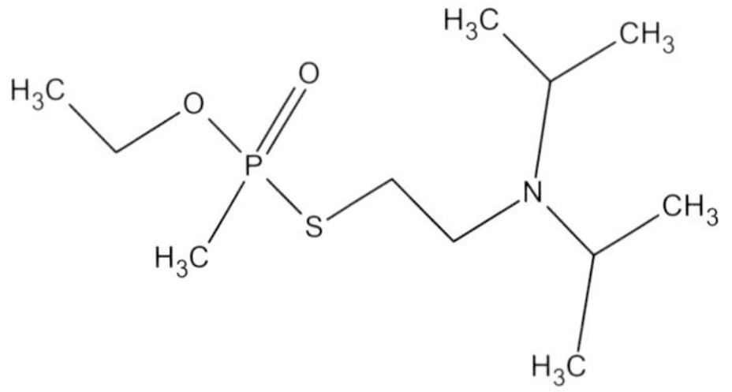

has a very low volatility. Liquid droplets on the skin do not evaporate quickly, thereby increasing absorption. VX by this percutaneous route is estimated to be more than 100 times as toxic as GB. VX by inhalation is estimated to be twice as toxic as GB [1, 46, 47].

Another V-agent of interest is Vx, called "V sub x." Information on this agent is limited. The properties of Vx are similar to those of VX. Vx is nearly ten times more volatile than VX but is very persistent in comparison to the G-agents. The physiological action, protection, and decontaminants for V_X are the same as for VX [47, 49]. Another designation for VX is "V-gas."

Hydrolysis

The first step in the hydrolysis of VX is the attack of hydroxide on the P-atom to form a phosphorus intermediate:

HO^- H_3C O P S O CH_3 CH_3 N H_3C CH_3 CH_3 → H_3C OH O P O^- H_3C S H_3C CH_3 N CH_3 H_3C

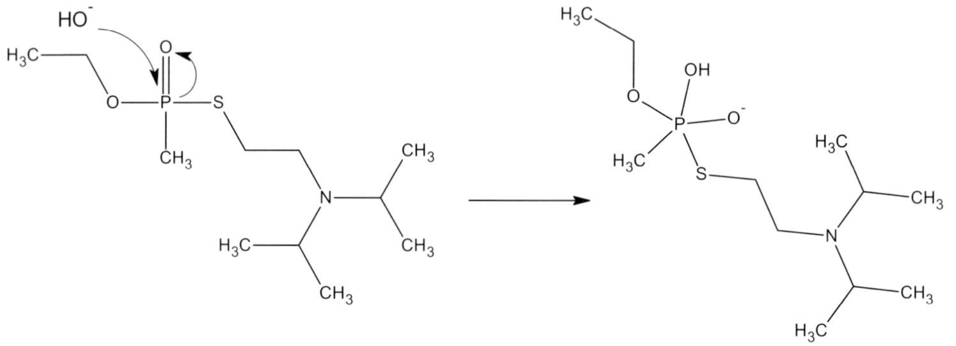

This phosphorus intermediate can decompose in two ways:

1.:

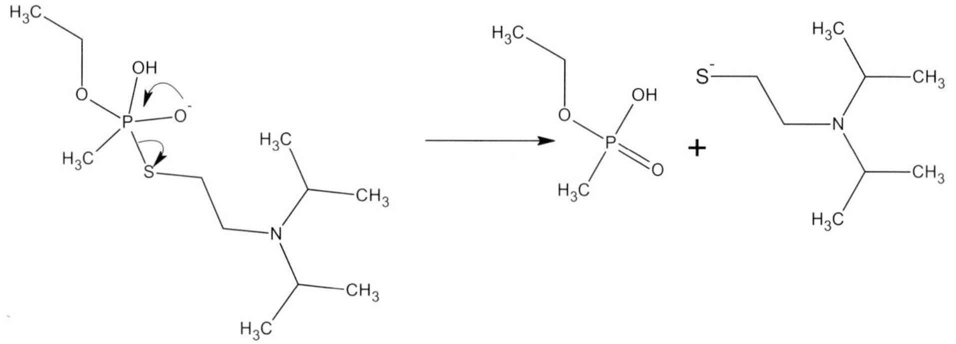

H_3C OH O P O^- H_3C S H_3C CH_3 N CH_3 H_3C → H_3C OH O P O H_3C + S^- H_3C CH_3 N CH_3 H_3C

The anion of diisopropylaminoethanethiol is expelled and as a result the ethyl ester of methylphosphonic acid is formed.

and 2.:

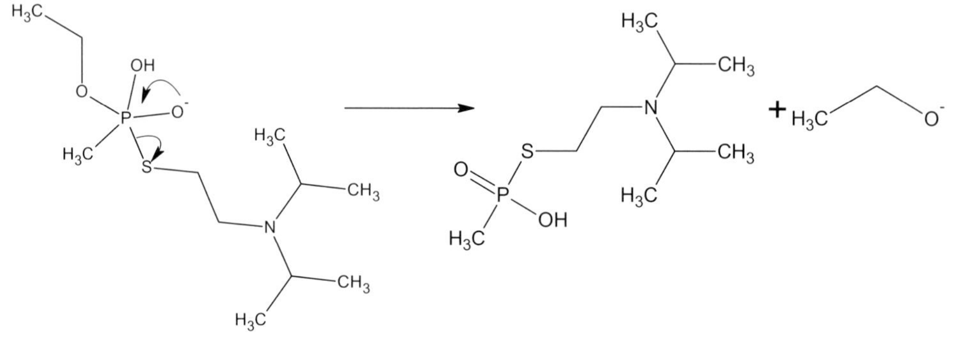

H_3C OH O P O^- H_3C S H_3C CH_3 N CH_3 H_3C → H_3C CH_3 N S CH_3 O P H_3C OH H_3C + H_3C O^-

The second reaction path shows that ethoxide is expelled to form a compound called EA 2192 [53, 54]. The ratio between EA 2192 and the reaction product of pathway 1 is 13% EA 2192 to 87% methylphosphonic ester. EA 2192 is stable against further hydrolysis (the hydrolysis rate of EA 2192 is about 1000 times slower than the hydrolysis of VX) and just as toxic as VX. Therefore, the formation of this compound has to be avoided by running the hydrolysis process during operations as shown in pathway 1. VX can also hydrolyze by displacement of the thiophosphonate anion from the carbon atom [53, 66]:

The hydrolysis of the Russian V-Gas can be expected to be similar.

Hydrolysis of VX is slow and, as stated, EA 2192 is formed [19]. Therefore, for the decontamination of VX, alkaline-catalyzed hydrolysis in combination with an oxidation reaction with commonly available bleach or HTH-solutions is better choice [54–56, 66].

7.4 Decontamination Procedures

The following paragraph gives an overview of decontamination procedures and a guide to equipment. The procedures and substances presented in this paragraph are not an exhaustive evaluation of all decontamination technologies and decontamination equipment available, but will give the reader an impression of the different procedures and why an overall concept, not only for decontamination, but also for the whole sector of NBC-defense, is necessary.

7.4.1 Generalities

Decontamination of personnel and equipment means in general dividing the decontamination side into two parts, a black (contaminated) and a white (clean) part.

In the black part of the site contaminated personnel and material is assigned to the decontamination procedures appropriate and the white part allows the reassembly of personal gear, vehicles and equipment without the need to wear

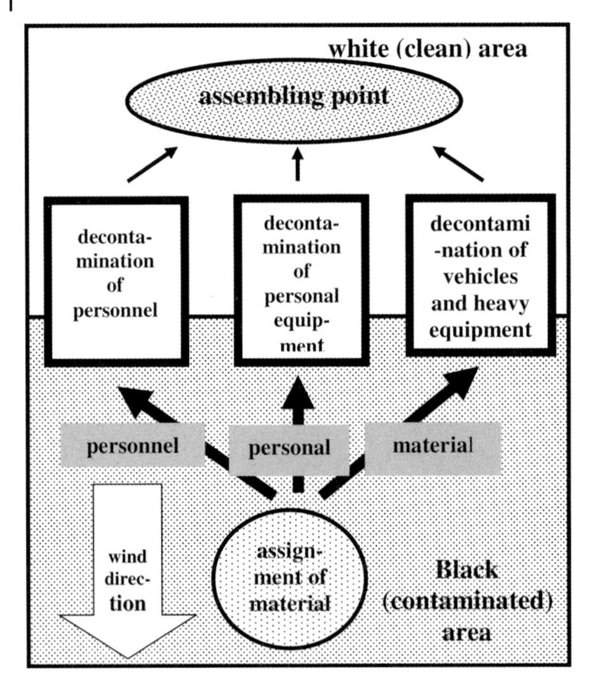

Figure 7.3 Schematic decont site.

NBC-protection. One of the most important parts is the design of a decontamination site under consideration of the wind direction (Figure 7.3).

7.4.2 Equipment Decontamination

The decontamination of equipment can be divided into two major processes:

- Wet procedures, using liquid decontaminants on an aqueous or non-aqueous base;
- Dry procedures using adsorbents.

Most of the decontamination procedures used by the armed forces all over the world are based on "wet procedures," which means most of them use aqueous or non aqueous decontaminants for the thorough decontamination of equipment.

7.4.2.1 Wet Procedures

"Wet procedures" use water based or solvent based decontaminants. Emulsions/micro emulsions are a combination of solvent and aqueous solutions. In general, they can be used with the same type of spray equipment used for the water-based systems.

7.4.2.1.1 **Aqueous Decontaminants** This group of decontaminants is still widely used by different armies of the world and well established since WW I. Most of the systems for large equipment decontamination use a mixing tank, where a solid decontamination product is dissolved in water. The ready to use solution is sprayed on the object to be decontaminated and after a dwell time of several minutes, the "film" is rinsed off by using water jets with elevated pressure. The Russian decontamination truck ARS 14 [57] and the US decontamination system M 12 A1, which is skid mounted and transported on a 5-ton truck [2, 58] are examples (Figures 7.4 and 7.5).

The German 7 to decontamination truck has additional capabilities. The vehicle is equipped with mixing equipment to make C8 – Emulsion, which is the standard decontaminant of the German Army. The vehicle has a tank capacity of $3\,m^3$ aqueous decontaminant or C8-Emulsion. The emulsion can be sprayed on the surfaces to be decontaminated by operating two hoses and special spray guns or using a fixed spray system to decontaminate road and infrastructure.

A smaller vehicle, the so-called TEP, is based on a 5 to truck and able to decontaminate large equipment, personal equipment, personnel and infrastructure. Figure 7.6 shows the German 7 to decontamination truck.

Figure 7.4 Skid mounted US decont system M12 A1 [58].

Figure 7.5 ARS 14, Russian decont system truck mounted [58].

Figure 7.6 German 7 to –decont truck.

Smaller and more mobile equipment, such as the **M**ulti **P**urpose **D**econtamination **S**ystem of the Kaercher Company or the Sanijet family of the Christanini s.p.a. are based on high pressure spray systems, operating with water soluble decontamination chemicals in combination with integrated injector pumps driven by the high pressure water jet or with prefabricated decontamination solutions. The integrated injector pumps supply the system with the amount of chemicals to produce a ready to use decontamination solution. The systems are more mobile than the decontamination trucks due to their small dimensions and jeeps or small trucks can transport them. Normally they are not used for the decontamination of great numbers of vehicles or bulky equipment due to the amount of time necessary to decontaminate a great number of vehicles. Figures 7.7 and 7.8 show an example.

Figure 7.7 MPDS, high pressure decont system, Kaercher, Germany [10].

Figure 7.8 Sanijet, Christanini, s.p.a., Italy [4].

For the decontamination of smaller numbers of vehicles, vehicle parts or equipment most armies use small size decontamination systems operating with a pressurized storage container for the decontaminationaminant. The decontamination solutions can be sprayed on the surface, sometimes a disposable brush is integrated to increase efficiency. One example for these smaller decontamination systems is the German DS 10 Sprayer, which is shown in Figure 7.9.

The German DS 10 sprayer is designed to be used with emulsions based on Tetrachloroethene (German C8–Emulsion) or Xylene and can be used to spray solutions of disinfectants and aqueous solutions of Fichlor or HTH. With additional equipment, for example foaming equipment, even decontamination foams for biological or nuclear decontamination are preparable [10]. The former East German Army used the principle of pressurized spray cans for the decontamination of large numbers of vehicles or equipment by loading several pressurized and decontamination solution filled containers on a transport truck, which brought the pressurized vessels to the places where they were needed (Figures 7.10 and 7.11).

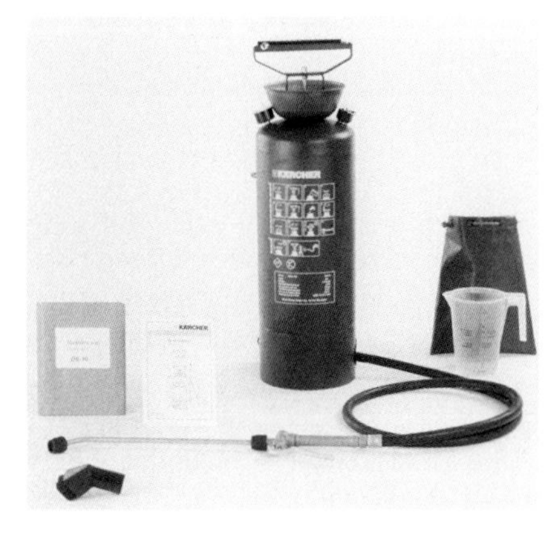

Figure 7.9 Decont sprayer DS 10 [10].

Figure 7.10 East German decont system EA 12 [59].

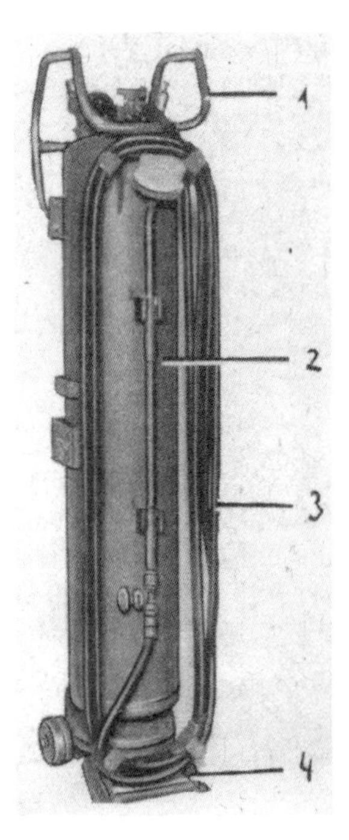

Figure 7.11 Single pressurized container EA 12 [59]. 1 = valve protection; 2 = pressure vessel; 3 = hose with spraygun; 4 = wheelbase.

Loading with pressurized air took place on the truck due to an integrated compressor- and pressurized air distribution system. Decontamination solutions used were mainly Hypochlorite solutions (10% by weight HTH [59]). The system was very flexible and able to supply troops anywhere in the theater with ready to use decontamination capacity.

7.4.1.1.2 **Non-Aqueous Decontaminants** The M 11 and M13 decontamination systems are designed to work with DS2, a non-aqueous decontaminant effective

Figure 7.12 US Decontamination apparatus M13 [58].

against all CW agents. Main parts are a storage container and a spray tube with integrated pump and brush. The decontaminant is pumped out of the storage vessel and simultaneously spread on the surface to be decontaminated by brushing. The decontamination of large surface areas is labor intensive. The user needs to climb on higher parts of a vehicle which makes decontamination operations due to DS2 on the surface somewhat dangerous. The decontaminant is introduced into several armies of the world and effective against all CW agents. Brushing is necessary for a successful decontamination operation [12] (Figure 7.12).

In Europe the M13 pendants are the DS 10 of the Kaercher Company (Figure 7.9) [10] and the OWR TRS 10 spray apparatus (Figure 7.13) [13]. They are both able to deal with several non-aqueous decontamination solutions including DS 2, GDS 2000 [10], GD 5 and GD 6 [13], aqueous solutions of HTH and emulsions of the Xylene- or Terachloroethene-type. Adaptation to the decontaminant can be

Figure 7.13 OWR TRS 10 decont sprayer [13].

Figure 7.14 US decont sprayer M11 [67].

made by simply changing the nozzle of the spray gun. Figure 7.13 shows the OWR-equipment.

For partly or hasty decontamination several spray cans are on the market to enable troops to clean mission essential parts of weapon systems and vehicles. The US M 11 is one of the well-known sprayers. The system is filled normally with DS2 and can be adapted to GDS 2002 or GD 6 and GD6. Figure 7.14 shows the device.

The aerosolization of non aqueous decontaminants, such as GD 5, GD6, DS 2 or GDS 2000 is another possible way to fulfill decontamination needs. The advantage is, that the aerosols can be deposited on any kind of surface, even surfaces, which are difficult to reach by using conventional techniques. There are three or more systems on the market and Figure 7.15 shows the OWR–device as an example.

The decontamination of large numbers of vehicles and heavy equipment with more or less automated decontamination systems using non aqueous decontaminants such as DS 2, GDS 2000 or GD 5/GD6 is under development in several countries. In Germany a new system (TEP 90) will be introduced into the German

Figure 7.15 OWR Turbofogger decont system generating GD5/GD6 aerosols [13].

Figure 7.16 TZ 74 gas turbine driven decont system.

army. This piece of equipment has decont capability in the area of large equipment, clothing, sensitive equipment and inner surfaces of vehicles or buildings. It consists of a truck as carrier and four modules for the different decontamination needs. The modules can work autonomously and be separated from the carrier. The manufacturer of this system is the Kaercher Company of Germany [10].

7.4.2.2 **Dry Procedures**

Dry decontamination procedures use the exhaust energy of combustion engines, such as gas turbines or pulse jet engines. The thermal energy in form of heat and high velocity gas flow evaporates and shears off CW agents from the surface. The principle was and still is in use in Russia, where a jet engine is truck mounted and used to decontaminate vehicles of any kind. The advantage of the system over wet processes is the dry surface after decontamination and the saving of large amounts of decontaminationamination chemicals. Most of the systems carried out have the additional possibility to inject aqueous solutions of decontaminants or water into the plume and use the increased reaction velocity of hot decontamination solutions. Many countries made developments and performed testing of gas turbine driven decontamination systems, but only Russia and the former Eastern Block Countries had systems in operation [57, 58]. Figure 7.16 shows the Czech TZ 74, which was used in Eastern Germany and other Warsaw Pact countries.

7.4.3 Clothing and Protective Clothing

Contaminated protective clothing clothing and must be discarded and replaced by clean material, which is costly and the logistical burden is very high. Therefore all armies of the world are trying to find procedures and equipment, which makes them able to clean up contaminated clothing in a minimum of time with a maximum of effectiveness. Several decontamination procedures and processes were introduced into the different armies and replaced by "new" procedures starting from simple washing processes over dry cleaning to treatment with adsorptive material. Especially when using protective suits, such as the well-known over- and undergarments, which have adsorptive capacity due to integrated charcoal layers, decontamination is difficult. Washing processes do harm to the adsorptive capacity

Figure 7.17 Field laundry [10, 67].

of the charcoal layers or remove at least a good part of the integrated charcoal. Additional problems are caused by the high heat of evaporation of water, which is responsible for the long drying times of wet garment material. Due to the limited drying temperatures water will be adsorbed on the charcoal and this decreases the protective properties of the garments down to a level, which cannot be tolerated for a protective suit. Figure 7.17 shows a military used field laundry integrated into a 20' Isocontainer [10]. The system is designed to disinfect and launder clothing of any kind and requires only water fuel and detergents. It is currently introduced into the German and Norwegian Army (Figure 7.17).

Solvent treatment or dry cleaning of protective suits based on integrated charcoal layers with solvents like FC113*, Tetrachloroethene or relatives cause problems due to solvent loads on the charcoal which reduce the adsorptive capacity for CW agents. These systems are much faster than washing machines due to the shorter drying times, but solvent residues within the garments can cause health and environmental problems. Another way to decontaminate garment material or protective clothing is the use of "dry processes." These dry processes are operating with hot air or mixtures of water vapor and hot air. The advantage over the washing process is, that there are no drying times have to be taken in consideration because the garment material comes dry out of the process. Figure 7.18 shows a unit developed by the German Army, which is ready for production and able to process 50 to 60 sets of protective suits and the additional personal gear per hour. The system works at 130 °C–170 °C, is operated by two soldiers and is able to decontaminate chemical, biological and nuclear contamination (Figure 7.18).

Another decontamination system is integrated into the Decocontain container of the Kaercher Company [10, 60–62]. It is introduced into the Portuguese Army and operates with a comparable process while using the exhaust of the generator gas turbine with into the plume injected water. Several companies offer combinations of clothing decontamination and equipment decontamination capability. Figure 7.19 shows a system of the OW.R. Company of Germany, where personal-, equipment and infrastructure decontamination capability is built in a 20-foot standard container.

* FC 113 = Fluoro Carbon 113.

Figure 7.18 Clothing and personal equipment unit in cooperation with a personnel decont unit. Prototypes of the German Army.

Figure 7.19 Personal-, equipment and infrastructure decontamination system MPD 100 [13, 60–62].

Several other major companies in the decontamination business offer different equipment and processes for the decontamination of clothing [60–62], so that the user must decide what type of decontamination process matches his needs.

7.4.4 Decontamination of Personnel

Persons suspected of being contaminated need immediate decontamination to minimize casualties. To achieve this goal, personnel decontamination can be done by two major processes:

1. Hasty decontamination to remove CW agents from skin, personal equipment and protective gear as soon as possible;
2. Thorough decontamination of personnel and personal equipment.

This paragraph describes some of the most important and available procedures and systems for the decontamination of personnel and personal equipment.

7.4.4.1 Hasty Decontamination of Personnel and Personal Gear

In a CW attack, soldiers and civilians are likely to become contaminated. Every soldier should be well educated in terms and methods of hasty decontamination. They should know how to react in dangerous situations and should be able to choose the right solutions. Civilians do not have this background and therefore the first time responders have to react in the right way. The first time responders have to follow the overarching concept for the decontamination of civilians. However, based on the military experience the civilian side could learn and could adept the military guide lines for their own purposes. The time it takes for a definitive (complete) decontamination is not given and the soldier based on his training should react as soon as possible to minimize agent penetration through protective or non-protective clothing causing massive casualties. One must remember, that the first ten minutes after the incident are critical to a favorable outcome. Under these circumstances hasty decontamination means to remove every, and the emphasis lies on every, reachable smear or droplet, which might be CW agent. Many armies have detailed standard operation procedures established to do so. Structural decontamination chemicals for hasty decontamination are introduced in many armies and in the following some examples are presented.

Some armies use fuller's earth, a solid absorbent that is a clay material characterized by the property of absorbing basic colors and removing them from oils. It is composed mainly of alumina, silica, iron oxides, lime, magnesia, and water, in extremely variable proportions, and is generally classified as sedimentary clay, common name for a number of fine-grained, earthy materials that become plastic when wet. Chemically, clays are hydrous aluminum silicates, ordinarily containing impurities, for example potassium, sodium, calcium, magnesium, or iron in small amounts [63, 64, 67, 68].

These absorbents are able to bind liquid CW agents and make them removable from cloth surfaces or personal equipment such as weapons, gas masks and so on. When solid absorbent materials like Fuller's Earth, soil, or diatomaceous earth are used, the contaminant is usually not altered. For example, petroleum products are readily absorbed but are not changed in their character. Thus, the sorbent material becomes toxic and so must be collected and disposed of afterwards. Caution needs to be taken during the collection process, as fine dust or particles can be inhaled or stuck to exposed skin.

Some NATO nations use adsorbent decontaminants in an attempt to reduce the quantity of chemical agent available for uptake through the skin. In emergency situations dry powders such as soap detergents, earth, and flour, may be useful. One example for the use of adsorbents combined with reactive capacity is the M 291 Skin decontamination kit of the U.S. Army. Figure 7.20 shows the kit in use.

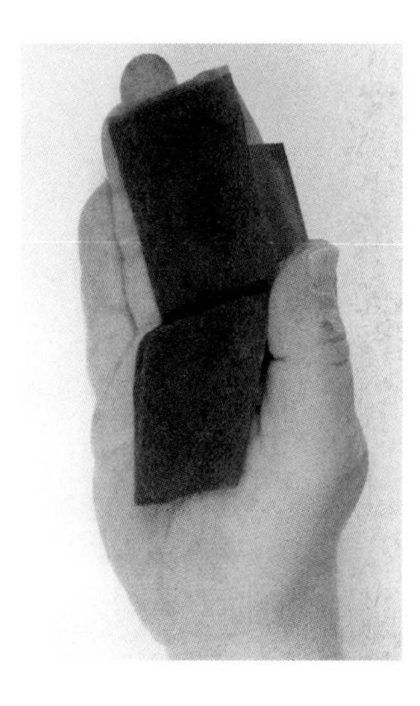

Figure 7.20 M 291 Skin Decontamination Kit [12, 65].

The kit operates with a mixture of so called reactive adsorbents (Ambersorb 348F carbonaceous adsorbent), made by Rohm&Haas [65] and charcoal. Another example of dealing with contamination of personnel and hasty decontamination is the German "Dutch Powder," a Hypochlorite containing powder for the hasty decontamination of personal gear and equipment. It is not suitable for the decontamination of human skin. Figure 7.21 shows the 60 g container ready for use.

Newer developments, such as the Canadian RSDL [14–16], an Oxime-based liquid decontaminant have FDA approval for the use on human skin.

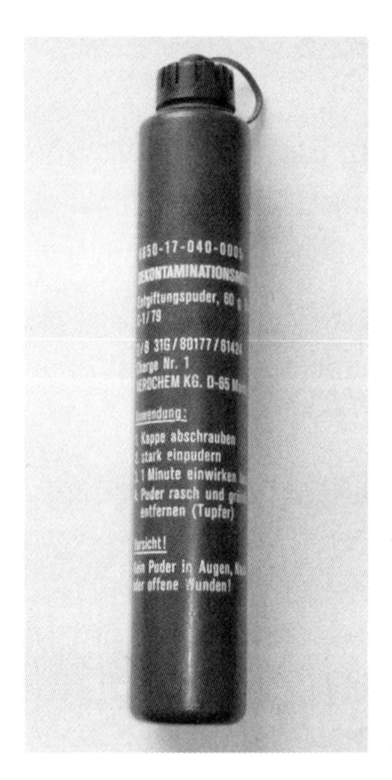

Figure 7.21 German decont powder.

RSDL is a topical decontamination solution that has been tested and shown to be able to reduce toxic effects from exposure to chemical warfare agents like VX and HD and T-2 toxin. RSDL contains Dekon 139 and a small amount of 2,3 butadiene monoxime (DAM). These compounds are dissolved in a solvent composed of polyethylene glycol monomethyl ether (MPEG) and water. This solvent system is particularly important as it promotes the decontamination reaction by actively desorbing, retaining and sequestering the chemical agent, while the active ingredient chemically reacts with, and rapidly neutralizes the vesicant chemical or the organo phosphorous nerve agent. This reaction starts immediately and neutralization is usually complete within seconds or a few minutes [14–17].

Future developments for hasty decontamination are clearly going into the direction micro emulsions with enzymes as a part of the active ingredients effective against CW agents. The advantage of these systems is the very low enzyme concentration necessary to destroy CW agents safely in combination with skin friendly micro emulsions [28, 30]. These systems could be usable on human skin without thinking about medical aspects such as burns due to the use of aggressive chemical compounds in the decontamination solution and they will be friendlier to the environment than established decontaminants are.

7.4.5.2 **Thorough Decontamination of Personnel**

Persons suspected being contaminated have to go through a thorough decontamination process even after hasty decontamination of protective gear and personal equipment. They are normally separated by sex and led into a decontamination system designed for the decontamination of personnel. The decont unit is strictly separated in a "black" or contaminated and a "white" clean area (see Chapter 7.1 generalities). After passing a strip down room, where the contaminated clothing is removed, they enter a wash down room, where they are showered. After have taken a shower they enter a re-dressing room, where new or freshly decontaminated clothing and personal gear is issued and after re-dressing they leave the decontamination unit on the "white" side: The decontamination unit has to be placed into the environment in a way, that no contact- or gas hazard can occur on the white side (Figure 7.22). Figure 7.20 shows the personnel decontamination unit introduced into the German Army.

Another decontamination system for personnel is integrated into the Decocontain container of the Kaercher Company [10, 60–62] and in the decontamination system MPD 100 of the OW.R. Company of Germany [13, 60–62].

Several other manufacturers are on the market with decontamination systems for personnel and can be found in the open literature [60–62]. They all have integrated showers in their decontamination process, due to the fact, that showering under use of specialized soap- or detergent systems are the most effective way to remove possible contamination on humans.

The "contaminated" personal equipment is lead into a specialized unit for the decontamination of personal equipment and protective suits/-clothing. Some of the units and processes are described in Section 7.4.3.

Figure 7.22 Personnel decontamination system of the German Army.

7.5 Summary and Conclusions

In this chapter a introduction into the world of C-decontamination was given. Several methods and procedures for decontamination under different conditions were explained. In the case of a contamination the deciders have to react fast to rescue most of the contaminated persons. Also, the need for new and better decontamination solutions was shown. One part of these solutions could be enzymes, which could play an important role in an overarching concept for decontamination and, therefore, for a better C-protection.

References

1 Franke, S. *et al.* (1977) *Lehrbuch der Militärchemie*, Vol. 2, Militärverlag der DDR.

2 Technical Manual TM 3-220, Department of the Army, US Army, 1953.

3 Industrial uses of ACL chlorinating Compositions ACL 56, ACL 59, ACL 60 and ACL 90 plus; Bulletin Nr. IC/WT-105, Monsanto Industrial Chemicals company, St. Louis, Mt.

4 NBC Defence, civil protection, FD 98, 1998, Christanini S.p.A. 37010 Rivoli, Italy.

5 Compilation of available data on building decontamination alternatives, EPA/600/R-05/036, March 2005.

6 Wagner, G.W. *et al.* (2002) *Decon green and other common decontaminants, efficacy tests.* Presentation at the 23th Army Scientific Conference, Orlando, FL, 2003.

7 ECBC undated, Decon Green Fact Sheet, Edgewood Chemical Biological Center, Aberdeen Proving Ground, Aberdeen, MD, USA.

8 Wagner, G.W. and Yang, Y.C. (2002) Rapid nucleophilic/oxidative decontamination of chemical warefare agents. *Industrial and Engineering Chemistry Research*, **41** (2002), 1925–28.

9 Katritzky, A.R. *et al.* (1989) Copper chloride-cyanopyridine complexes as catalysts for the decomposition of

fluorophosphonate esters. *Journal of Fluorine Chemistry*, **44** (1), 121–31.

10 Kärcher Future Tech, Waterpurification-Protective- and Decontamination-Equipment, Winnenden, Germany.

11 Kirk Othmer, Vol. 5, 3rd edn.

12 US Army Field Manual 3-5, Decontamination.

13 Odenwaldwerke Rittersbach (OWR), Decontamination- and Protective-Equipment, Rittersbach, Germany.

14 Today Science Tomorrow defence: Meeting the Challenge defence Research at Suffield CDN, Ralston ALTA (CAN) (1994).

15 Clewley, R.G., Purdon, J.G. and Chenier, C.L. (1992) Reactive Skin Decontaminant Reactivity Studies: The Effect of O-Acetyl 2,3 Butandion Monoxim on the stability of 2,3,Butandione Monoximat, DRES, Report No. 1382.

16 Zimmermann, Y. (1998) Diplomarbeit "Entwicklung eines Systems zur Zwischenlagerung und Sicherung von beschädigten, tox. Stoffe enthaltenden Wirkkörpern", Universität Dortmund, Physikalisch – Chemische Verfahrenstechnik.

17 RSDL Safety Information, http://www.rsdecon.com/safety.htm (2006).

18 Kappa, H.W. (1980) Hydrolysestudien an Organophosphaten, Thesis, Universität Mainz 1980.

19 Yang, Y.C., Baker, J.A. and Ward J.R. (1992) Decontamination of chemical warfare agents. *Chemical Reviews*, **92**, 1729–43.

20 CRC Handbook, 59th edn (1978) CRC Press Inc, C 726 ff.

21 Lagaly G. (1984) Stabilität kolloidaler Dispersionen, University of Kiel.

22 Jones, R.A. (1976) Application of Phase Transfer Catalysis in Organic Synthesis, School of Chemical Sciences, University of East Anglia, Norwich, GB.

23 Anna, H. and Alberti, J. (1978) Herkunft und Verwendung von Organohalogenverbindungen und ihre Verbreitung im Wasser und Abwasser, Kongressvorträge Wasser, Berlin 1977.

24 McConnel, G., Ferguson, D.M. and Pearson, C.R. (1975) Chlorinated hydrocarbons in the environment, *Endeavor*, **121**, 13–8.

25 German Patent, DBP DE 3815753.

26 German Patent, DBP DE 3638625 C2.

27 Bidyut, K.P. and Satya, P.M. (2001) Uses and applications of microemulsions, *Current Science*, **80** (8), 990–1001.

28 Richardt, A., Mitchell, S. and Blum, M.M. (2006) Chemie unserer Zeit 4.

29 Menger, F.M. and Rourk, M.J. (1999) Deactivation of mustard and nerve agents via low temperature microemulsions. *Langmuir*, **15**, 309–13.

30 Blum, M.M., Löhr, F., Richardt, A., Rüterjahns, H. and Chen, J.C-H. (2006) Binding of a Designated Substrate Analog to Diisopropyl Fluorophosphatase (DFPase): Implications for the Phophotriesterase Mechanism, JACD .

31 Richardt, A. and Mitchell, S. (2006) Enzymes for environmentally friendly decontamination of sensitive equipment. *Journal of Defense Science*, **10**, 261–5.

32 Larsson, K.M., Adlercreutz, P., Mattiasson, B. and Olsson, U. (1990) Enzymatic catalysis in microemulsions: enzyme reuse and product recovery. *Biotechnology and Bioengineering*, **36**, 135–41.

33 Bretschneider W. Prävention 02/2003, 76–78.

34 Gimaex-Schmitz Fire and Rescue GmbH. www.gimaex-schmitz.com (May 2007).

35 Spence, M., Ho, J. and Ogston, J. (2001) Decontamination of Vehicles Using CASCAD after Exposure to Biological Hazard.

36 National institute of Justice (2001) *Guide for the Selection of Chemical and Biological Decontamination Equipment for Emergency First Responders*, Vol 1. NIJ Guide 103-00, Vol. 1 US Department of Justice, October 2001.

37 NBC Team Ltd, P.O. Box 11040, 921 Barton St., Stoney Creek, Ontario, Canada, L8E 5P9.

38 Sandia (undated draft report) DF 200, An enhanced formulation for decontamination and mitigation of CBW agents and biological pathogens.

39 Sandia (2002) Sandia Decon Formulation, Publication Nr.: SAND2000-0625, Sandia National Laboratory.

40 Sandia (2003) Field demonstration for biological agent decon, http//www.sandia.gov/sandiaDecon/demos/demo5.htm (23 May 2007).

41 DuPont (1998) DuPont Oxone Monopersulfate Compound, technical information, Nov. 2004.

42 DuPont Oxone compound applications, Nov. 2004.

43 Meyer, V. (1886) Universitätslaboratorium Göttingen, *Ueber Thiodiglykolverbindungen, Berichte*, **19**, 3256–66.

44 Stockholm International Peace Research Institute (1971) *The Problem of Chemical and Biological Warfare*, Vol. 1, *The Rise of CB Weapons*, Humanities Press, NY, pp. 47–50.

45 Wojtowicz, J.A. (1979) *Kirk Othmer Encyclopedia of Science and Technology*, 3rd edn, Vol. 5, Olin Corp., pp. 795–802.

46 US Army Field Manual 3-9, Potential Military Chemical/Biological Agents and compounds.

47 Chemical agent data sheet, Vol. I, EO-SR-74001, Edgewood Arsenal, 1974.

48 Bartlett, P.D and Swain, C.G. (1949) Kinetics of hydrolysis and displacement reactions of b,b'-Dichlorodiethyl Sulfide and of b Chloro-b-hydroxydiethyl Sulfide. *Journal of the American Chemical Society*, **71**, 1406–15.

49 http://www.mitretek.org/ChemistryOfMustard/Sarin/Soman/VX.htm (10 May 2007).

50 Yang, Y.C., Szafraniec L.L., Beaudry, W.T. and Ward, R.J. (1988) Kinetics and mechanism of the hydrolysis of 2-Chloroethylsulfides. *The Journal of Organic Chemistry*, **53** (14), 3293–7.

51 Albrizo, J.M. and Ward, J.R. (1991) Soman hydrolysis catalyzed by HEPES buffers. *Journal of Molecular Catalysis*, **66** (2), 191–4.

52 Ward, J.R., Yang, Y.C., Wilson, R.B., Burrows, W.D. and Ackerman, L.L. (1988) Base-catalysed hydrolysis of 1.2.2-trimethypropyl methylphosphonofluoridate an examination of the saturation effect. *Bioorganic Chemistry*, **16** (1), 12–6.

53 Yang, Y.C., Szafraniec, L.L., Beaudry, W.T. and Bunton, C.A. (1993) Perhydrolysis of nerve agent VX. *The Journal of Organic Chemistry*, **58**, 6964–5.

54 Yang, Y.C. (1999) Detoxification of nerve agent VX. *Accounts of Chemical Research*, **32**, 109–15.

55 Epstein, X. *et al.* (1974) The kinetics and mechanism of hydrolysis of phosphonothiolates in dilute aqueous solutions. *Phosphorus, Sulfur, and Silicon and the Related Elements*, **4**, 157–63.

56 Szanfraniec, L.L. *et al.* (1990) On the Stoichiomety of Phosphonothiolate Ester Hydrolysis, CRDEC-TR.212, AD-A250773.

57 Handbuch für Schirrmeister Ch und Instrukteure Ch, 3rd edn, Militärverlag der DDR1986.

58 History of decontamination, US Army Soldier and Biological Chemical Command.

59 AZS-EA 12, Anlage zur Spezialbehandlung, Field manual A053/1/214, NVA, 1982.

60 http://www.ncjrs.gov/pdffiles1/nij/189725d.pdf#search=%22Decontamination%20Equipment%22 (15 February 2007).

61 http://www.epatechbit.org/pdf/103-00_vol2_part3.pdf22 (14 February 2007).

62 http://www.studysphere.com/Site/Sphere_13783.html22 (14 February 2007).

63 Fuller's earth. Encyclopædia Britannica (2006) Encyclopædia Britannica Online. 5 October 2006.

64 Wikipedia, The Free Encyclopedia.

65 www.rohmhaas.com (22 March 2007).

66 Yang, Y.C. *et al.* (1990) Oxidative detoxification of phosphonothiolates. *Journal of the American Chemical Society*, **112** (18), 6621–7.

67 http://www.ncjrs.gov/NIJ Guide103-00 (14 February 2007).

68 http://en.wikipedia.org/wiki/Fuller's_earth (14 February 2007).

8
A Short Introduction to Enzyme Catalysis

Marc-Michael Blum

8.1
Introduction

The principles of protein chemistry and enzyme catalysis have been the subject of many textbooks, review articles, specialized volumes and numerous contributions in the scientific literature. This chapter cannot and shall not be an all-embracing introduction into the field, but will try to give those readers of this volume who are not specialists in biochemistry and enzymology (who can of course skip this chapter if they whish) the necessary toolkit at hand to enable them to steer their way through the chapters dealing with enzymes for decontamination in more detail.

Those readers who wish to dig more deeply into the field are referred to textbooks on biochemistry and many more detailed monographs, some of which are listed in the section on recommended reading at the end of this chapter.

8.2
Thermodynamic Equilibrium, Reaction Velocities and the Need for Catalysis

If we want to find out whether a chemical reaction will proceed spontaneously we have to turn to the field of thermodynamics. Two equations are of fundamental importance. The first relates the Gibbs free energy G to the equilibrium constant k of a reaction.

$$\Delta G = -RT \ln k \tag{8.1}$$

R is the gas constant ($8.314\,\mathrm{J\,mol^{-1}\,K^{-1}}$) and T the absolute temperature in Kelvin. The second equation relates the change in Gibbs free energy with changes in enthalpy H and entropy S. A negative ΔG will result in a positive value for the equilibrium constant indicating that the equilibrium will be located on the side of the products.

Decontamination of Warefare Agents.
Edited by André Richardt and Marc-Michael Blum

ISBN: 978-3-527-31756-1

$$\Delta G = \Delta H - T\Delta S \tag{8.2}$$

This second relation also explains why certain reactions proceed spontaneously even though they draw energy from the surrounding. The gain in entropy will compensate the endothermic enthalpy part rendering ΔG negative and therefore making the reaction exergonic. The mixing of ice and salt to produce temperatures well below 0°C is an example of such a process. If ΔG is positive the process is endergonic and will not proceed spontaneously and if ΔG is 0 the system is in equilibrium. As the Gibbs free energy is dependent on pressure and temperature it is important that these quantities are known for the process of interest.

If the chemical reaction of interest does posses a negative ΔG at the specified reaction conditions we know that the reaction will proceed spontaneously but we do not know the reaction velocity. The Equations 8.1 and 8.2 will tell us about the location of the equilibrium and the direction a system not in equilibrium will head to (towards $\Delta G = 0$).

Several theories exist to account for chemical kinetics. The simple collision theory is normally used to introduce the topic in physical chemistry courses using ideal gases as a model. If we are interested in biocatalysis a more demanding description is required and transition state theory is a suitable model for our purposes. The theory is best described using a diagram of a fictive reaction as depicted in Figure 8.1. In this diagram the Gibbs free energy is plotted against a reaction coordinate that proceeds from the substrates to the final products of the reaction. The difference between G for the substrate and G for the product is the ΔG we have already met in Equations 8.1 and 8.2. As the reaction proceeds it will go uphill in terms of G. The highest peak represents the transition state in which bonds are being made and broken. As the reaction approaches the product it might do so in a direct way or through one or more saddle points that represent intermediates. In order to be able to cross the barrier represented by the transition state the necessary Gibbs free energy of activation $\Delta G^{\ddagger}$ has to be provided which is the difference between G for the substrates and G for the transition state.

It is assumed that the ground state of the substrate and the transition state are thermodynamic equilibrium and the concentration of the transition state $[X^{\ddagger}]$ can be calculated from $\Delta G^{\ddagger}$ according to (Equation 8.3).

$$[X^{\ddagger}] = [X]\exp\left(-\frac{\Delta G^{\ddagger}}{RT}\right) \tag{8.3}$$

It can be shown that the rate of consumption of X is given by (Equation 8.4).

$$-\frac{d[X]}{dt} = [X]\left(\frac{kT}{h}\right)\exp\left(\frac{-\Delta G^{\ddagger}}{RT}\right) \tag{8.4}$$

The first-order rate constant for the consumption of X is therefore (Equation 8.5).

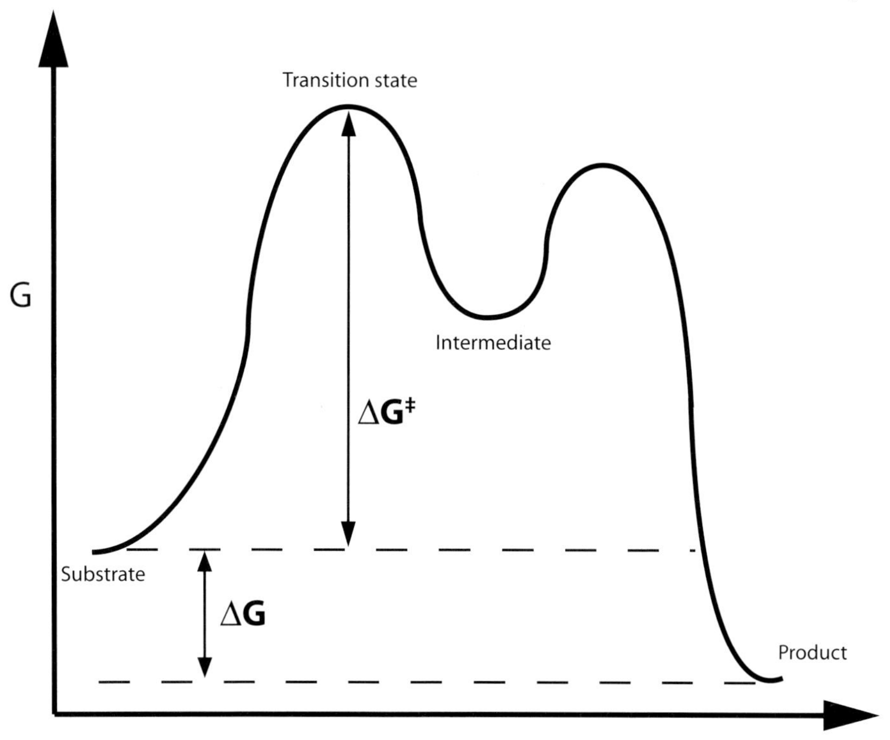

Figure 8.1 Fictive reaction of a substrate along the reaction coordinate processing via a high energy transition state and an intermediate to the product.

$$k_1 = \left(\frac{kT}{h}\right)\exp\left(\frac{-\Delta G^{\ddagger}}{RT}\right) \tag{8.5}$$

Many reactions that show a negative ΔG but proceed with reaction velocities so slow that the reaction half-times are in the range of thousands or even millions of years. One example with great industrial importance is the reaction of nitrogen and hydrogen to produce ammonia as employed by the Haber-Bosch process. At ambient temperature ΔG is negative but the reaction does not proceed at measurable speed.

Catalysts will enhance reaction rates without being consumed in the reaction. The catalyst can form intermediates with the substrate but will be recycled to be available for the next turnover cycle. It is important however to note that the catalyst in unable to change the thermodynamic equilibrium as defined by Equation 8.1. The rate enhancement due to the presence of the catalyst will be achieved by lowering the free energy of the transition state, the highest peak on the reaction coordinate which is rate limiting. This can be realized by stabilizing the transition state due to favourable interaction with the catalyst or by providing an alternative reaction pathway that will proceed through a different transition state with lower energy. Chemical industry and nature both use catalysts and nature is often far more efficient than industrial chemistry, a good reason why industry employs more and more enzymes for industrial processes. A good example is the production of ammonia from hydrogen and nitrogen mentioned above. Industry employs reaction temperatures of 400°C and a pressure of 300 bar in addition to an iron-based catalyst being responsible for more than 1% of the world energy consumption. Nature employs the enzyme nitrogenase that is able to produce ammonia from nitrogen at ambient temperatures and normal pressure. It should be noted however that between 16 and 24 molecules of the energy rich compound ATP is consumed to turn one molecule of N_2 into two molecules of NH_3.

It is not only important for biochemical processes in nature to enhance very slow reactions, but sometimes these reactions must be completed within a very short timeframe. Nerve conduction is just one example, but a very relevant one as one of the prime enzymes involved is also the main target for nerve agents as discussed in other chapters in this book. The necessity to rapidly hydrolyse acetylcholine in the synaptic cleft to turn off the nerve signal requires a very efficient enzyme. Indeed, the enzyme acetylcholinesterase acting as the catalyst for this reaction is a perfect enzyme enhancing the reaction velocity to a value where the limiting factor of the reaction is no longer the activation barrier of the transition state but the substrate diffusion rate.

8.3
Proteins and Structure – A Prerequisite for Enzyme Function

Enzymes are proteins acting as catalysts. Other biocatalysts include catalytic antibodies and catalytically active RNA (Ribozymes). Proteins are formed by the 20 known natural alpha-amino acids (rare exceptions like Selenocysteine exist) shown in Table 8.1 by formation of a continuous polypeptide chain. The sequence of amino acids of a protein is known as the primary structure and is encoded by the standard genetic code (a codon of three nucleic acids per amino acid). This primary structure contains in principle all the information that will finally determine the three dimensional shape of the protein. As certain amino acid sidechains might be subject to posttranslational modification and cofactors like metal ions might be present reality can be a bit more complicated. In fact even though great advances have been made to understand protein folding it is still almost impossible to predict the three dimensional shape of a protein just from the primary structure without additional information.

Certain sequences of amino acids lead to the formation of three dimensional structures in local segments of the biopolymer chain known as the secondary structure. Among these local structure elements are α-helices and β-sheets. α-helices are right-handed coiled formations resembling a spring and are formed by specific hydrogen bonding interactions between residues. Figure 8.2 depicts the

Table 8.1 The 20 common natural amino acids.

Amino acid name	*Three letter code*	*One letter code*	*Chemical structure*
Alanine	Ala	A	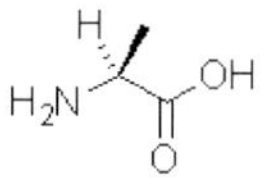
Arginine	Arg	R	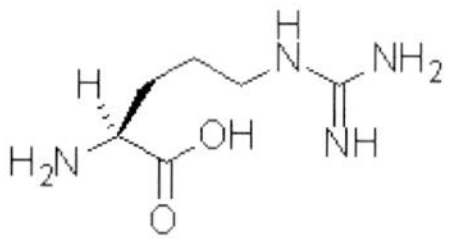
Asparagine	Asn	N	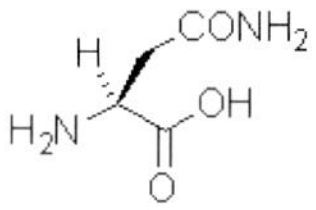
Aspartic acid	Asp	D	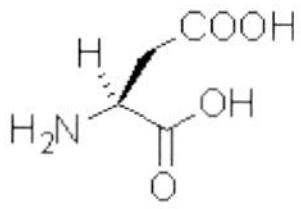
Cysteine	Cys	C	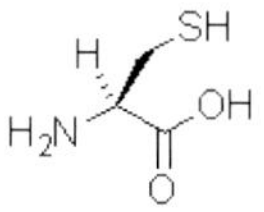
Glutamic acid	Glu	E	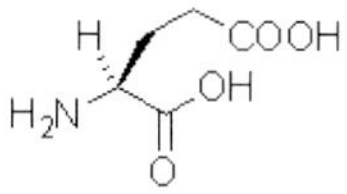
Glutamine	Gln	Q	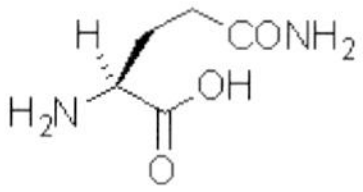
Glycine	Gyl	G	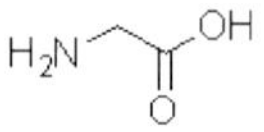
Histidine	His	H	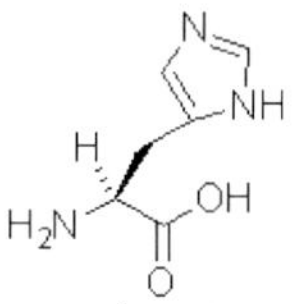
Isoleucine	Ile	I	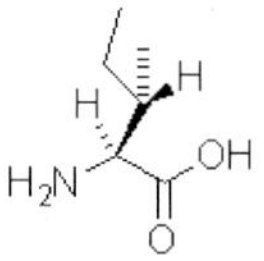
Leucine	Leu	L	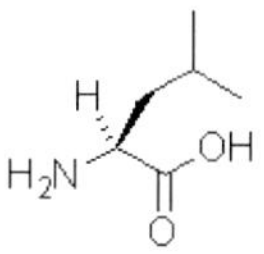
Lysine	Lys	K	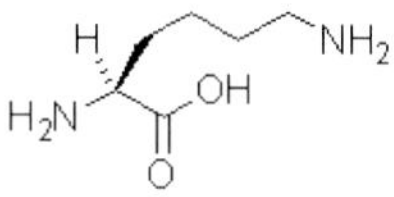

Table 8.1 continued

Amino acid name	*Three letter code*	*One letter code*	*Chemical structure*
Methionine	Met	M	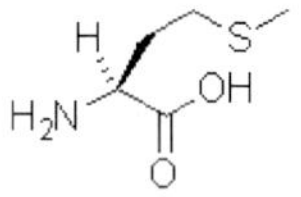
Phenylalanine	Phe	F	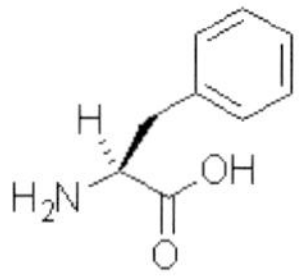
Proline	Pro	P	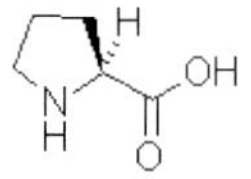
Serine	Ser	S	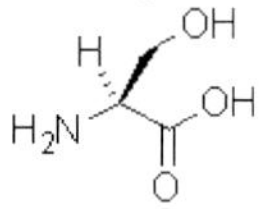
Threonine	Thr	T	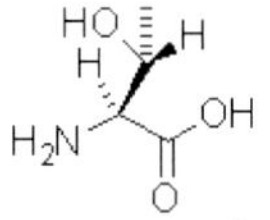
Tryptophan	Trp	W	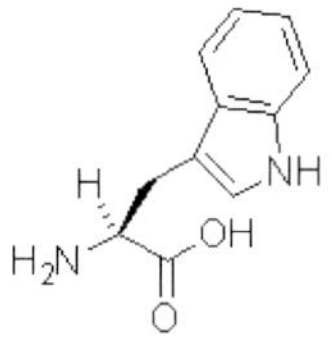
Tyrosine	Tyr	Y	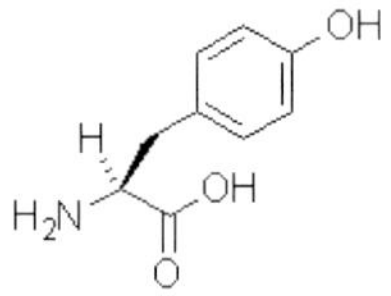
Valine	Val	V	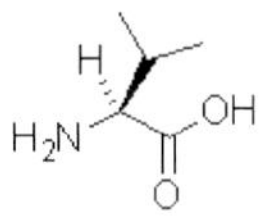

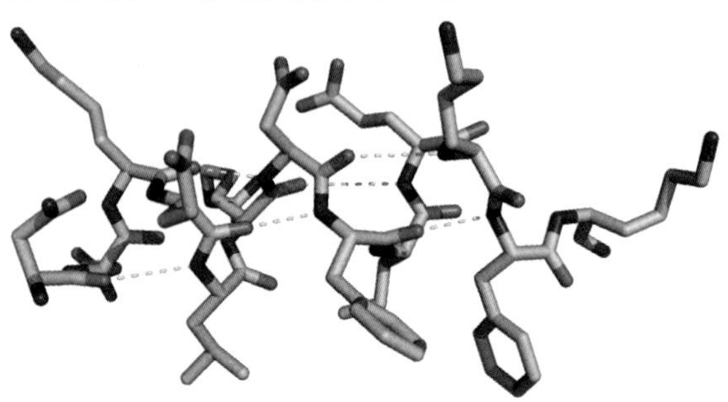

Figure 8.2 Stick representation of an α-helical peptide segment. Important hydrogen bonding interactions are represented as dashed lines.

structure of a simple small polypeptide forming a single helix in a stick representation while Figure 8.3 shows the same structure in the common cartoon representation in which only the backbone of the polypeptide chain is shown as well as secondary structure elements.

β-sheets consist of β-strands connected by hydrogen bonds. The β-strand itself is a stretch of amino acids with almost fully extended protein backbones. Figure 8.4 shows the structure of a small polypeptide folding into a β-sheet in stick representation and Figure 8.5 is the corresponding cartoon representation clearly showing the individual β-strands.

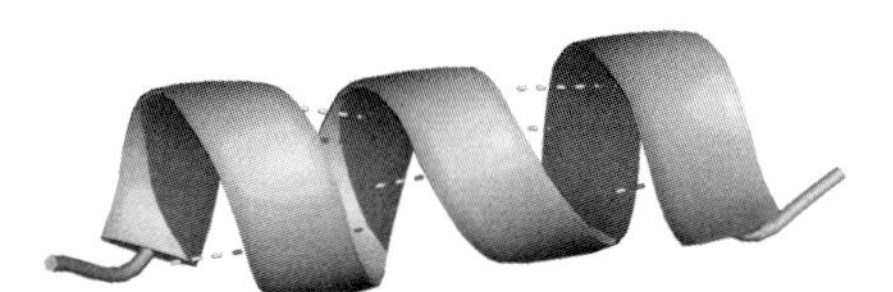

Figure 8.3 Cartoon representation of the α-helix from Figure 8.2. Hydrogen bonds are shown for orientational purposes.

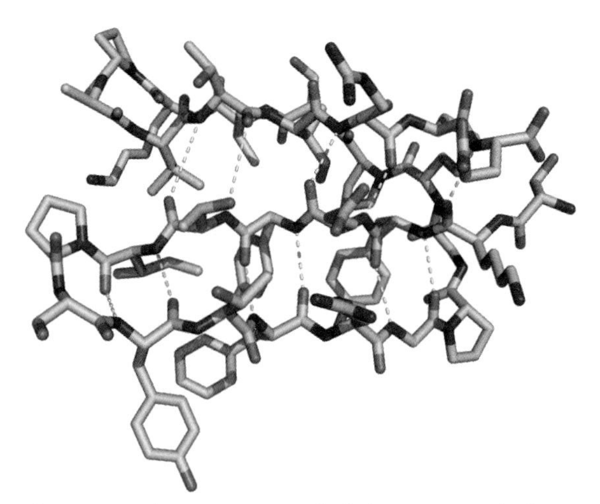

Figure 8.4 Stick representation of a β-sheet composed of three β-strands. Important hydrogen bonds are indicated by dashed lines.

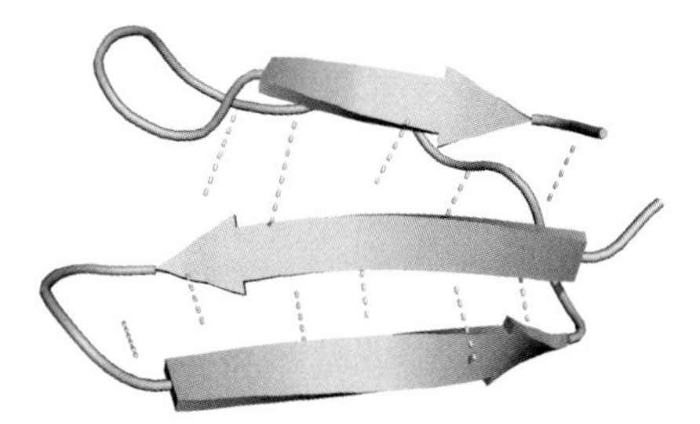

Figure 8.5 Cartoon representation of the β-sheet from Figure 8.4.

Other parts of a protein might form turns or show no specific secondary structure at all (also known as random coil). The overall structure of the protein in space is then known as the tertiary structure. If a protein is assembled from more than one unit the structure of the multi-subunit complex is known as the quaternary structure. The tertiary structure of the nerve agent hydrolysing enzyme diisopropyl fluorophosphatase (DFPase) from the squid *Loligo vulgaris* is shown in Figure 8.6 as an example for a protein containing mainly β-sheets and the sulfur mustard hydrolysing enzyme haloalkane dehalogenase LinB from *Sphingobium japonicum UT26* is shown in Figure 8.7 as an example for a protein with a considerable content of α-helices. Both enzymes are treated in more detail in Chapter 9.

The folded state of the protein is in a delicate balance with its unfolded state. For most proteins the Gibbs free energy of the folded compared to the unfolded state is only in the range of about −50 kJ/mol although quite a few proteins are known that are very stable in solution even at high temperatures like those found in thermophile bacteria. Hydrogen bonding is not the major driving force for protein folding. As the potential hydrogen bond donors and acceptors in the polypeptide can form hydrogen bonds with the surrounding water molecules in the unfolded state. The water molecules are very flexible in their orientation in order to ensure optimal interactions. The importance of hydrogen bonds within folded proteins is rather that the potential donors and acceptors are satisfied in order not to destabilize the folded state. On the other hand it is often found that proteins contain a hydrophobic core. Hydrophobic side chains like those of the amino acids phenylalamine, methionine, valine or leucine are packed together inside of the protein stabilizing the folded state by hydrophobic interactions and avoidance of

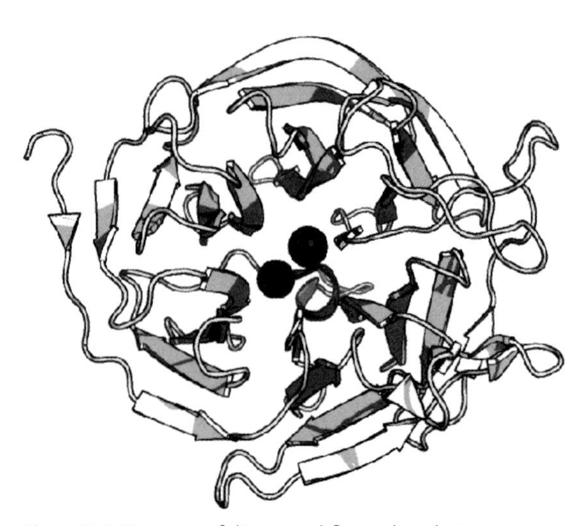

Figure 8.6 Structure of diisopropyl fluorophosphatase, a nerve agent hydrolysing enzyme almost exclusively composed of β-sheets. The two black spheres represent calcium ions.

Figure 8.7 Structure of haloalkane dehalogenase LinB, a sulfur mustard hydrolysing enzme with significant α-helix content.

the exposure of these side chains towards the solvent. This is known as the hydrophobic effect and often also plays an important role in ligand binding to proteins.

The ability to bind the substrate is the first step in enzyme catalysis. The three dimensional structure of the folded protein provides binding sites for ligands and substrates with specific spatial characteristics. It is important to note that proteins in the folded state are not rigid bodies. The structures are fluctuating and motions of certain parts of an enzyme can be important for catalysis but also for the binding of a substrate or the release of a product. Two special cases need to be mentioned. Sometimes a substrate can only bind to an enzyme after some other ligand (the effector) is bound to a different binding site resulting in some rearrangements in the structure. This is known as allosteric regulation. The other case is related to the biding of the ligand itself. According to the classical lock and key model, formulated by Emil Fischer in 1894, both the enzyme and the binding substrate possess complementary geometric shapes that fit exactly into one another like lock and key. While this explains the specificity of certain enzymes for specific substrate molecules it does not account for enzyme promiscuity and it also fails to explain the stabilization of the transition state that will be discussed below. Daniel Koshland modified this model in 1958 by the introduction of the induced fit model. According to Koshland the interaction of the substrate with the enzyme will reshape the binding site to some extent. As a result the substrate will not only bind to a rigid cavity but will reorient the amino acid sidechains for optimal binding and catalytic efficiency.

The two most dominant methods for the determination of protein structure at present are X-ray diffraction on protein crystals and NMR spectroscopy on proteins in solution. While X-ray crystallography, the most prominent method, determines atom positions of a protein in a crystallized state and therefore generates a snapshot image of the structure, NMR structures are reported as a bundle of structures satisfying the distance constraints between atoms obtained from different NMR experiments. The Protein Data Bank (PDB) stores these protein structures in electronic form and can bee accessed freely on the web (www.rcsb.org). It was shown that both methods basically generate the same information and that X-ray structures and NMR structures of the same protein are compatible. In Figure 8.8 the protein backbone of the reduced form of human thioredoxin is shown in a ribbon representation as an example. The black ribbon is the protein backbone of the structure obtained by X-ray crystallography (PDB-code: 1ERT), the gray ribbons represent the bundle of NMR structures (PDB-code: 4TRX).

8.4 From Ligand Binding to Enzyme Activity

Binding of a potential substrate to a protein is only one step in enzyme catalysis. Antibodies for example are proteins with very high binding affinities for their specific binding partner (the antigen) but still they do not catalyze any reactions. Many drugs bind strongly to enzymes but they are not turned into products and

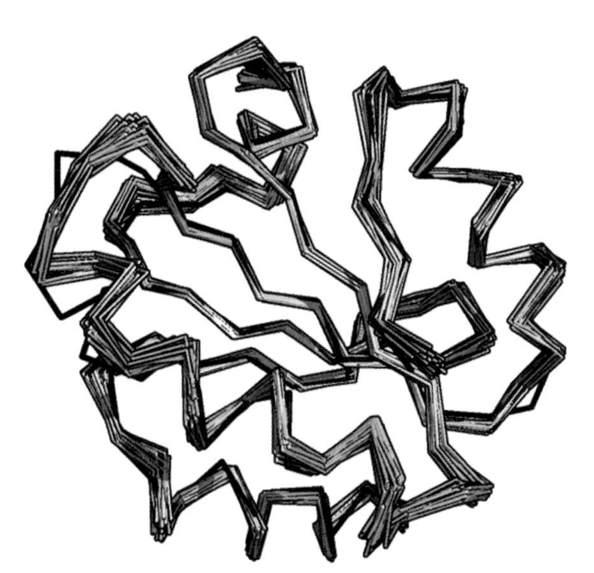

Figure 8.8 Ribbon diagram of the protein backbone of human thioredoxin. The protein backbone obtained from X-ray crystallography is in black and the structure bundle obtained by NMR spectroscopy is shown in grey.

rather inhibit the turnover of the original substrate molecules. So what differentiates an enzyme from a protein with good binding properties for a substrate?

One fundamental observation is that for example antibodies have great binding affinities for the ground state of a ligand molecule while enzymes have exceptional affinity for the transition state of the reaction it catalyses. As we saw earlier rate enhancements for the reaction depend on lowering the Gibbs free energy of activation. This can be achieved by tight binding of the transition state that results in its stabilization. An equation for the enhancement of a reaction rate by a catalyst related to the relative strength of the binding of the transition state vs. the ground state was formulated by Kurz in 1963 based on a combination of transition state theory and the thermodynamic (Born-Haber) cycle.

$$\frac{k_{cat}}{k_u} \approx \frac{K_S}{K_T} \tag{8.6}$$

In this equation k_{cat} is the rate constant for the catalyzed reaction and k_u for the uncatalyzed reaction, k_{cat}/k_u is therefore the rate acceleration due to catalysis. K_S is the dissociation constant for the enzyme substrate complex and K_T for the enzyme transition state complex. From Equation 8.6 follows that the tighter the binding of the transition state by the enzyme compared to the enzyme substrate complex the higher the rate acceleration.

Classic examples for transition state stabilization are serine proteases. These are digestive enzymes like trypsin that catalyze the hydrolysis of a peptide (amide) bond. A serine residue attacks the carbon of the peptide bond and generates a tetrahedral transition state as depicted in Figure 8.9. In the transition state negative charge accumulates on the former carbonyl carbon atom. To stabilize this transition state, serine proteases posses a structural feature called an oxyanion hole. This hole is created by two backbone amides that function as hydrogen bond donors

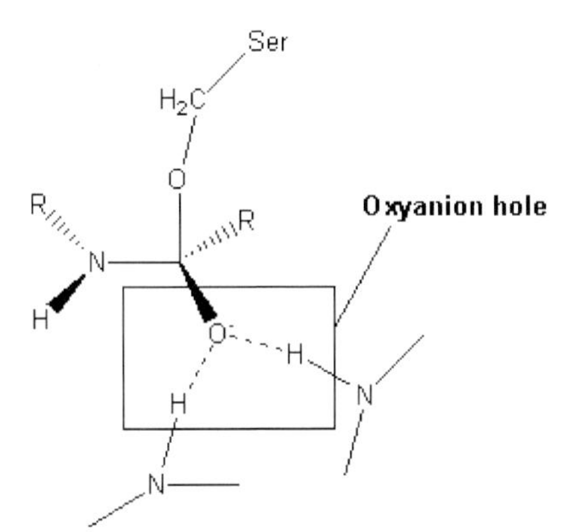

Figure 8.9 Transition state in serine proteases. The developing negative charge on the oxygen is stabilizes in the oxyanion hole by hydrogen bonding to two backbone amides.

for the oxyanion generated in the transition state. These hydrogen bonds stabilize the transition state, lower the free activation energy and enhance the reaction rate.

Another possibility to stabilize a developing negative charge is the presence of a metal ion. Metal ions in proteins function as electrophiles withdrawing electron density from bound ligands. This can be a carbonyl group of an ester or amide or a phosphoryl group of a phosphorus ester. In this case the metal ion functions as an oxyanion hole. Another way enzymes utilize metal ions is for the production of hydroxide ions at neutral pH. Water coordinated to metal ions with a charge greater than one shows a significantly reduced pK_a and hydroxide bound to a metal is an effective nucleophile. A prime example is the enzyme carbonic anhydrase that contains a hydroxyl bound to a zinc ion that acts on the substrate carbon dioxide. Also amino acid sidechains can act as nucleophiles apart from metal bound hydroxide ions. In case of the serine proteases the active nucleophile is a serine residue. Also the acidic amino acids glutamic and aspartic acid can function as nucleophiles. Aspartate D229 is proposed as the active nucleophile in the phosphotriesterase DFPase (Figure 8.6 and Chapter 9).

One important aspect for the rate enhancement in enzymes catalyzing ligation or addition reactions is that of proximity and pre-orientation. By binding two substrates in the proper orientation for reaction an originally intermolecular reaction becomes a quasi-intramolecular reaction. This leads to an enhanced effective concentration of the reagents and is also favourable in entropic terms. The condensation of two molecules to a larger one leads to a loss in translational and rotational entropy, which is unfavourable in terms of the Gibbs free energy of activation for the reaction. If the two substrates are already bound to the enzyme and the reaction becomes quasi-intramolecular this loss in entropy is significantly reduced and the activation barrier is lowered.

Many reactions also benefit from the different environment of an enzyme active site compared to the bulk solvent. While the dielectricity constant of water is about $\varepsilon = 80$ at room temperature it is significantly reduced inside of enzymes and easily reach values of $\varepsilon < 10$. In the hydrophobic cores of proteins a value of $\varepsilon = 4$ is assumed. The high dielectric constant of water reduces the impact of electrostatic interactions. This is apparent from the coulomb's law (Equation 8.7) describing the electrostatic force between two charged particles.

$$F = \frac{1}{4\pi\varepsilon} \frac{|q_1||q_2|}{r^2} \tag{8.7}$$

The interaction becomes stronger as the charges q of the two particles increase and is reduced by an increasing distance r between the particles and an increasing dielectric constant ε. Interactions like the stabilization of a negative charge by a metal cation are therefore stronger in an enzyme cavity compared to the bulk solvent.

Finally the importance of acid/base catalysis has to be mentioned. Many reactions involve proton transfer. Protonation or deprotonation are typical steps in hydrolytic reactions but they can also be found as steps in the mechanisms of various other enzyme catalyzed reactions. Specific and general acid/base have to be separated. Specific means that catalysis occurs specifically by hydroxide or protons while general means that the nature of the acid/base is not restricted to H^+/OH^-. As an example we can return to the mechanism of serine proteases. The role of the oxyanion hole and the role of a serine residue as the attacking nucleophile were discussed. As the normal hydroxyl group on serine is not nucleophilic it is necessary that a proton acceptor removes the proton of the hydroxyl group. A neighbouring histidine residue fulfils this task and accepts the proton therefore acting as a general base catalyst as depicted in Figure 8.10.

This protonated histidine will also act as an acid catalyst. After the transition state as shown in Figure 8.9 is created, the amine moiety is the potential leaving group that will generate to peptide fragments. This moiety is a rather bad leaving group but protonation of the nitrogen atom turns it into a good leaving group. This proton is transferred from the protonated histidine as shown in Figure 8.11.

Several amino acid sidechains contain functional groups that can be protonated and deprotonated and can therefore serve in acid/base catalysis. Mosten often aspartic and glutamic acid as well as histidine is involved but also lysine, arginine, threonine and serine are found. The ability of a group to accept or donate a proton is determined by its pK_a. In order to be able to act as proton acceptors and donors the pK_a of these functional groups should not be too different from the pH of the surrounding solvent. In case of Glu and Asp the pK_a is slightly above 3 while the pH of the solvent around an enzyme in its native environment is often close to a neutral pH of 7 so both should be present as aspartate and glutamate anions. While this is true for solvent exposed residues this is not necessarily so for internal groups inside the protein or in the active site. The local electrostatic environment, dielectric constant as well a specific interactions with neighbouring groups can

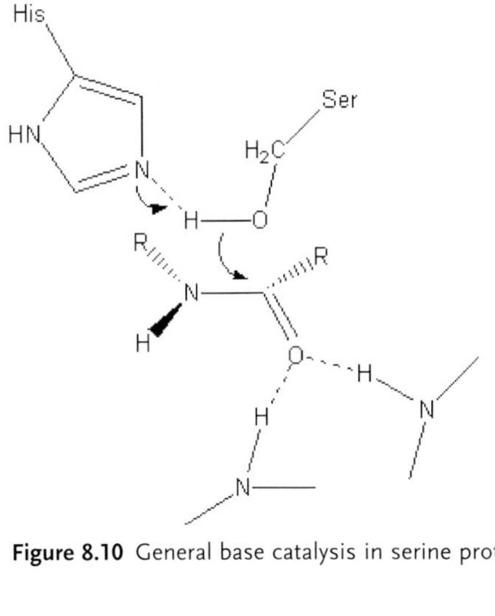

Figure 8.10 General base catalysis in serine proteases by a histidine.

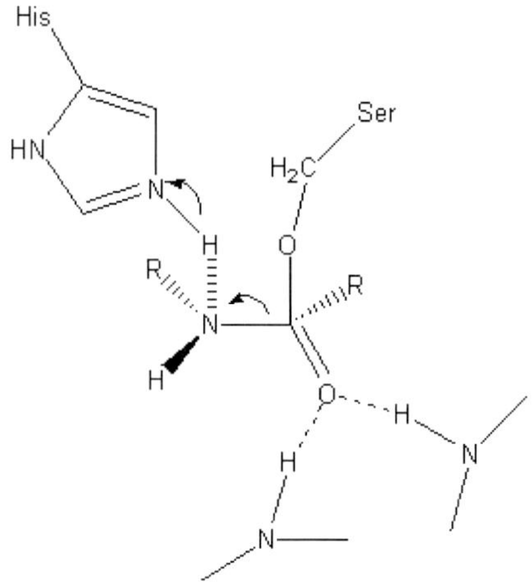

Figure 8.11 General acid catalyis in serine proteases by a protonated histidine.

alter the pK_a dramatically and it should not be assumed that a residue is protonated or not by simply referring to the pKa of the isolated amino acid.

8.5
Enzyme Classification

A clear indication for an enzyme is the ending of its name on –ase, although some older names are still in common use and do not follow this convention. The serine proteases trypsin, chemotrypsin and subtilisin are good examples for this. Normally the name of the enzyme gives information on the chemical reaction that is catalyzed. Names like diisopropyl fluorophosphatase, D-xylose isomerase or dihydrofulate reductase already indicate the reactions involved.

The International Union of Biochemistry and Molecular Biology (IUBMB) publishes an enzyme nomenclature and classification scheme currently in its sixth edition. The Enzyme Commission (EC) number classifies enzymes by the chemical reaction they catalyze. The code consists of the letters EC followed by four numbers separated by periods. Each of the four numbers describes the chemistry catalyzed by the enzyme in more detail from left to right. The first number is the top-level code and separates enzymes into six main categories that can be found in Table 8.2.

To illustrate the systematics of the system we will investigate the EC code for the enzymes DFPase (Figure 8.6) and Haloalkane Dehalogenase LinB (Figure 8.7). DFPase is a hydrolase (EC 3) that hydrolyses ester bonds (EC 3.1). The substrates of DFPase are phosphorus triesters (EC 3.1.8) and it is a diisopropyl fluorophosphatase (**EC 3.1.8.2**). LinB is also a hydrolase (EC 3) but it acts on halide bonds (EC 3.8). The substrates of LinB are carbon-halide compounds (EC 3.8.1) and LinB is a haloalkane dehalogenase (**EC 3.8.1.5**).

Table 8.2 Top-level EC codes.

Group	Class	Reaction catalyzed
EC 1	Oxidoreductases	Oxidation/reduction reactions
EC 2	Transferases	Transfer of a functional group from one substance to another
EC 3	Hydrolases	Formation of two products from one substrate by hydrolysis
EC 4	Lyases	Non-hydrolytic and non-oxidative cleavage of bonds
EC 5	Isomerases	Intramolecular rearrangements
EC 6	Ligases	Joining of two molecules by forming a new covalent bond

8.6 Kinetics of Enzyme Reactions

The kinetics of enzyme catalyzed reaction can be complicated but even when the reaction mechanism is complex many of these reactions follow the well known Michaelis–Menten equation proposed in 1913 by Leonor Michaelis and Maud Menten. The model assumes a simple mechanism (Equation 8.8) in which substrate and enzyme first form a complex in a reversible reaction, which then proceeds to the product and the free enzyme in an irreversible reaction step.

$$\mathrm{E+S} \underset{k_{-1}}{\overset{k_1}{\rightleftarrows}} \mathrm{ES} \xrightarrow{\mathrm{k_2}} \mathrm{E+P} \tag{8.8}$$

Although this very simple mechanism is in most cases insufficient to describe the mechanism of an enzyme, many more complex treatments can yield the Michaelis–Menten mechanism. In these cases the rate constants like k_2 are composed of many individual rate constants. It should be noted however that some enzyme kinetics cannot be described using the Michaelis–Menten equation.

Two values are of special importance in enzyme kinetics. The first is V_{max} which is the maximum reaction rate of the enzyme and is closely related to k_2. As V_{max} depends on the enzyme concentration a useful derivative is $k_{cat} = V_{max}/[E_0]$. The other value is the Michaelis constant Km. In the simple case of the original mechanism (Equation 8.8) Km is a value indicating the affinity of the substrate S to the enzyme E. In the case that $k_2 << k_1$ km is simply k_{-1}/k_1. The Km value is also equal to the substrate concentration at half-maximal rate. The full Michaelis–Menten equation relates V_{max} and Km to the overall rate of reaction depending on the substrate concentration (Equation 8.9).

$$v = \frac{V_{max}[S]}{K_m + [S]} \tag{8.9}$$

This leads to a saturation curve depicted in Figure 8.12.

The higher Km, the higher the necessary substrate concentration to reach V_{max}. As substrate concentration are often limited by effects like limited solubility it

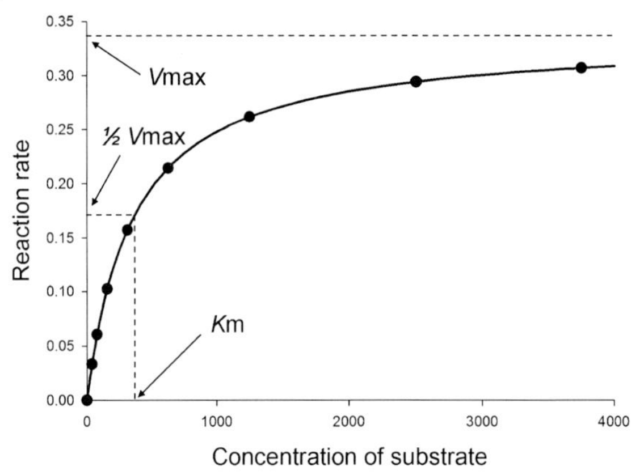

Figure 8.12 Realtion between substrate concentration and reaction rate in Michaelis–Menten kinetics.

might be impossible to operate an enzyme at its V_{max}. One warning at this point: as many reaction mechanisms are far more complicated it is advisable to be very cautious and preferably not to refer to Km as the affinity of the substrate to the enzyme but to use the term apparent affinity. The same is true for k_{cat}, which is an apparent first-order rate constant.

Not all small molecules are either substrates of an enzyme or non-interacting with the enzyme at all. Some molecules might interact with the enzyme without being turned into products. The compounds are termed inhibitors if they lower the overall rate of the reaction. In rare cases they might also be beneficial for the reaction. One case that is easy to visualize is a compound that competes with the substrate for binding in the active site of the enzyme. Whenever the inhibitor blocks the active site no substrate can enter and the overall rate is reduced. Raising the substrate concentration can compensate this. Therefore a competitive inhibitor changes the Km as a higher substrate concentration is needed to reach V_{max} but V_{max} itself is unchanged. In case of non-competitive inhibitors the binding to the inhibitor does not influence the binding of the substrate (by binding to an allosteric site and not to the active site) but does inhibit the release of product. In this case V_{max} is reduced but Km remains unchanged. The third and final possibility is an inhibitor that changes both V_{max} and Km. All these types of inhibition are reversible but also irreversible inhibition exists. A prime example is the irreversible inhibition of acetylcholinesterase (AChE) by organophosphorus nerve agents. The agent, forming a covalent bond, residue, phosphorylates the important serine residue in AChE. Compared to the acetylated intermediate that is formed in the normal reaction with acetylcholine, the phosphorylated form is not hydrolyzed spontaneously and inactivates the enzyme that will remain inactivated even if the agent is removed from the surrounding of the enzyme.

8.7 Enzymes for Industrial Applications

Without the possibility to produce enzymes in bulk amounts enzyme catalysis would be an interesting field but without the possibility for application in industry, or for the main topic of this volume – the decontamination of chemical and biological warfare agents. While the use of micro organisms and their enzymes by humans is dating back many centuries and was mainly employed in the production of food (e.g. beer, vinegar, cheese), the real advent of biocatalysis for industrial purposes came with the arrival of recombinant techniques for protein expression. The ability to isolate a gene coding for an enzyme of interest and to clone it into an expression host able to overexpress the protein in great amounts opened the way to bulk enzyme production. Fermentation of a wide variety of expression hosts including a variety of bacteria, fungi or yeasts, in fermenters of different sizes up to more than $100\,m^3$ is common today and a large variety of enzymes are produced and commercially exploited. Examples for enzymes produced in multi-ton scale include phytase as an additive to animal food, xylanases and cellulases in the paper and pulp industry, enzymes for washing detergents (proteases, amylases, lipases and cellulases) and glucose isomerase in the starch industry. More specialized applications include the production of enzymes used as catalysts in the production of fine chemicals and pharmaceuticals.

Compared to the isolation and production of enzymes in the laboratory in mg quantities industrial processes vary considerably, especially concerning the isolation and purification of the product. Chromatographic techniques normally employed in the lab like affinity chromatography, ion exchange chromatography or gel filtration can normally not be used for the preparation of enzymes in large scale due to the costs and capacity problems. Older techniques like salt precipitation are often used instead due to their scalability and cost effectiveness. The separation problem is simplified if the enzyme is secreted into the fermentation broth by the microorganisms because the intact cells can be separated easily by centrifugation and compared to cell lysis less alien protein has to be separated from the product.

Stability is an important issue for industrial use of enzymes but two types of stability must be distinguished. The first is operational stability that is the stability of the enzyme in the reaction mixture while actively catalysing reactions. This kind of stability depends on intrinsic properties of the enzyme as well as the reaction conditions and the solvents. The second kind of stability is storage stability that can be influenced by additives. The normal way to prepare enzyme formulations for long-term storage is either freeze or spray drying with the latter being the preferred technique, as it is more economical.

Recommended Further Reading

Many general biochemistry textbooks that also cover protein science and enzyme catalysis exist. For those who want to find a point to start into the field these textbooks are a good choice.

When asking colleagues for recommendations for a specific book many different suggestions were made and indeed a lot of high quality books exist and it is more a question of personal taste which one to choose. Therefore no specific recommendation is made at this point but the reader is advised to check with their local academic bookstore and to find a book of their liking. The following recommendations cover works that are beyond these very introductory texts but still far more general than reviews and papers in the scientific literature:

Fersht, A. (1999) *Structure and Mechanism in Protein Science – A Guide to Enzyme Catalysis and Protein Folding*, W.H. Freeman and Company.

Although dating from 1999 this book is still very commendable and provides a detailed description of enzyme catalysis, protein folding and structure as well as protein engineering.

Böhm, H.J. and Schneider, G. (eds) (2003) *Protein-Ligand Interactions – From Molecular Recognition to Drug Design*, Wiley-VCH Verlag GmbH, Weinheim.

This volume contains contributions from several authors focusing on different aspects of protein-ligand interaction. Although the book is more focused on drug design and molecular recognition, it is a very good introduction into the major non-bonding interactions governing the binding of small molecules to macromolecular receptors. Some basic knowledge in protein science and structural biology is recommended.

Bomarius, A.S. and Riebel, B.R. (2004) *Biocatalysis*, Wiley-VCH Verlag GmbH, Weinheim.

The book is well worth reading for anybody who is interested in an in depth introduction to the production of enzymes in technical scale and their application in industry. This does include a theoretical treatment of enzyme mechanism, protein engineering and bioanalytical techniques.

Rhodes, G. (2006) *Crystallography Made Crystal Clear – A Guide for Users of Macromolecular Models*, 3rd Edition, Academic Press – Elsevier, Burlington – San Diego – London.

A very good introduction to protein X-ray crystallography for the beginner, that also contains a brief introduction to other methods like NMR. Valuable not only for those who want to determine protein structures but also for those using structural models obtained by these methods by enabling the reader to critically asses the available data.

Marangoni, A.G. (2003) *Enzyme Kinetics – A Modern Approach*, Wiley & Sons, Inc., Hoboken.

A good introduction to enzyme kinetics. The book focuses on modern methods linked with the availability of computers for calculations and contains practical examples making a thorough understanding easier. For example the focus is on nonlinear least-squares regression to solve the Mechaelis-Menten equation compared to older methods using linearization (like Linewaever-Burke plots or Eadie-Hofstee plots).

9
Hydrolytic Enzymes for Chemical Warfare Agent Decontamination

Marc-Michael Blum and André Richardt

9.1
Problematic Warfare Agents and Pathways for Detoxification

Many agents with very different chemistries exist. Not all of these agents represent a problem when it comes to their decontamination and detoxification. Very volatile agents like chlorine or hydrogen cyanide are rapidly diluted with air or can be washed down using water sprays. Other compounds like phosgene are highly reactive and rapidly react with other available compounds in the environment. Those agents that are thought to require decontamination are persistent agents or those agents with a very high toxicity that pose a risk even in very small concentrations.

This is true for the organophosphorus (OP) nerve agents and sulfur mustard. Mustard is a highly persistent agent that is relatively easy to produce and, although the lethality is rather low, mustard injuries are painful and require extensive medical treatment and time to heal. The nerve agents are highly toxic even in low quantities. While VX is a highly persistent agent that is also easily absorbed through the human skin, the G-agents are more volatile but can be turned into persistent agents when mixed with polymers that are used as thickeners, and even the volatile agent sarin might require decontamination when used inside of buildings or underground, as during the attack on the Tokyo subway.

Decontamination is defined as the removal of the agent from a contaminated surface and this does not necessarily include detoxification. Due to the hazards posed by these agents, a decontamination fluid that contains active agent after use is not desirable. Preferably, the agent should be converted chemically to non-hazardous products that pose no further risk [1]. Detoxification is possible by oxidation, which is relatively easy to achieve with mustard and VX, but the G-agents are rather resistant to oxidation [2]. Also, the reaction with nucleophiles is possible. Hypochlorite for example, that is used in the current German Army's decontamination formulation, reacts with the G-agents not by an oxidative mechanism, but

Decontamination of Warefare Agents.
Edited by André Richardt and Marc-Michael Blum

ISBN: 978-3-527-31756-1

the hypochlorite anion acts as a nucleophile attacking the phosphorus atom of the agent molecule [3].

9.2
Hydrolysis of Warfare Agents

This chapter will discuss enzymatic ways to catalyze the reaction of the G- and V-type nerve agents, as well as sulfur mustard, with water. Hydrolysis will detoxify the nerve agents by producing the corresponding phosphon acids (phosohonic or phosphoric) and the anions of the leaving groups (Figure 9.1a). Hydrolysis of the C-Cl bonds in sulfur mustard will finally produce thiodiglycol and chloride ions (Figure 9.1b). Both reactions release protons and pH control using buffers is required to avoid highly acidic conditions.

Inspection of the chemical structure of nerve agents in Figure 9.1a will reveal that these compounds (with the exception of DFP) contain four different groups bound to the phosphorus atom. Therefore, these compounds exist as optical stereoisomers [4]. These stereoisomers have different inhibition potencies towards acetylcholinesterase and as a consequence different toxicities. Soman even exists in four stereoisomers as a second stereocenter is found in the pinakolyl sidechain. In contrast to carbon, phosphorus stereocenters racemize in aqueous solution. In case of Sarin, Soman and Cyclosarin, this process is catalyzed by fluoride ions in solution [5]. The different hydrolytic enzymes discussed below show different degrees of stereoselectivity. In general the less toxic stereoisomer is hydrolyzed faster by the wild-type enzymes.

9.3
Early Examples of Enzymatic Decontamination

The first article published in the scientific literature was an article by Abraham Mazur in 1946 [6]. Mazur, who worked at the Medical Division of the Chemical

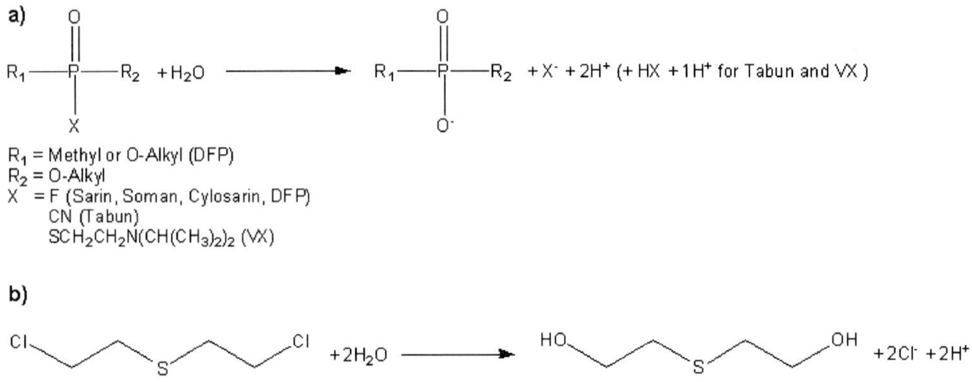

a)

$$R_1\text{-}P(=O)(X)\text{-}R_2 + H_2O \longrightarrow R_1\text{-}P(=O)(O^-)\text{-}R_2 + X^- + 2H^+ \text{ (+ HX + 1H}^+ \text{ for Tabun and VX)}$$

R_1 = Methyl or O-Alkyl (DFP)
R_2 = O-Alkyl
X = F (Sarin, Soman, Cylosarin, DFP)
CN (Tabun)
$SCH_2CH_2N(CH(CH_3)_2)_2$ (VX)

b)

$$Cl\text{-}CH_2CH_2\text{-}S\text{-}CH_2CH_2\text{-}Cl + 2H_2O \longrightarrow HO\text{-}CH_2CH_2\text{-}S\text{-}CH_2CH_2\text{-}OH + 2Cl^- + 2H^+$$

Figure 9.1 (a) Hydrolysis reaction of organophosphorus nerve agents (b) Hydrolysis reaction of sulfur mustard.

Warfare Service at Edgewood Arsenal in Maryland during the war holding the rank of a captain, reported that human and rabbit tissues, red blood cells and plasma contained an enzyme that accelerated the hydrolysis of dialkyl fluorophosphates. He was able to partially characterize the enzyme from rabbit kidneys and concluded that the fate of dialkyl fluorophosphates is either the irreversible inactivation of cholinesterases or detoxification by hydrolysis, enhanced by enzymatic catalysis, especially in the liver.

Although Mazur's report was the first report in the open literature, work on enzymatic methods for the detoxification of organophosphorus nerve agents began during World War II. The nerve agent Soman (GD) was discovered in Richard Kuhn's group at the Kaiser Wilhelm Institute (since 1950: Max Planck Institute) for Medical Research in Heidelberg. A subgroup formed by Günter Quadbeck, Konrad Henkel and Helmut Beinert investigated the effects of several enzymes on the hydrolysis of Soman and Sarin [7]. For these experiments extracts of mammalian liver were used as well as extracts of maggots of the fly *Phormia regina*.

9.4 Current Hydrolytic Enzyme Systems

Over the years, many enzymes that show some ability to hydrolyze toxic organophosphorus compounds have been described. Most of these enzymes have poor enzymatic activities, suffer from stability issues and have not been purified or properly characterized. At the time of writing, three enzymes have entered a stage of development beyond the laboratory. This means they show the required activity and stability and can be produced in large quantities. The first is the enzyme diisopropyl fluorophosphatase (DFPase) from the squid *Loligo vulgaris*, the second is organophosphorus hydrolase (OPH) from *Pseudomonas diminuta* and the third is organophophorus acid anhydrolase (OPAA) from *Sphingomonas*. In addition to these three enzymes acting against organophosphorus nerve agents, this chapter will also discuss a haloalkane dehalogenase LinB from *Sphingobium japonicum*. This enzyme is active against sulfur mustard and is subject of ongoing development work to make it useable for technical decontamination purposes. Finally, due do close similarities with DFPase and because of its medical importance, this chapter will also discuss mammalian serum paraoxonase 1 (PON1).

Drawing 3D-structures in 2D-space is always demanding. Due to restrictions regarding the use of color this task becomes even more difficult. The proteins discussed are only presented in cartoon representation to give an overview of their structural characteristics. The reader is highly encouraged to retrieve the structures from the Protein Data Bank (PDB) at www.rcsb.org and to investigate them on a computer screen with the ability to move the molecule and zoom into interesting areas. The PDB-codes for the wild-type structures are given in the text and several free viewers for PDB-files are available on the Internet.

9.5 Squid DFPase

The oldest known enzyme that is still relevant for large-scale decontamination is diisopropyl fluorophosphatase (DFPase) from squid. While the screening of microorganisms for enzymes with a desired catalytic activity becomes more and more a routine operation, it is quite remarkable that an enzyme hydrolyzing highly toxic organophosphorus compounds was found in a higher organism well before modern techniques like the polymerase chain reaction (PCR) were available.

The history of DFPase is closely linked to the pioneering work of David Nachmannsohn. Nachmannsohn worked on the theory of axonal conduction including the role of the cholinergic synaptic transmission system [8]. Electrophysiology was still limited in the 1950s and 1960s, with the modern patch clamp technique not available. To investigate axonal conduction with the available electrodes it was necessary to work with a model species that contained an axon with a size suitable for electrode insertion. Nachmannsohn's group used the calmar *Loligo pealei* for their experiments and one of these experiments tried to block axonal conduction by irreversibly inhibiting the cholinesterases using the potent inhibitor diisopropyl fluorophosphate (DFP). The DFP concentration needed to block conduction turned out to be exceptionally high. A concentration of 10^{-2} M was required while cholinesterase in solution is completely inhibited employing a concentration of 10^{-5} M. Francis C.G. Hoskin, at that time assistant professor at Columbia University NY, tried to investigate this surprising behavior by using ^{14}C labeled DFP. He was able to show that radioactivity rapidly accumulated in the interior of the axon. But the compound found was not DFP but diisopropyl phosphoric acid. He concluded that the axonal envelope contains a potent hydrolytic enzyme that accounts for the high concentration of DFP required to block conduction [9].

Francis Hoskin continued his work on the remarkable enzyme from squid after becoming full professor at the Illinois Institute of Technology, Chicago. He spent his summer months at the Marine Biological Laboratory in Woodshole on the Cape Cod peninsula preparing the enzyme from squid tissue. Apart from the axon, the enzyme is also found in the hepatopancreas, saliva and head ganglion [10, 11]. He demonstrated that the enzyme is not only able to hydrolyze DFP, but also the nerve agents Sarin, Soman and Tabun [12, 13].

The major problem with DFPase was the very low amount of protein that could be isolated from squid. Efforts to isolate the complete gene of DFPase from *Loligo pealei* were unfortunately not successful [14]. At this point, we have to cross the Atlantic to continue the DFPase story. Prof. Heinz Rüterjans at the University of Frankfurt in Germany showed great interest in the squid enzyme after conducting some research on an OP hydrolase from hog kidney. While *Loligo pealei* was readily available in the waters around Woodshole, the Frankfurt lab worked with the Mediterranean squid *Loligo vulgaris*. Experiments with monoclonal antibodies revealed that the enzymes from both species are very similar although not identical [15]. Becoming regular customers at the fish market in Frankfurt the group of Prof. Rüterjans was able to create a cDNA bank of the squid and to finally isolate the gene of DFPase. Subsequently, the gene was cloned into *E. coli* and enzyme yields from a liter culture reached more than 100 mg/l after optimal growth and expression conditions were found [16]. Purification of the protein was further simplified by the introduction of a polyhistidine tag that could be removed by enzymatic digestion. From that point on, enough protein for biochemical as well as structural studies was available.

In parallel to kinetic studies and studies of the effects of chemical modifications on the protein [17], structural investigations were initiated. After suitable crystals of DFPase were successfully obtained, the first X-ray structure was solved at a resolution of 1.8 Å (PDB-code 1E1A) [18]. Subsequent determinations of the DFPase structure were able to push the resolution to 0.85 Å (PDB-code 1PJX) [19]. At this resolution individual atoms are visible in the electron density maps as well as some normally invisible hydrogen atoms in certain parts of the protein. The structure of DFPase from *Loligo vulgaris* is shown in Figure 9.2.

DFPase consists of 314 amino acids and contains two calcium ions. The overall structure resembles a sixfold β-propeller with a central water filled tunnel. A high affinity calcium ion is located in the center of the molecule and is important to maintain the structural integrity of the protein. The second low-affinity calcium ion is located at the base of the active site, sealing the water filled tunnel, and is important for catalysis. The importance of the active site-bound calcium ion for catalysis was demonstrated by the removal of the ion that resulted in a folded, yet inactive enzyme [20]. Inspection of the active site shows the calcium ion coordinated by four amino acid residues at the bottom of the active site and a total coordination number of seven. The three remaining ligands are water molecules. Two of them are located below the metal ion forming the "dead end" of the central water filled tunnel and one is located on top of the metal ion in the active site. DFPase efficiently hydrolyzes DFP and G-type nerve agents including Sarin, Soman, Cyclosarin and Tabun. DFPase does not catalyze the hydrolysis of VX and Paraoxon. A certain degree of stereoselectivity was noticed, but complete detoxification can still be achieved using the wild type enzyme [21].

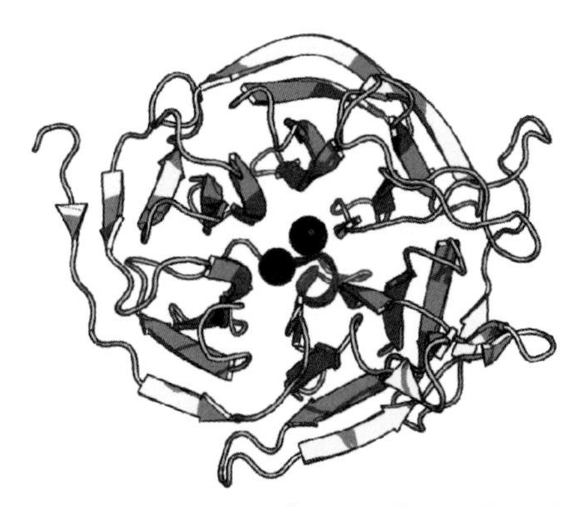

Figure 9.2 Structure of DFPase from *Loligo vulgaris*.

The first reaction mechanism of DFPase was proposed at the same time the X-ray structure was published [18]. An obvious candidate for a water-activating residue was histidine 287. In fact mutant H287N turned out to retain only a small residual activity. So the first proposed mechanism that is shown in Figure 9.3 argued that the incoming substrate DFP would replace the calcium coordinating water molecule in the active site and the metal would function as an electrophile making the phosphorus atom of DFP more susceptible for nucleophilic attack by water activated by H287.

This mechanism was challenged when mutations of DFPase (H287F and H287L) were generated that still maintained 65–80% of the wild-type activity [22]. Computational docking of DFP in the DFPase active site also revealed that the orientation of the DFP molecule with the fluoride-leaving group pointing away from H287 as required for an inline attack of the activated water was energetically unfavourable. Other point mutations did not reveal any other amino acid residue responsible for water activation. A new mechanism was proposed based on new experimental findings [23]. In this mechanism, calcium coordinating residue D229 is the active nucleophile that attacks the phosphorus atom of the substrate forming a phosphoenzmye intermediate. This intermediate is then hydrolyzed to regenerate the enzyme and release the product. One might argue that a metal coordinating aspartate should not be able to function as an active nucleophile as the metal will withdraw electron density from the carboxylate group. While this is definitely a valid point looking at isolated species one has to take into account the protein environment at the metal binding site. For other enzymes like the phosphatase activity in human epoxide hydrolase such a mechanism has been proposed [24] and detailed computational calculation show that the step forming the phosphoenzmye intermediate is indeed possible [25]. Isotope labeling using single and multiple turnover experiments in ^{18}O labeled water supported the

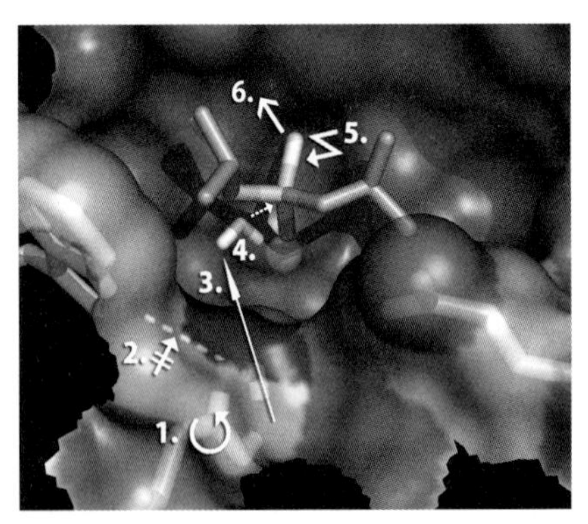

Figure 9.3 Old reaction mechanism of DFPase 1. The imidazole ring of H287 turns around 2. Breaking the hydrogen bond to W244 3. Activating a water molecule 4. Activated water performs nucleophilic attack on the phosphorus atom of the substrate DFP 5. The bond between the phosphorus atom and the fluoride leaving group is broken 6. Fluoride is released an diisopropyl phosphoric acid generated.

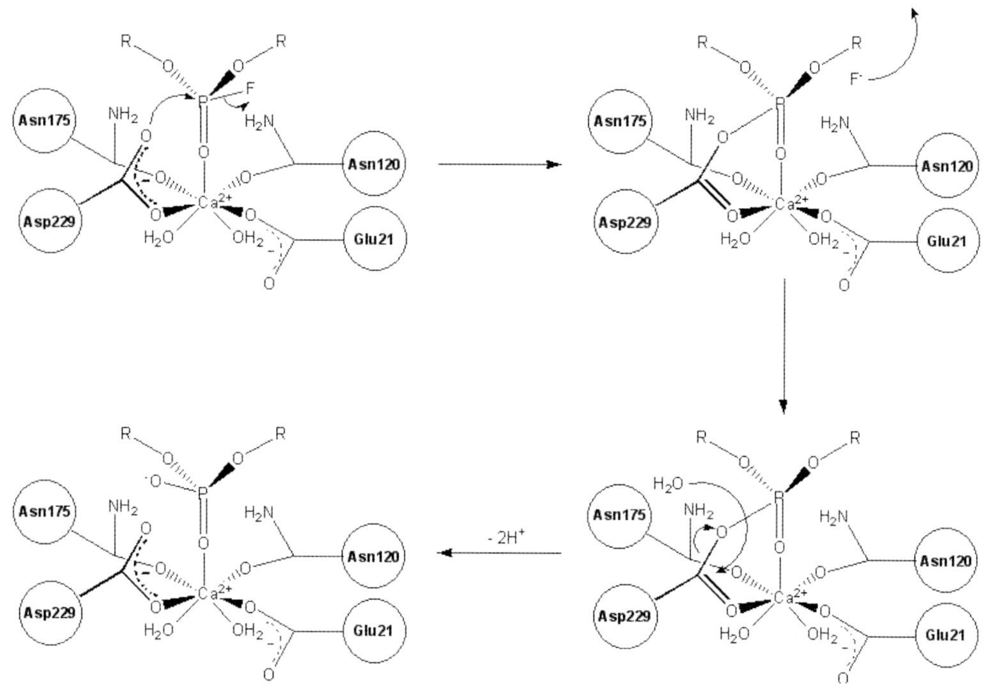

Figure 9.4 New reaction mechanism of DFPase. D229 performs a nucleophilic attack on the phosphorus atom of the substrate generating a phosphoenzyme intermediate. This is subsequently hydrolyzed by water attacking the carbonxylate caron atom of D229 regenerating the enzyme and releasing the product.

proposed mechanism for DFPase, which is depicted in Figure 9.4. In case of the multiple turnover experiment ^{18}O was incorporated into the product while in case of the single turnover experiment ^{16}O was found in the product. This is indicative of an oxygen atom transferred from the protein to the product by the formation of a covalent intermediate. Limited digestion of DFPase also revealed a peptide fragment that was labeled with ^{18}O and contained D229.

In order to allow an inline attack on the phosphorus atom by D229 the DFP molecule must bind to the active site with the correct orientation. Structural characterization of DFPase in complex with the organophosphorus inhibitor *O,O*-dicylopentylphosphoroamidate, in which the fluoride leaving group of the substrate is replaced by an amino group, revealed that the amino group was pointing towards H287 [23]. While this was in very good agreement with computational docking experiments it was the wrong orientation for an attack by D229 but detailed computational docking studies with DFP revealed that the correct orientation (Figure 9.5) was only 0.5 kcal/mol higher in energy than the conformation lowest in energy so that both states should be well populated at room temperature.

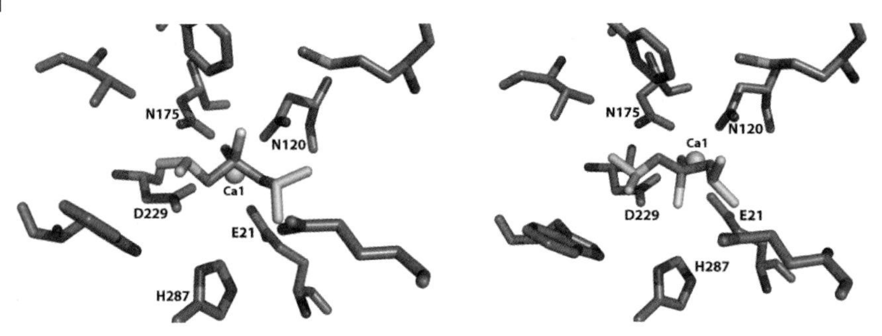

Figure 9.5 Reactive (left) ad non-reactive binding of DFP into the DFPase active site.

Further insights into the mechanistic aspects of DFPase can be expected from a neutron diffraction structure awaiting it release at the time of print. A dataset of 2.2 Å resolution was recorded and the structure will reveal more details about the protonation states of amino acid side chains and water molecules in the active site and the central tunnel [26].

DFPase is not only of academic interest but also of practical relevance as bulk amounts of the enzyme are readily available from fermentation with the yeast *Picha pastoris.* DFPase is also a highly stable protein. It can be used over a wide temperature range and is compatible with large amounts of different organic solvents. It is also compatible with microemulsions as carrier systems, thus also enabling the system to dissolve thickened agent [27]. The application of DFPase for technical decontamination will be discussed towards the end of this chapter.

9.6 Paraoxonase (PON)

The human enzyme paraoxonase 1 (PON1) is one of the major enzymes in the human body that is able to hydrolyze organophosphorus nerve agents and the serum level of this enzyme is thus of great importance for the protection of humans against these highly toxic agents. PON1 is a high-density lipoprotein (HDL) component, which may function to inactivate toxic byproducts of lipid oxidation by the low-density lipoprotein (LDL) complex [28]. PON1 is normally co-purified from human serum with a human phosphate binding protein (HPBP) that adds to the stability of PON1 in solution but even then the presence of detergents in the solution is required [29]. This does not speak in favor of PON1 as an enzyme for decontamination, at least not for decontamination in the field. The reasons for mentioning PON1 in this chapter are threefold. The first is that PON is a quite well characterized protein that efficiently hydrolyses G-type nerve agents and even exhibits a small activity against VX [30]. The second reason is that PON1 is an interesting protein for medical use after organophosphorus intoxication, as the human enzyme will not cause a reaction of the immune system compared to enzymes from other organisms [31]. Finally, the structure of PON1 is very similar to that of squid DFPase making it an interesting case for comparison.

PON1 is a tricky candidate for expression into common microorganisms like *E. coli*. Early trials resulted in aggregated protein. Directed evolution of PON1 led to variants that could be expressed in *E. coli* in a functional and soluble form [32]. The newly evolved PON1 rapidly converged towards the sequence of rabbit PON1, which seems to encode the most stable and soluble PON1. This resulted in the availability of PON1 in sufficient amounts for crystallization trails and subsequent structural characterization of PON1.

The publication of the structure of an engineered variant of PON1 (in the following just PON) was a major breakthrough on the way to understanding the function and activity of this enzyme (PDB-code 1V04) [33]. Striking similarities to the DFPase structure are apparent. PON also has the structure of a sixfold β-propeller and also contains two calcium atoms, of which one has a structural and the other a catalytic role. In contrast to DFPase, three hydrophobic helices can be found on one side of PON. These helices are thought to act as anchors of the protein in the HDL particle and add to the instability of the native protein in solution where it tends to aggregate without the presence of detergents. The structure of PON is shown in Figure 9.6.

Despite the structural similarities of PON and DFPase there are also major differences regarding the enzymatic activity and the range of substrates. In addition to the phosphotriesterase activity, PON also possesses activity against lactones and esters. PON was characterized as a native lactonase with promiscuous activities against the other substrates [34]. A first reaction mechanism based on the insight gained from the structure of the PON variant proposed a catalytic His-His dyad as the crucial residues for enzymatic activity [33], but at least for phosphorus triesters, this is assumption had to be modified as several mutants changing the histidines of the dyad to other amino acids still maintained substantial activity [35]. H115W for example did retain activity against paraoxon, but almost completely lost activity against phenylacetate and lactones like δ-valerolactone and

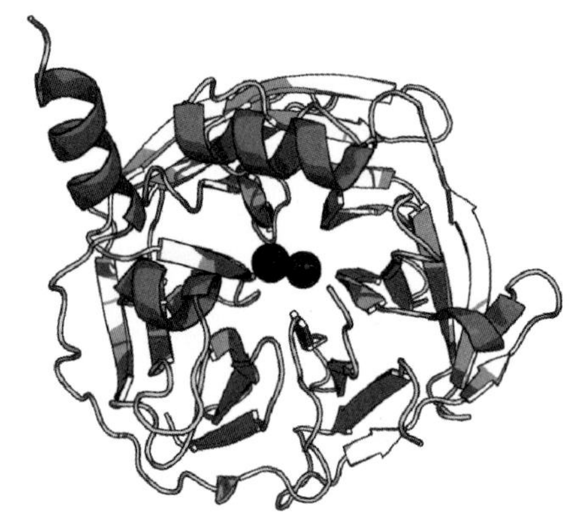

Figure 9.6 Structure of an engineered variant of PON1.

Table 9.1 Activity of PON1 in U/mg for different substrates [36].

	Paraoxon	*Dihydrocoumarin*	*Phenylacetate*	*δ-Valerolactone*
Wild-type	1.62 ± 0.06	183 ± 10	633 ± 60	142 ± 6
H115A	2.70 ± 0.02	670 ± 15	1.893 ± 0.006	4.3 ± 0.1
H115W	3.2 ± 0.3	477 ± 18	0.35 ± 0.02	0.9 ± 0.2
H115Q	0.52 ± 0.07	490 ± 24	3.67 ± 0.03	6.9 ± 0.1
H134Q	9.7 ± 0.3	141 ± 5	62.98 ± 0.03	18 ± 3

γ-nonalactone. At the same time, the mutant was even more active against the aromatic lactone dihydrocoumarin compared to the wild-type. Khersonsky and Tawfik concluded that the His-His dyad mediates the lactonase activity of PON (with exception of aromatic lactones like dihydrocoumarin) by activating a water molecule for attack on the carbonyl carbon of the lactone or ester that is activated itself by coordination to the catalytic calcium ion, but the dyad does not play a crucial role in the phosphotriesterase mechanism, which remains unknown as no apparent residues for the activation of water could be identified apart from the His-His dyad [36]. Some activities of PON1 for different substrates are displayed in Table 9.1.

When the current reaction mechanism for DFPase was published, it was noted that the calcium binding site of PON would allow a reaction mechanism similar to that of DFPase. In this case aspartate 269 would act as the nucleophile [23].

PON variants and mutants were not only generated to accomplish soluble and functional expression of the protein in *E. coli*. Enhanced activity against organophosphorus compounds and also enhanced stereoselctivity towards the more toxic isomers were also in the focus of research [37, 38]. While site directed mutagenesis was mainly used to gain further understanding of structure/activity relationships [39], directed evolution was the method of choice to enhance the catalytic proficiency of PON [32].

9.7 Organophosphorus Hydrolase (OPH) from *Pseudomonas diminuta*

Organophosphorus hydrolase (OPH), also named Phosphotriesterase (PTE) from *Pseudomonas diminuta* and *Flavobacterium sp.* is the second hydrolytic enzyme of practical relevance for large-scale decontamination. It is also among the best-investigated enzymes with more than 100 papers found in the literature. The enzyme was first discovered from *Flavobacterium sp.* isolated from a Phillippine rice paddy that was treated with the pesticide diazinon [40]. Identification of the enzyme in the soil bacterium *Pseudomonas diminuta* occurred around the same time due to its ability to hydrolyze parathion [41]. The *Opd* gene encoding for the enzyme was located in extrachromosomal plasmids in both cases and the gene sequence for the enzyme was also identical [42]. The gene from *Pseudomonas* was cloned into different expression vector including *E. coli* [43] providing sufficient amounts of protein for experiments.

The pesticide Paraoxon is among the best substrates of OPH with a k_{cat} of $1500\,s^{-1}$ and k_{cat}/K_M of $3.5 \times 107\,M^{-1}s^{-1}$ for the native zinc containing enzyme [44] and it was found that the reaction occurs via inversion of the configuration using an enantiomer of a chiral insecticide [45]. The substrate specificity of OPH is relatively broad and values for substrate specificity are given in Table 9.2.

A thiophosphoryl group can even replace the phosphoryl group depicted in Figure 9.1a. The enzyme also tolerates a variety of leaving groups ranging from small ions like fluoride in sarin or soman, via *p*-nitrophenolate in Paraoxon to the large leaving group of VX. Rate of hydrolysis is dependent on the pK_a of these leaving groups [46]. Among the enzymes discussed in this chapter OPH is the only one with an appreciable activity against compounds containing a P-S bond to the leaving group [47]. V-type nerve agents like VX fall into this group, although turnover rates are relatively slow with $v_{max} = 0.56\,mol\,min^{-1}\,mg^{-1}$ for VX [48]. OPH is inactive against esters like phenylacetate and phosphodiesters are very poor substrates [49].

The recombinant enzyme was crystallized and high-resolution X-ray structures are available (PDB-code 1HZY for the zinc containing enzyme) [50]. The protein displays a $(\alpha\beta)_8$-barrel motif, also known as a TIM-barrel, as depicted in Figure 9.7, which is a common fold for proteins. OPH belongs to a superfamily of metal-containing amidohydrolases. Other family members include urease, dihydroorotase and adenosine deaminase. The active site is located at the C-terminal end of the central β-sheet. The active site of OPH contains a binuclear metal center with two zinc ions bound within a cluster of histidine residues. The two metal ions are bridged by a hydroxide ion and a carbamate function resulting from posttranslational modification of a lysine residue. The native zinc ions can be removed from the enzyme without destroying the overall structure [51]. The metal ions can also be replaced by other metal ions like Co^{2+}, Cd^{2+}, Ni^{2+} or Mn^{2+} retaining the enzymatic function or even enhancing it [44].

The structure of OPH was also determined in complex with non-hydrolysable substrate mimics that allowed the identification of potential enzyme–substrate

Table 9.2 Substrate specificity of OPH [60].

Substrate	k_{cat}/K_m ($M^{-1}s^{-1}$)
Paraoxon	1×10^8
Diisopropyl fluorophosphate (DFP)	1×10^7
Sarin	8×10^4
Soman	1×10^4
Demeton-S	7×10^2
VX	7×10^2
Acephate	2×10^1

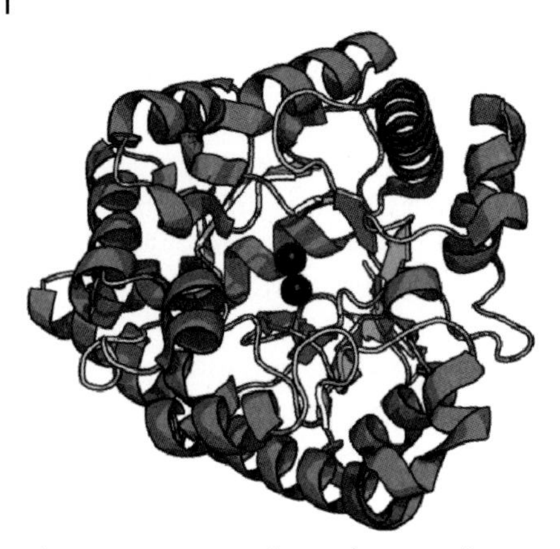

Figure 9.7 Structure of OPH from *Pseudomonas diminuta.*

interactions [52, 53]. Together with detailed kinetic studies, determination of isotope effects [54] and Brønstedt analyses [55] a reaction mechanism for OPH was proposed, which is shown in Figure 9.8 [56]. The mechanism was recently supported by detailed theoretical calculations employing Density Functional Theory (DFT) [57].

The incoming substrate replaces a water molecule on the β-metal (the more solvent exposed) by coordinating via the phosphoryl oxygen. This interaction weakens the binding of the bridging hydroxide to the β-metal as shown by the larger metal-oxygen distance in complexes with substrate mimics. The coordination to the metal ion polarizes the phosphoryl group rendering the phosphorus atom more electrophilic. The bridging hydroxide is positioned in a favourable near attack conformation and nucleophilic attack is facilitated by proton abstraction from D301. H254 is thought to accept the proton from D301 and enhance transfer to the bulk solvent.

Apart from mutating OPH to gain insights into the major forces guiding substrate binding and turnover, another aspect came into focus, namely how to increase OPH's activity against the nerve agents, especially VX and Soman for which wild-type activity is rather low, and to tweak the stereoselectivity of the enzyme towards the more toxic stereoisomers of the agents. The first report about efforts to enhance the activity of OHP was published shortly after the structure of the protein was determined by X-ray crystallography [58]. Two phenylalanine residues (F132, F306) contributing to the shape and size of the substrate binding site were targeted by site directed mutagenesis based on rational design, drawing conclusions from the binding mode of substrate analogs. V_{max} tested with DFP as a substrate increased as expected. Following work continued these efforts by mutating further residues in the substrate binding site and focusing on the substrate specificities of the enzyme [59, 60]. It was apparent from the crystal structure that several subsites in the binding pocket were guiding the stereoselectivity, and it was found that only a few residues had to be changed in order to alter this

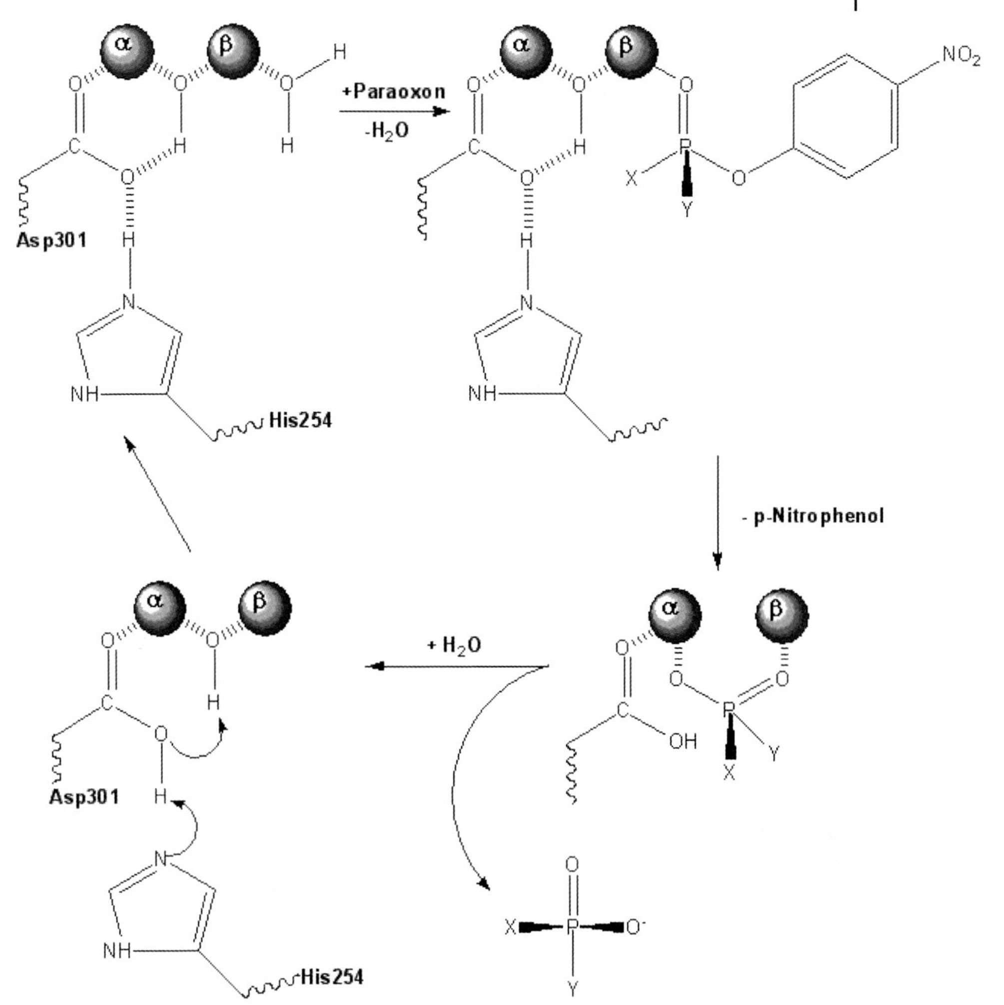

Figure 9.8 Reaction mechanism fort the hydrolysis of toxic organophosphorus compounds by OPH. Mechanism shown with the pesticide Paraoxon.

selectivity [61, 62]. One striking example is the introduction of a single methyl group in the mutant G60A. The stereoselectivity for the test substrate ethyl phenyl *p*-nitrophenylphosphate is about 21 in favor of the S_P-enantiomer, but in the mutant this selectivity is increased to 11 000 : 1 [63]. It was also possible to reverse the stereoselectivity [64]. Some recent work even proved that it is possible to use a set of OPH mutants to isolate both enantiomers of several nerve agent mimics from a racemic mixture by using the right OPH mutant from the set therefore using the enzyme for the synthesiz of chiral organophosphorus compounds that are valuable probes both for enzymatic studies but also in other areas of research like toxicology [65, 66].

As OPH is among the small group of enzymes showing activity against the nerve agent VX it was also tried to enhance activity specifically for this substrates as the wild-type activity was not sufficient for practical use. This turned out to be a more demanding task than expected. Early work only resulted in an increase in activity of 33% found for the mutant L136Y [67]. It should be noted however that VX is a problematic substrate for several reasons [68]. The chemistry and polarization of the P-S bond in VX is different to the P-F bond found in sarin and soman. VX also contains an amino function in the leaving group that can be protonated. This occurs at low pH and enhances the otherwise very limited solubility of the agent in water. OPH on the other hand shows best activities above neutral pH. If VX analogs are used, these are not sufficient to mimic the agent's properties. In fact, it turns out that one has to test with VX itself. This is, of course, impossible in public university labs and requires collaborative work with those labs allowed to handle life agents.

Finally, it should be noted that a phosphotriesterase very similar to OPH was found in *Agrobacterium radiobacter*. The gene coding for this enzyme termed OpdA has a sequence identity of 88.4% with the *Opd* gene in *Pseudomonas diminuta* encoding for OPH [69]. The amino acid sequence identity is 90.0%. One striking difference between the two proteins is the C-terminus where OpdA has an additional 20 residues. The X-ray structure of OpdA was solved at a resolution of 1.8 Å (PDB-code 2D2J) [70]. The structure of the protein was also determined in complex with several non-hydrolysable organophosphorus substrate mimics [71]. Despite the close similarity to OPH both enzymes exhibit different substrate activities and also exhibit slightly different metal coordination environments although it is believed that the reaction mechanism of both enzymes is virtually identical [72, 73]. Unfortunately, no activity of OpdA against nerve agents has been reported yet, although it seems likely that this activity exists. So far, only data for pesticides was published.

9.8 Organophosphorus Acid Anhydrolase (OPAA) from *Alteromonas*

The third and final enzyme hydrolyzing nerve agents for which the possibility of industrial production currently exists is organophosphorus acid anhydrase (OPAA) isolated from a halophilic *Alteromonas* strain *(A. sp. JD6.5)* by the laboratory of Joseph DeFrank. Originating from salt springs near the Great Salt Lake in Utah [74], OPAA is composed of 517 amino acids. Another very similar protein was isolated from *Alteromonas haloplanktis* containing 440 amino acids [75]. In contrast to OPH and DFPase, for which the natural substrate is still unknown, OPAA has been identified as a proline dipeptidase [76]. Compared to the other enzymes, only a few articles are available from the literature and the structure of OPAA has not been published yet, but the X-ray structure of the prolidase from *Pyrococcus furiosus* was determined (PDB-code 1PV9) [77] and is closely related to OPAA, although not an exact match (J.J. DeFrank, personal communication). The structure of the

Pyrococcus enzyme is depicted in Figure 9.9 and reveals a binuclear metal center formed by two cobalt ions bridged by a hydroxide ion that is somewhat similar to that in OPH [78]. OPAA activity is stimulated by the addition of cobalt, but also manganese ions.

OPAA does not hydrolyze VX, but is highly effective against the fluoride containing G-type nerve agents like Sarin and Soman, and also against the cyanide containing agent Tabun. The k_{cat} values for different nerve agents are given in Table 9.3.

The stereochemical preference of OPAA was tested using a similar set of substrates also employed to test OPH. The stereoselectivities are comparable showing a preference for the less toxic stereoisomer [79, 80]. To test the practical usability of OPAA for decontamination, several fire fighting foams and wetting agents were tested as suitable carrier systems for the enzyme [76]. In some of these systems, the enzyme turned out to be highly stable with an enzymatic activity even greater than in normal buffered solution. Although the first publication on OPAA appeared in the literature in 1991, only a few papers on this interesting enzyme have been published since then and details about the actual structure and the catalytic mechanism remain unknown.

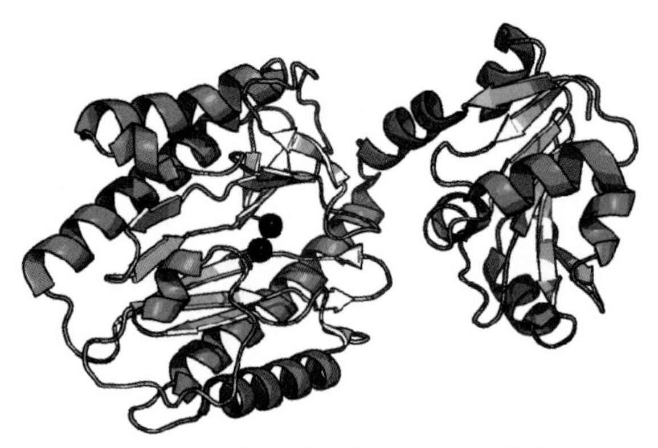

Figure 9.9 Structure of Prolidase from *Pyococcus furiosus*, a close relative of OPAA. For clarity only one monomeric unit of the dimeric structure is shown.

Table 9.3 Comparison of enzymatic activities of OPAA from *A. sp.JD6.5* [76].

Substrate	k_{cat} (s^{-1})
DFP	1650
Tabun	85
Sarin	611
Soman	3145
Cyclosarin	1650

9.9
Haloalkane Dehalogenase from LinB from Sphingobium Japonicum

The hydrolytic detoxification of sulfur mustard (HD) is not a trivial problem as HD is poorly soluble in water. If neat agent is added to an aqueous solution, the droplet will sink to the bottom of the reaction vessel and remain there even under stirring. At the interface between the aqueous phase and the organic agent phase some hydrolysis occurs, but the hydrolysis products will form a protective layer stopping any further reaction. This phenomenon is of practical relevance as large amounts of mustard filled ammunition were dumped in the Baltic Sea after the World Wars. Agent is leaking from corroded shells and forms cake-like structures that are sometimes retrieved by fishermen or washed ashore. On the outside, there is a skin a hydrolyzed products, but, below this layer, neat agent is still present. If mustard is pre-dissolved in organic solvent like alcohols that can be mixed with water, the agent readily dissolves in the mixture and hydrolyses rapidly. Just finding a suitable enzyme to speed up hydrolysis is, therefore, only part of a possible solution for mustard decontamination.

Such an enzyme is haloalkane dehalogenase, capable of hydrolyzing carbon-halogen bonds. This class of enzymes is also relevant for bioremediation of chlorinated pesticides like Lindane (hexachlorocyclohexane). One such enzyme from *Sphingobium japonicum* (formerly: *Sphingomonas paucimobilis*) UT26 called LinB is capable of detoxifying both Lindane and HD [81]. The protein was isolated by a group in Japan from bacteria retrieved from a test field that was sprayed with Lindane for 12 years. The enzyme was crystallized and the structure determined by X-ray diffraction [82]. An atomic resolution structure at 0.95 Å is available (PDB-code 1MJ5) [83] and the structure is shown in Figure 9.10.

A catalytical triad could be identified consisting of the three amino acids D108, H272, E132 [84]. The reaction mechanism of LinB was proposed as follows: D108

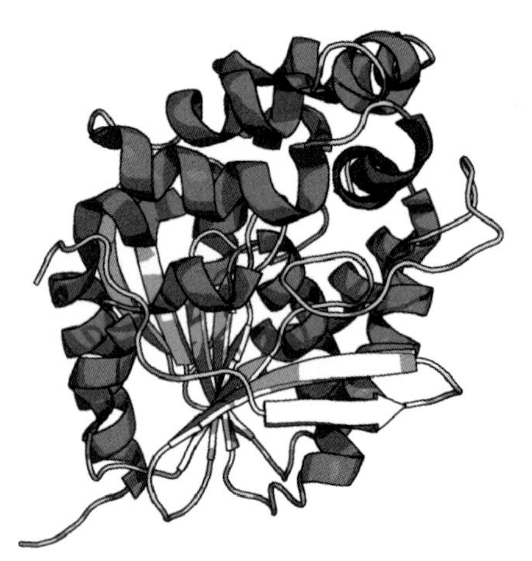

Figure 9.10 Structure of Haloalkane Dehalogenase LinB from *Shingobium Japonicum*.

is thought to act as a nucleophile causing release of a halide ion and formation of an acyl-enzyme intermediate. A hydrolytic water molecule cleaves the covalent ester intermediate. This water molecule is activated by H272 acting as a base catalyst prior to hydrolysis. E132 keeps H272 in the proper orientation and stabilizes a positive charge on the histidine imidazole ring during the reaction. This mechanism is drawn in Figure 9.11. The substrate binding site is found in the inside of the protein but the crystal structure reveals a tunnel connecting this cavity to the surface of the molecule. Extensive computational simulations have shown that the resulting alcohol does not leave the cavity through the tunnel, but through a slot that opens wide enough during normal low frequency motions of the protein, as revealed by normal mode analysis (NMA) on the X-ray crystal structure [85].

An interesting feature of LinB is that substrate hydrolysis occurs in a hydrophobic cavity inside of the protein. Spontaneous hydrolysis of mustard proceeds via a sulfonium ion intermediate that is a potent alkylating compound and causes massive cellular damage [86]. This makes LinB an interesting enzyme for topical use as a skin decontaminant. If hydrolysis of mustard can be achieved inside of a

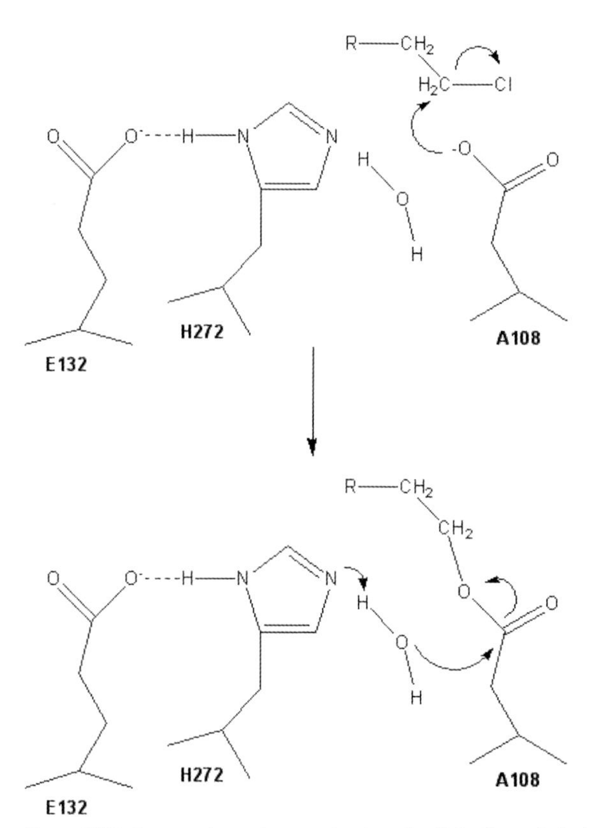

Figure 9.11 Proposed reaction mechanism for Haloalkane Dehalogenase LinB.

protein cavity, proceeding via non-toxic intermediates, the amount of unwanted reaction intermediate can be significantly reduced as well as skin damage.

9.10 Carrier Systems for Hydrolytic Enzymes

It was already mentioned that enzymes like OPAA are stable and active in fire fighting foams [76] and that DFPase was tested in microemulsions [27]. But why is it necessary to use special carrier systems for these enzymes and not simply use a buffered aqueous solution? The limited solubility of the agents in water, pointed out in this chapter, is only one aspect. Some agents show limited water solubility themselves (like VX and sulfur mustard). Others can become less water-soluble by the addition of polymer additives. These so called "thickened agents" are used to increase the persistency of otherwise volatile agents. They are normally used to contaminate terrain for a long time, therefore denying access or the use of contaminated equipment, while volatile agents like Sarin are used to inflict high casualties within a short period of time.

Apart from simply reporting the amount of agent soluble in one liter of water, one can also employ two other parameters. One is the partition coefficient between water and n-octanol that is frequently employed in environmental analytical chemistry to measure the hydrophobicity and persistency of compounds. Often this partition coefficient is reported in a logarithmic scale as *log* P [87, 88]. The other parameter is the so-called Hildebrand number (H) frequently employed in technical and polymer chemistry [89, 90]. This solubility parameter, used in a dimensionless form, is an indicator of the intermolecular forces between the molecules of a chemical compound and can be calculated from the enthalpy of evaporation and the molar volume of the compound. Values for chemical agents can be calculated and fall in the region between $H = 8$ and $H = 11$. To effectively dissolve these agents, the Hildebrand number of the solvent should be similar to the value for the agents and should not differ by more than two integer units. Water is not a suitable solvent as its Hildebrand number is too high ($H = 23.4$). Organic solvents like acetone ($H = 9.9$) are more appropriate.

At this point, a look at the other aspects of a suitable carrier system is necessary. The enzymes used must be stable and active in the solvent system used. Pure organic solvents do not fulfil these requirements, as hydrolytic enzymes require water for the reactions they catalyze and pure organic solvents normally affect protein stability adversely. The third and final aspect that has to be accommodated is the requirement for a carrier system to stay in close contact with the surface to be decontaminated. Adhesive forces must be strong enough to keep enough of the decontaminant sticking to the surface, even if the surface is vertical.

Systems that contain a mixture of water with organic solvents are possible solutions. If additives turn this mixture into foam it will stick to a surface. Fire fighting foams were therefore tested with OPH [91] and OPAA. Another possibility is the use of emulsions. They combine the beneficial properties of organic solvents (solubility of the agents) and water (solubility of the enzymes and substrate for hydrolytic reactions). Such an emulsion that is currently used by the German

Armed Forces is the so called "German Emulsion" using tetrachloroethylene as the organic solvent and calcium hypochlorite (C8) as the reactive component for decontamination [1]. The emulsion has very good wetting properties even on vertical surfaces. DFPase was tested in this emulsion that contained no C8, but a buffer to control pH. It was shown that DFPase was active in this emulsion over extended periods of time and effectively decontaminated metal plates coated with polyurethane paint and contaminated with sarin according to NATO standards (10 g Sarin per m^2). The agent was allowed to diffuse into the surface of the plate for three hours prior to decontamination [92].

Even though the German Emulsion is an established carrier system for decontaminants and currently fielded, certain drawbacks exist. Tetrachloroethylene is a harmful and persistent pollutant in the environment. The emulsion itself is a macroemulsion that is thermodynamically unstable and has to be prepared by vigorous stirring before use. In the case of improper preparation, the emulsion might not form, or will break after a short period of time. Therefore, current research efforts search for alternative emulsions. Microemulsions are among the prime candidates for new carrier systems. In contrast to classic macroemulsions, they are thermodynamically stable and form spontaneously when the components are mixed. They contain at least two immiscible fluids and an amphiphile component [93]. They also display a much smaller droplet size (10–200 nm compared to 1–90 μm), greater interphase surface and good solubility properties for hydrophilic and hydrophobic compounds [94, 95]. Microemulsions and the activity of DFPase in them are described in detail in a separate chapter of this volume. Figure 9.12 depicts the action of a microemulsion for decontamination [96].

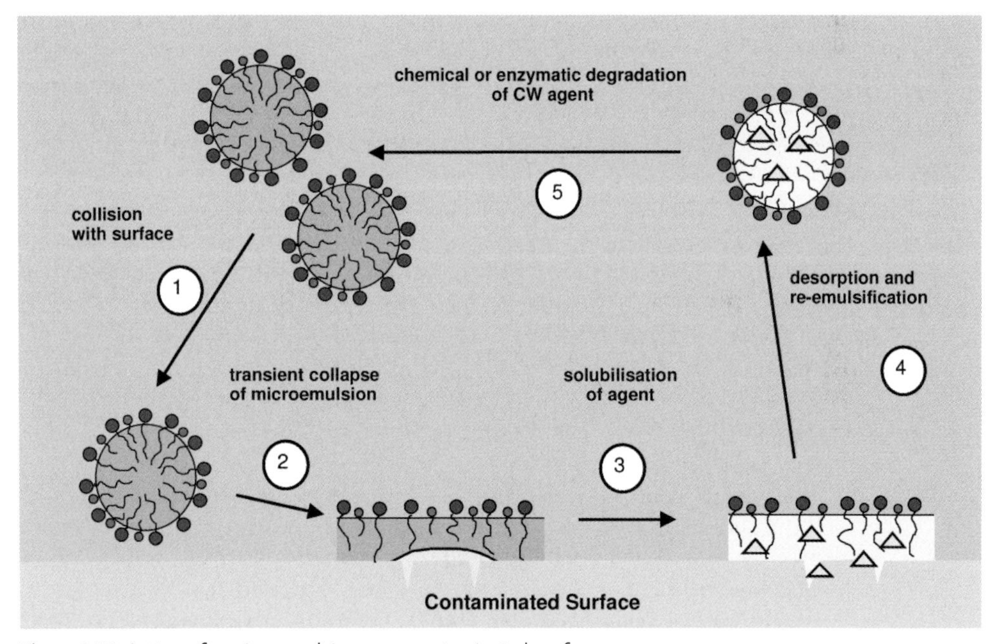

Figure 9.12 Action of a microemulsion on a contaminated surface.

9.11 Technical Decontamination with DFPase–An Example

The ability to use enzymes to detoxify chemical warfare agents in the laboratory and the ability to use them in the field are very different. Decontamination in the lab normally includes the use of dilute agent solutions in a small reaction volume, employing very small amounts of enzyme in a controlled environment. These conditions are required for enzyme characterization and determination of kinetic constants. Technical decontamination, on the other hand, takes place in a variable environment regarding temperature and humidity, includes large amounts of toxic agents spread on a variety of surfaces and requires the efficient dispersal of large amounts of enzyme and decontamination fluid. Due to the high toxicity of the warfare agents, even technical testing has to be carried in a highly controlled environment. The dispersal of chemical agents in the open is not allowed in most countries, even when this takes place in very remote and highly secured areas.

One major step required for the usability of enzymes for technical decontamination is the production of bulk quantities of enzyme and stabilization of the enzyme in a storable form. Preferably, the enzyme should be stored in a dry form to avoid microbial contamination and to ensure storability at elevated temperatures. Freeze- or spray-drying are established methods for this purpose. DFPase can be produced in bulk amounts using the methylotrophic yeast *Pichia pastoris* fermentation [92]. Simple lyophilization of a DFPase solution in the standard Tris-Buffer on the contrary was not successful, as a significant amount of protein could not be resolubilized and enzyme activity rapidly declined during storage. As it turned out, the Tris-Buffer is unsuitable for freeze-drying as the pK_a of this buffer is highly dependent on temperature and the pH of a buffered solution can change dramatically when cooled below 0 °C [97, 98]. As a solution, the buffer was changed to 3-(N-morpholino)propanesulfonic acid (MOPS) or N-2-hydroxyethylpiperazine-N′-2-ethanesulfonic acid (HEPES). MOPS belongs to the original "Good Buffers" showing a significantly reduced temperature dependence and HEPES was developed shortly afterwards [99]. This led to vastly improved results regarding the resolubility of the enzyme after freeze-drying, but long term stability was still a problem. The final solution was to use polyethylene glycol (PEG) with a molecular weigh of about 10 kDa as a lyoprotectant. PEG is used for protein stabilization but also as a precipitant as it competes with water binding to the protein [100]. The final enzyme powder retained its activity for month and les than 10% of the enzyme was lost when resolubilized [101]. The current work-up procedure allows the preparation of a stabilized enzyme power without the use of column chromatography in high yields. This enzyme powder was mixed with sodium hydrocarbonate as the buffer component and stored at three different temperatures (4 °C, RT and 65 °C) and activity was monitored over a period of 9 months. The powder retained almost complete activity at RT and 4 °C but even at 65 °C substantial activity was found after 9 months, as can be seen in Figure 9.13.

With the availability of stabilized enzyme power in great amounts it was possible to test DFPase in a setting as closely resembling a realistic case as possible. As a tool

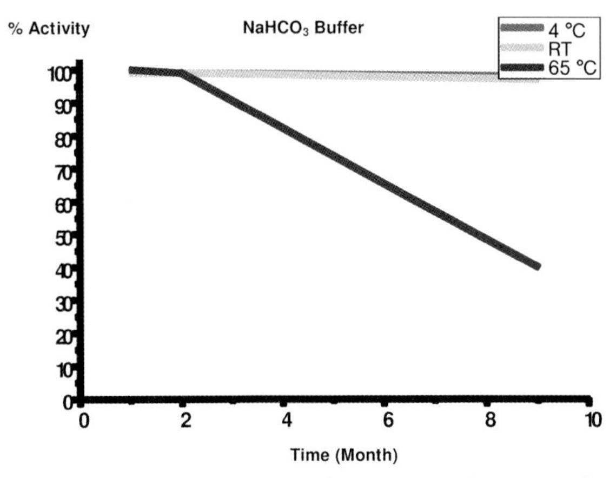

Figure 9.13 Long term stability of DFPase over the course of nine month.

Figure 9.14 Enzymatic decontamination of paint coated test plates contaminated with Soman.

for dispersal, a modified fire extinguisher [102] was used. In addition to a gas cylinder containing CO_2, the extinguisher also contained a special compartment for the storage of the enzyme powder. The decontamination fluid used consisted of 4 litres of water, 6 litres of a non-aqueous concentrate, 400 mM ammonium hydrocarbonate and 2 mM calcium chloride. A total of 1 million units of DFPase were used (5 g of enzyme). Testing was conducted at the Armed Forces Institute for Protection Technologies/NBC-Protection in Munster using metal plates (10 cm × 10 cm) coated with polyurethane paint contaminated with soman according to NATO standards. Some plates were immediately washed with isopropanol as control plates to determine the amount of agent. Plates were allowed so rest with the contamination for 30 minutes to allow penetration of the paint by the agent. Some more plates were removed at this point to determine the remaining agent on the plate by GC analysis and correct for any loss by evaporation. The remaining plates were arranged at different angles to simulate horizontal, vertical and slanted surfaces and the decontamination fluid containing the enzyme was sprayed onto the remaining plates using the modified fire extinguisher as can be seen in Figure 9.14.

Table 9.4 Results of decontamination trials with DFPase (Bdl = below detection limit).

	Agent applied per plate (mg)	*Agent after incubation (mg)*	*Evapo-ration (%)*	*Agent in run-off (µg/ml)*		*Agent on surface (µg/cm²)*		*Residual agent in paint (µg/cm²)*		*Efficacy (%)*
				15 min	*30 min*	*15 min*	*30 min*	*15 min*	*30 min*	
Horizontal	9.60	6.50	33	Bdl	Bdl	Bdl	Bdl	Bdl	Bdl	>99.9
Slanted	9.60	6.50	33	Bdl	Bdl	Bdl	Bdl	Bdl	Bdl	>99.9
Vertical	9.60	6.50	33	0.07	Bdl	Bdl	Bdl	Bdl	Bdl	>99.9

After 15 and 30 minutes, half of the plates were removed and the residual agent was determined by GC analysis. The hydrolysis of Soman was completed in a short period of time, as can be seen from Table 9.4.

Also, there was no residual contamination in the paint after 15 minutes and there was no detectable soman in the run-off after 30 minutes. These results show clearly that a technical decontamination based on enzymes can be used for an environmentally benign decontamination system [103].

9.12 Conclusions and Outlook

Enzymatic decontamination has come a long way since its beginnings in the 1940s. A significant number of effective enzymes have been identified and characterized. Some of them are available in bulk amounts and can be stored for years. But the road does not end here. It is unlikely that a single enzyme will be able to decontaminate all warfare agents. Therefore, the right mixture of enzymes has to be identified that are stable in the same carrier system and can coexist in solution without affecting each other's activities. While the decontamination of G-type nerve agents on a technical scale is possible today, it has to be demonstrated that the same is possible for V-type nerve agents and sulfur mustard. Research is continuing in several research groups and companies throughout the world. Novel enzymes might be discovered and known ones might be improved, regarding both their activity and substrate diversity.

At the same time, it is clear that enzymatic decontamination might be based on complicated science, but the products originating from this science are by no means complicated. In the end, there is no need for a thorough understanding of enzymatic reaction mechanisms for the people using the enzymes. The same is true for the use of enzymes in other areas of daily life like washing detergents. It is important that enzymes can achieve the desired result in a fast and efficient manner following easy protocols. Examples like the decontamination extinguisher show that this can be almost as easy as pressing a single button.

References

1 Yang, Y.-C., Baker, J.A. and Ward, J.R. (1992) Decontamination of chemical warfare agents. *Chemical Reviews*, **92**, 1729–43.

2 Franke, S., Koehler, K.F., Kostyanovsky, R.G. and Kuntsevich, A.D. (1994) *Chemie der Kampfstoffe*, Dr. Koehler GmbH, Munster.

3 Epstein, J., Bauer, V.E., Saxe, M. and Demek, M.M. (1956) The chlorine-catalyzed hydrolysis of isopropyl mehtylphosphonofluoridate (sarin) in aqueous solution. *Journal of the American Chemical Society*, **78**, 4068–71.

4 Benschop, H.P. and De Jong, L.P.A. (1988) Nerve agent stereoisomers: analysis isolation and toxicology, *Accounts of Chemical Research*, **21**, 368–74.

5 Christen, P.J. and van den Muysenberg, J.A. (1965) The enzymatic isolation and fluoride catalyzed racemisation of optically active sarin. *Biochimica et Biophysica Acta*, **110**, 217–20.

6 Mazur, A. (1946) An enzyme in animal tissue capable of hydrolyzing the phosphorus-fluorine bond of alkyl fluorophosphates. *The Journal of Biological Chemistry*, **164**, 271–89.

7 Schmaltz, F. (2005) *Kampfstoff-Forschung im Nationalsozialismus*, Wallstein Verlag, Göttingen, pp. 506–10.

8 Nachmansohn, D. (1959) *Chemical and Molecular and Basis of Nerve Activity*, Academic Press, New York.

9 Hoskin, F.C.G., Rosenberg, P. and Brzin, M. (1966) Re-examination of the effect of DFP on electrical and cholinesterase activity of squid giant axon. *Proceedings of the National Academy of Sciences of the United States of America*, **55**, 1231–5.

10 Hoskin, F.C.G. and Long, R.J. (1972) Purification of a DFP-hydrolyzing enzyme from squid head ganglion. *Archives of Biochemistry and Biophysics*, **150**, 548–55.

11 Garden, J.M., Hause, S.K., Hoskin, F.C.G. and Roush, A.H. (1975) Comparison of DFP-hydrolyzing enzyme purified from head ganglion and hepatopancreas of squid (*Loligo pealei*) by means of isoelectric focusing. *Comparative Biochemistry and Physiology*, **52C**, 95–8.

12 Hoskin, F.C.G. and Roush, A.H. (1982) Hydrolysis of nerve gas by squid-type diisopropyl phosphorofluoridate hydrolyzing enzyme on agarose resin. *Science*, **215**, 1255–7.

13 Hoskin, F.C.G. (1971) Diisopropylphosphorofluoridate and tabun: enzymatic hydrolysis and nerve function. *Science*, **172**, 1243–5.

14 Kopec-Smyth, K., Deschamps, J.R., Loomis, L.D. and Ward, K.B. (1993) A partial primary structure of quid hepatopancreas organophosphorus acid anhydrolase. *Chemico-Biological Interactions*, **87**, 49–54.

15 Deschamps, J.R., Kopec-Smyth, K., Poppino, J.L., Futrovsky, S.L. and Ward, K.B. (1993) Comparison of organophosphorus acid anhydrolases from different species using monoclonal antibodies. *Comparative Biochemistry and Physiology C: Comparative Pharmacology*, **106**, 765–8.

16 Hartleib, J. and Rüterjans, H. (2001) High-yield expression, purification and characterization of the recombinant diisopropylfluorophosphatase from *Loligo vulgaris*. *Protein Expression and Purification*, **21**, 210–9.

17 Hartleib, J. and Rüterjans, H. (2001) Insights into the reaction mechanism of the diisopropylfluorophosphatase from *Loligo vulgaris* by means of kinetic studies, chemical modification and site-directed mutagenesis. *Biochimica et Biophysica Acta*, **1546**, 312–24.

18 Scharff, E.I., Koepke, J., Fritzsch, G., Lücke, C. and Rüterjans, H. (2001) Crystal structure of diisopropylfluorophosphatase from *Loligo vulgaris*. *Structure*, **9**, 493–502.

19 Koepke, J., Scharff, E.I., Lücke, C., Rüterjans, H. and Fritzsch, G. (2003) Statistical analysis of crystallographic data obtained from squid ganglion DFPase at 0.85 Å resolution. *Acta Crystallographica. Section D, Biological Crystallography*, **59**, 1744–54.

20 Hartleib, J., Gschwindner, S., Scharff, E.I. and Rüterjans, H. (2001) Role of Calcium

Ions in the Structure and Function of the Di-isopropylfluorophosphatase from *Loligo vulgaris*. *The Biochemical Journal*, **353**, 579–89.

21 Schulz, W., Schäfer, B., Stroop, G. and Rüterjans, H. (1987) Enzymatische hydrolyse von hochtoxischen phophororganischen verbindungen. *Forschungsbericht aus der Wehrtechnik*, Auftragsnummer T/R 770/D 00001/D 1750.

22 Katsemi, V., Lücke, C., Koepke, J., Löhr, F., Maurer, S., Fritzsch, G. and Rüterjans, H. (2005) Mutational and structural studies of the diisopropylfluorophosphatase from *Loligo vulgaris* shed new light on the catalytic mechanism of the enzyme. *Biochemistry*, **44**, 9022–33.

23 Blum, M.M., Löhr, F., Richardt, A., Rüerjans, H. and Chen, J.C.-H. (2006) Binding of a designed substrate analogue to diisopropyl fluorophosphatase: implications for the phosphotriesterase mechanism. *Journal of the American Chemical Society*, **128**, 12750–7.

24 Gomez, G.A., Morisseau, C., Hammock, B.D. and Chrsitianson, D. W. (2004) Structure of human epoxide hydrolase reveals mechanistic inferences on bifunctional catalysis in epoxide and phosphate ester hydrolysis. *Biochemistry*, **43**, 4716–23.

25 De Vivo, M., Ensing, B., Dal Peraro, M., Gomez, G.A., Christianson, D.W. and Klein, M.L. (2007) Proton shuttles and phosphatase activity in soluble epoxide hydrolase. *Journal of the American Chemical Society*, **129**, 387–94.

26 Blum, M.M., Koglin, A., Rüterjans, H., Schoenborn, B.P., Langan, P. and Chen, J.C.-H. (2007) Preliminary time-of-flight neutron diffraction study on diisopropyl fluorophosphatase (DFPase) from *Loligo vulgaris*. *Acta Crystallographica. Section F, Structural Biology and Crystallization Communications*, **63**, 42–5.

27 Richardt, A., Blum, M.M. and Mitchell, S. (2006) Was wissen calamari über sarin. *Chemie in unserer Zeit*, **40**, 252–9.

28 Li, H.L., Liu, D.P. and Liang, C.C. (2003) Paraoxonase gene polymorphisms, oxidative stress, and diseases, *Journal of Molecular Medicine*, **81**, 766–79.

29 Morales, R., Berna, A., Carpentier, P., Contreras-Martel, C., Renault, F., Nicodeme, M., Chesne-Seck, M.L., Bernier, F., Dupuy, J., Schaeffer, C., Diemer, H., Van-Dorsselaer, A., Fontecilla-Camps, J.C., Masson, P., Rochu, D. and Chabriere, E. (2006) Serendipitous discovery and X-ray structure of a human phosphate binding apolipoprotein. *Structure*, **14**, 601–9.

30 Broomfield, C.A., Morris, B.C., Anderson, R., Josse, D., Masson, P. (2000), *Conference Proceedings CBMTS III*, Spiez, Switzerland, May 2000, (Available from www.jmcdcbr.org).

31 Masson, P., Josse, D., Lockridge, O., Viquié, N., Taupin, C. and Buhler, C. (1998) Enzymes hydrolyzing organophosphates as potential catalytic scavengers against organophosphate poisoning. *Journal of Physiology, Paris*, **92**, 357–62.

32 Aharoni, A., Gaidukov, L., Yagur, S., Toker, L., Silman, I. and Tawfik, D.S. (2004) Directed evolution of mammalian paraoxonases PON1 and PON3 for bacterial expression and catalytic specialization, *Proceedings of the National Academy of Sciences of the United States of America*, **101**, 482–7.

33 Harel, M., Aharoni, A., Gaidukov, L., Brumshtein, B., Khersonsky, O., Meged, R., Dvir, H., Ravelli, R.B.G., McCarthy, A., Toker, L., Silman, I., Sussman, J.L. and Tawfik, D.S. (2004) Structure and evolution of the serum paraoxonase family of detoxifying and anti-atherosclerotic enzymes. *Nature Structural and Molecular Biology*, **11**, 412–9.

34 Khersonsky, O. and Tawfik, D.S. (2005) Structure-reactivity studies of serum paraoxonase pon1 suggest that its native activity is lactonase. *Biochemistry*, **44**, 6371–82.

35 Yeung, D.T., Josse, D., Nicholson, J.D., Khanal, A., McAndrew, C.W., Bahnson, B.J., Lenz, D.E. and Cerasoli, D.M. (2004) Structure/function analyses of human serum paraoxonase (Hupon1) mutants designed from A Dfpase-like homology model. *Biochimica et Biophysica Acta*, **1702**, 67–77.

36 Khersonsky, O. and Tawfik, D.S. (2006) The histidine 115-histidine 134 dyad mediates the lactonase activity of mammalian serum paraoxonases. *The Journal of Biological Chemistry*, **281**, 7649–56.

37 Amitai, G., Gaidukov, L., Adani, R., Yishay, S., Yacov, G., Kushnir, M., Teitlboim, S., Lindenbaum, M., Bel, P., Khersonsky, O., Tawfik, D.S. and Meshulam, H. (2006) Enhanced stereoselective hydrolysis of toxic organophosphates by directly evolved variants of mammalian serum paraoxonase. *The FEBS Journal*, **273**, 1906–19.

38 Amitai, G., Adani, R., Yacov, G., Yishay, S., Teitlboim, S., Tveria, L., Limanovich, O., Kushnir, M. and Meshulam, H. (2007) Asymmetric fluorogenic organophosphates for the development of active organophosphate hydrolases with reversed stereoselectivity. *Toxicology*, **233**, 187–98.

39 Josse, D., Lockridge, O., Xie, W., Bartels, C.F., Schopfer, L.M. and Masson, P. (2001) The active site of human paraoxonase (PON1). *Journal of Applied Toxicology*, **21**, S7–11.Suppl. 1

40 Sethunathan, N. and Yoshida, T. (1973) A flavobacterium that degrades diazinon and parathion. *Canadian Journal of Microbiology*, **19**, 873–5.

41 Munnecke, D.M. (1976) Enzymatic hydrolysis of organophosphate insecticides, a possible pesticide disposal method. *Applied and Environmental Microbiology*, **32**, 7–13.

42 Harper, L.L., McDaniel, C.S., Miller, C.E. and Wild, J.R. (1988) Dissimilar plasmids isolated from *Pseudomonas diminuta MG* and a *Flavobacterium sp.* (ATCC 27551) contain identical opd genes. *Applied and Environmental Microbiology*, **54**, 2586–9.

43 McDaniel, C.S., Harper, L.L. and Wild, J.R. (1988) Cloning and sequencing of a plasmid-borne gene (opd) encoding a phosphotriesterase. *Journal of Bacteriology*, **170**, 2306–11.

44 Omburo, G.A., Kuo, J.M., Mullins, L.S. and Raushel, F.M. (1992) Characterization of the zinc binding site of bacterial phosphotriesterase. *The Journal of Biological Chemistry*, **267**, 13278–83.

45 Lewis, V.E., Donarski, W.J., Wild, J.R. and Raushel, F.M. (1988) Mechanism and stereochemical course at phosphorus of the reaction catalyzed by a bacterial phosphotriesterase. *Biochemistry*, **27**, 1591–7.

46 Hong, S.B. and Raushel, F.M. (1996) Metal-substrate interactions facilitate the catalytic activity of the bacterial phosphotriesterase. *Biochemistry*, **35**, 10904–12.

47 Hoskin, F.C.G., Walker, J.E., Dettbarn, W.D. and Wild, J.R. (1995) Hydrolysis of tetriso by an enzyme derived from *Pseudomonas diminuta* as a model for the detoxication of O-ethyl S-(2-diisopropylaminoethyl) methylphosphonothiolate (VX). *Biochemical Pharmacology*, **49**, 711–5.

48 Rastogi, V.K., DeFrank, J.J., Cheng, T.C. and Wild, J.R. (1998) Enzymatic hydrolysis of Russian-VX by organophosphorus hydrolase. *Biochemical and Biophysical Research Communications*, **241**, 294–6.

49 Shim, H., Hong, S.B. and Raushel, F.M. (1998) Hydrolysis of phosphodiesters through transformation of the bacterial phosphotriesterase. *The Journal of Biological Chemistry*, **272**, 17445–50.

50 Benning, M.M., Shim, H., Raushel, F.M. and Holden, H.M. (2001) High resolution X-ray structures of different metal-substituted forms of phosphotriesterase from *Pseudomonas diminuta*. *Biochemistry*, **40**, 2712–22.

51 Benning, M.M., Kuo, J.M., Raushel, F.M. and Holden, H.M. (1994) Three-dimensional structure of phosphotriesterase: an enzyme capable of detoxifying organophosphate nerve agents. *Biochemistry*, **33**, 15001–7.

52 Vanhooke, J.L., Benning, M.M., Raushel, F.M. and Holden, H.M. (1996) Three-dimensional structure of the zinc-containing phosphotriesterase with the bound substrate analog diethyl 4-methylbenzylphosphonate. *Biochemistry*, **35**, 6020–5.

53 Benning, M.M., Hong, S.B., Raushel, F.M. and Holden, H.M. (2000) The binding of substrate analogs to phosphotriesterase. *The Journal of Biological Chemistry*, **275**, 30556–5560.

54 Caldwell, S.R., Raushel, F.M., Weiss, P.M. and Cleland, W.W. (1991) Transition-state structures for enzymatic and alkaline phosphotriester hydrolysis. *Biochemistry*, **30**, 7444–50.

55 Caldwell, S.R., Newcomb, J.R., Schlecht, K.A. and Raushel, F.M. (1991) Limits of diffusion in the hydrolysisof substrates by the phosphotriesterase from *Pseudomonas diminuta*. *Biochemistry*, **30**, 7438–44.

56 Aubert, S.D., Li, Y. and Raushel, F.M. (2004) Mechanism for the hydrolysis of organophosphates by the bacterial phosphotriesterase. *Biochemistry*, **43**, 5707–15.

57 Chen, S.L., Fang, W.H. and Himo, F. (2007) Theoretical study of the phosphotriesterase reaction mechanism. *The Journal of Physical Chemistry. B*, **111**, 1253–5.

58 Watkins, L.M., Mohoney, H.J., McCulloch, J.K. and Raushel, F.M. (1997) Augmented hydrolysis of diisopropyl fluorophosphate in engineered mutants of phosphotriesterase. *The Journal of Biological Chemistry*, **272**, 25596–601.

59 Shim, H., Hong, S.B. and Raushel, F.M. (1998) Hydrolysis of phosphodiesters through transformation of the bacterial phosphotriesterase. *The Journal of Biological Chemistry*, **273**, 17445–50.

60 Di Sioudi, B.D., Miller, C.E., Lai, K., Grimsley, J.K. and Wild, J.R. (1999) Rational design of organophosphorus hydrolase for altered substrate specificities. *Chemico-Biological Interactions*, **119–120**, 211–23.

61 Hill, C.M., Li, W.S., Thoden, J.B., Holden, H.M. and Raushel, F.M. (2004) Enhanced degradation of chemical warfare agents through molecular engineering of the phosphotriesterase active site. *Journal of the American Chemical Society*, **125**, 8990–1.

62 Chen-Goodspeed, M., Sogorb, M.A., Wu, F., Hong, S.B. and Raushel, F.M. (2001) Structural determinants of the substrate and stereochemical specificity of phosphotriesterase. *Biochemistry*, **40**, 1325–31.

63 Chen-Goodspeed, M., Sogorb, M., Wu, F. and Raushel, F.M. (2001) Enhancement, relaxation, and reversal of the stereoselectivity for phosphotriesterase by rational evolution of active site residues. *Biochemistry*, **40**, 1332–9.

64 Li, W.S., Lum, K.T., Chen-Goodspeed, M., Sogorb, M. and Raushel, F.M. (2001) Stereoselective detoxification of chiral sarin and Soman analogues by phosphotriesterase. *Bioorganic and Medicinal Chemistry*, **9**, 2083–91.

65 Li, Y., Aubert, S.D., Maes, E.G. and Raushel, F.M. (2004) Enzymatic resolution of chiral phosphinate esters. *Journal of the American Chemical Society*, **126**, 8888–9.

66 Nowlan, C., Li, Y., Hermann, J.C., Evans, T., Carpenter, J., Gahanem, E., Shoichet, B.K. and Raushel, F.R. (2006) Resolution of chiral phosphate, phosphonate, and phosphinate esters by an enantioselective enzyme library. *Journal of the American Chemical Society*, **128**, 15892–902.

67 Gopal, S., Rastogi, V., Ashman, W. and Mulbery, W. (2000) Mutagenesis of organophosphorus hydrolase to enhance hydrolysis of the nerve agent VX. *Biochemical And Biophysical Research Communications*, **279**, 516–9.

68 Yang, Y.C. (1999) Chemical detoxification of nerve agent VX. *Accounts of Chemical Research*, **32**, 109–15.

69 Horne, I., Sutherland, T.D., Harcourt, R.L., Russell, R.J. and Oakeshott, J.G. (2002) Identification of an opd (organophosphate degradation) gene in an *Agrobacterium* isolate. *Applied and Environmental Microbiology*, **68**, 3371–6.

70 Yang, H., Carr, P.D., McLoughlin, S.Y., Liu, J.W., Horne, I., Qui, X., Jeffries, C. M.J., Russell, R.J., Oakeshott, J.G. and Ollis, D.L. (2003) Evolution of an organophosphate-degrading enzyme: a comparison of natural and directed evolution, *Protein Engineering*, 135–45.

71 Jackson, C., Kim, H.K., Carr, P.D., Liu, J.W. and Ollis, D.L. (2005) The structure of an enzyme-product complex reveals the critical role of a terminal hydroxide nucleophile in the bacterial phosphotriesterase mechanism. *Biochimica et Biophysica Acta*, **1752**, 56–64.

72 Jackson, C.J., Liu, J.W., Coote, M.L. and Ollis, D.L. (2005) The effects of substrate orientation on the mechanism of a phosphotriesterase. *Organic and Biomolecular Chemistry*, **3**, 4343–50.

73 Horne, I., Qui, X., Ollis, D.L., Russell, R.J. and Oakeshott, J.G. (2006) Functional effects of amino acid substitutions within the large binding pocket of the phosphotriesterase OpdA from *Agrobacterium sp. P230. FEMS Microbiology Letters*, **259**, 187–94.

74 DeFrank, J.J. and Cheng, T.C. (1991) Purification and properties of an organophosphorus acid anhydrase from a halophilic bacterial isolate. *Journal of Bacteriology*, **173**, 1938–43.

75 Cheng, T.C., Liu, L., Wang, B., Wu, J., DeFrank, J.J., Anderson, D.M., Rastogi, V.K. and Hamilton, A.B. (1997) Nucleotide sequence of a gene encoding an organophosphorus nerve agent degrading enzyme from *Alteromonas haloplanktis. Journal of Industrial Microbiology and Biotechnology*, **18**, 49–55.

76 Cheng, T.C., DeFrank, J.J. and Rastogi, V.K. (1999) Alteromonas prolidase for organophosphorus G-agent decontamination. *Chemistry and Biology*, **119–120**, 455–62.Interact

77 Maher, M.J., Gosh, M., Grunden, A.M., Menon, A.L., Adams, M.W., Freeman, H.C. and Guss, J.M. (2004) Structure of the prolidase from *Pyrococcus furiosus. Biochemistry*, **43**, 2771–83.

78 Du, X., Tove, S., Kast-Hutcheson, K. and Grunden, A.M. (2005) Characterization of the dinuclear metal center of *Pyrococcus furiosus* prolidase by analysis of targeted mutants. *FEBS Letters*, **579**, 6140–6.

79 Hill, C.M., Wu, F., Cheng, T.C., DeFrank, J.J. and Raushel, F.M. (2000) Substrate and stereochemical specificity of the organophosphorus acid anhydrolase from *Alteromonas sp. JD6.5* toward *p*-nitrophenyl phosphotriesters. *Bioorganic and Medicinal Chemistry Letters*, **10**, 1285–8.

80 Hill, C.M., Li, W.S., Cheng, T.C., DeFrank, J.J. and Raushel, F.M. (2001) Stereochemical specificity of organophosphorus acid anhydrolase toward *p*-nitrophenyl analogs of soxman and sarin. *Bioorganic Chemistry*, **29**, 27–35.

81 Nagata, Y., Nariya, T., Ohtomo, R., Fukuda, M., Yano, K. and Takagi, M. (1993) Cloning and sequencing of a dehalogenase gene encoding an enzyme with hydrolase activity involved in the degradation of gamma-hexachlorocyclohexane in *Pseudomonas paucimobilis. Journal of Bacteriology*, **175**, 6403–10.

82 Marek, J., Vévodová, J., Smatanová, I.K., Nagata, Y., Svensson, L.A., Newman, J., Takagi, M. and Damborsky, J. (2000) Crystal structure of the haloalkane dehalogenase from *Sphingomonas paucimobilis UT26. Biochemistry*, **39**, 14082–6.

83 Oakley, A., Klvana, M., Otypeka, M., Nagata, Y., Wilce, M.C.J. and Damborsky, J. (2004) Crystal structure of haloalkane dehalogenase LinB from *Sphingomonas paucimobilis UT26* at 0.95 A resolution: dynamics of catalytic residues. *Biochemistry*, **43**, 870–8.

84 Hynková, K., Nagata, Y., Takagi, M. and Damborsky, J. (1999) Identification of the catalytic triad in the haloalkane dehalogenase from *Sphingomonas paucimobilis UT26. FEBS Letters*, **446**, 177–81.

85 Negri, A., Marco, E., Damborsky, J. and Gago, F. (2007) Stepwise dissection and visualization of the catalytic mechanism of haloalkane dehalogenase LinB using molecular dynamics simulations and computer graphics. *Journal of Molecular Graphics and Modelling*, **26**, 643–651.

86 Prokop, Z., Oplustik, F., DeFrank, J. and Damborsky, J. (2006) Enzymes fight chemical weapons. *Biotechnology Journal*, **1**, 1370–80.

87 Sangster, J. (1997) *Octanol-Water Partition Coefficients: Fundamentals and Physical Chemistry*, Vol. **2**, John Wiley & Sons, Inc., Chichester.

88 Eugene Kellogg, G. and Abraham, D.J. (2000) Hydrophobicity: is logP(o/w) more than the sum of its parts? *European Journal of Medicinal Chemistry*, **35**, 651–61.

89 Hildebrand, J.H. (1936) *The Solubility of Non-electrolytes*, ACS Monograph No. 17, Reinhold, New York.

90 Brandrup, J. and Immergut, E.H. (1989) *Polymer Handbook*, John Wiley & Sons, Inc., Chichester.

91 LeJeune, K.E. and Russell, A.J. (1999) Biocatalytic nerve agent detoxification in fire fighting foams. *Biotechnology and Bioengineering*, **62**, 659–65.

92 Dierstein, R., Grabowski, A., Klein, N., Kremer, J. and Richardt, A. (2001) Enzymxxatischer Abbau von Chemischen Kampfstoffen, *Wehrtechnik*, 106–9.

93 Hoar, T.P. and Schulmann, J.H. (1943) Transparent water-in-oil dispersions: the oleopathic hydromicelle. *Nature*, **52**, 102–3.

94 Bonkhoff, K., Schwuger, M.J. and Subklew, G. (1997) Use of Microemulsions for the extraction of contaminated solids, in *Industrial Applications of Microemulsions, Surface Science Series* 66 (eds C. Solans and H. Kuineda), Chap. 17, Marcel Dekker, New York.

95 Myers, D. (ed.) (1991) *Surfaces, Interfaces and Colloids*, Chap. 16, VCH Verlagsgesellschaft, Weinheim.

96 Richardt, A. and Mitchell, S. (2006) Enzymes for environmentally friendly decontamination of sensitive equipment. *Journal of Defence Science*, **10**, 261–5.

97 Orii, Y. and Morita, M. (1977) Measurement of the pH of frozen buffer solutions by using pH indicators. *Journal of Biochemistry*, **81**, 163–8.

98 Chang, B. and Randall, C. (1992) Use of a subambient thermal analysis to optimize protein lyophilization. *Cryobiology*, **29**, 632–56.

99 Good, N.E., Winget, G.D., Winter, W., Conolly, T.N., Izawa, S. and Singh, R.M.M. (1966) Hydrogen ion buffers for biological research. *Biochemistry*, **5**, 467–77.

100 Morita, T., SakamuraY., Horikiri, Y., Suzuki, T. and Yoshino, H. (2000) Protein encapsulation into biodegradable microspheres by a novel S/O/W emulsion method using poly (ethylene glycol) as a protein micronization adjxuvant. *Journal of Controlled Release : Official Journal of the Controlled Release Society*, **69**, 435–44.

101 Rüterjans, H. (2004) *Abschlussbericht zum Forschungsvorhaben E/E590/2 X041/ M5137 Stabilisierung der DFPase*, J.W. Goethe University, Frankfurt.

102 Richardt, A., Fischer, G., Hoff, E. and Teichmann, J. (2004) Neuartiger Dekontlöscher, German Patent 10231740.

103 Blum, M.M., Mitchell, S., Danielsen, S. and Richardt, A. (2007) *Enzymatic Decontamination of Nerve Agents by the Enzyme DFPase – From Concept to Product*, Conference Proceedings, 9th Symposium on Protection Against Chemical and Biological Warfare Agents, Gothenburg.

10
Laccases – Oxidative Enzymes for Bioremediation of Xenotics and Inactivation of Bacillus Spores

Bärbel Niederwöhrmeier, Lars Ostergaard, André Richardt and Steffen Danielsen

10.1
Introduction

Oxidoreductases with oxidative reaction mechanisms and broad substrate specificity have the potential to destroy a wide range of toxic substances such as chemical warfare agents simultaneously. By incorporating only one type of enzyme in a formulation instead of a mixture of different types of enzymes it is possible to avoid compatibility problems between the different enzymes such as differences in pH-optimum, mutual inhibition by different metal ligands, and interferences by using different buffer systems.

Laccases (EC 1.10.3.2) are oxidoreductases that are widely distributed among microbes, plants and insects. They have received much attention due to their ability to oxidize both phenolic compounds as well as highly toxic environmental pollutants. Therefore, their substrate versatility makes them highly interesting for different biotechnology based applications, including pulp bleaching and bioremediation. They are able to catalyze the exchange of electrons or redox equivalents between donor and acceptor molecules. As a result of the growing potential of oxidative enzymes like laccases for decontamination of chemical and biological warfare agents, we would like to give a detailed description of the reaction mechanism, biochemistry, potential mediators and applications in the biotechnical and medical industry for one type of oxidoreductases. A comprehensive understanding of the fundamentals of these enzymes is necessary to find new approaches for technical deontamination systems. Furthermore, first results of oxidoreductases for the inactivation of *Bacillus anthracis* spores will be described.

10.2
Nomenclature

Laccases [EC 1.10.3.2; p-diphenol: dioxygen oxidoreductase] are defined by the Nomenclature Committee of the International Union of Biochemistry and

Decontamination of Warefare Agents.
Edited by André Richardt and Marc-Michael Blum

ISBN: 978-3-527-31756-1

Molecular Biology as oxidoreductases. They are mostly extracellular glycoproteins with molecular weight between 60 and 80 kDa. Together with the enzymes plant ascorbate oxidase [EC 1.10.3.3] and ceruloplasmin [EC 1.16.3.1] they belong to the group of enzymes called blue copper proteins (blue copper oxidases or multicopper oxidases) [1, 2]. The enzymes phenoxazinon-synthase, bilirubin-oxidase [EC 1.3.3.5], dihydrogeodin-oxidase, sulochrin-oxidase [EC 2.10.3.8] and FET3 are also integrated into this group [3]. They catalyze the four–electron reduction of molecular oxygen to water with concomitant one-electron reduction of the substrate [4]. Compared to enzymes with very high substrate specificity, laccases are able to oxidase a broad range of substrates. These include phenols, diphenols, polyphenols, substituted phenols, benzenthiols, diamines, and aromatic amines [5]. Artificial substrates such as ABTS (2,2′-azino-bis-3-ethylbenzthiazoline-6-sulfonic acid) are able to act as mediators enabling the oxidation of non-phenolic compounds which normally cannot be oxidized by laccases. With these substrates (mediators), the range of applications of laccases can be considerably expanded. Surprisingly, even some inorganic compounds like iodine can be oxidized and will also work as mediators [5]. The oxidized mediator loses a single electron and the normally formed free radical is able to catalyze further reactions [6, 7].

10.3 Distribution of Laccases

Laccases are widely distributed and can be found in fungi, plants and insects [8] and also in some prokaryotes [9]. The laccase found in the sap of *Toxicodendron vernicifluum*, also called Chinese or Japanese lacquer trees (*Rhus* species), was the first to be reported, in 1883 [10]. The designation laccase was derived from this tree [11].

10.3.1 Distribution in Plants

Laccases have been discovered in many other plants [12–16]. It has been shown that plant laccases are able to polymerize monolignols within the plant cell wall matrix [17–19]. These studies indicate that laccases are involved in the early stages of lignification [16]. However, the role of laccases in the lignification process remains an unsolved matter [16]. There are also speculations and some evidence, that laccases are involved in the healing process of wounded leaves [14]. In investigations of the general mechanism of laccases, *Rhus vernicifera* laccase has been used [20].

10.3.2 Distribution in Fungi

As efficient lignin degraders, white-rot basidiomycetes are an excellent source of laccases. Therefore, the majority of laccases have been found in fungi (Table 10.1).

Table 10.1 Examples for laccases in fungi.

Fungi	*References*
Boltrytis cinera	[21]
Chaetomium thermophilum	[22]
Coprinus cinereus	[23]
Neurospora crassa	[24]
Phlebia radiata	[25]
Pleurotus ostreatus	[26]
Pycnoporus cinnabarinus	[27]
Trametes (Coriolus, Polyporus) versicolor	[28]

Laccase from white-rot fungi, such as *Trametes versicolor* and *Pycnoporus cinnabarinus*, reduce lignification activities by the host, where they mainly oxidize the phenolic subunits of lignin [27, 29–31]. In plant-pathogenic fungi, laccases are important factors in virulence. They can protect the fungal pathogen from the toxic phytoalexins and tannins in the host environment [32–34]. Some fungal secreted laccases act as a detoxifying enzyme to protect the fungus from toxic metabolites [35–37]. For example, the laccase in grapevine gray mold, *Botrytis cinerea*, is necessary for pathogenesis. The assumed role of the laccase could be related to detoxification of toxic defence metabolites produced by the plant [38].

10.3.3 Distribution in Prokaryotes

Also, there is increasing evidence for the existence of the multi-copper oxidase enzyme family in prokaryotes [9, 39]. The first bacterial laccase was detected and described in the plant root-associated bacterium *Azospirillum lipoferum* [40, 41]. Further corresponding genes have been found in other gram-negative and gram-positive bacteria, including species living in extreme habitats [40, 42–49]. A laccase-like enzyme activity was found in thermostable spores of different Bacillus strains. The production of a thermostable CotA laccase by *Bacillus subtilis* involved in pigment production in the endospore coat was described [50–52].

10.4 General Structure, Biochemical and Active Site Properties of Laccases

In order to optimize enzymes for their intended use in a technical system, a comprehensive understanding of the enzymes is often an advantage. For laccases, the structure has been studied in great detail [53–56], and the electrochemical properties of laccases [20, 57] have also been studied intensely. The mechanism of the internal electron transfer and the reduction of oxygen to water has been clarified

[58]. The overall structure of laccases has been studied intensively. Three cupredoxin-like domains A, B and C can be found as the common structure among copper proteins [55, 56, 59, 60]. This structure has also been found in other copper proteins (Table 10.2).

Such domains in copper containing proteins have different names in the literature (Table 10.3).

All three domains are important for the catalytic activity of laccases:

- The substrate-binding site is located in a cleft between domains B and C.
- The mononuclear copper center with one type-1 copper atom is located in domain C.
- The trinuclear copper center with one type-2 copper atom and a pair of type-3 coppers is located at the interface between domains A and C [2].

As an active holoenzyme form the laccase is a dimeric or tetradimeric glycoprotein. For the catalytic activity a minimum of four copper atoms per active protein unit is needed. Bound to three redox sites (T1, T2 and T3 Cu pair) usually four copper atoms can be found per monomer (Figures 10.1, 2).

Each of the monomers has a molecular mass between 50 and 100 kDa. A covalently linked carbohydrate moiety (10–45%) could be responsible for the high stability of laccases [67]. Electron paramagnetic resonance (EPR) spectrum can be used to distinguish the four copper atoms. One strong signal in the spectrum belongs to the paramagnetic "blue" T1 copper site. The covalent coordination bond between T1 and S_{Cys} has a strong electronic absorption at 610 nm and causes the typical blue color of laccases [3]. Another signal belongs to the T2 paramagnetic

Table 10.2 Cupredoxin-like domains in proteins.

Proteins with cupredoxin-like domains	*References*
Plant plastocyanin	[61, 62]
Bacterial azurin	[63]
Ascorbate oxidase	[64]
Ceruloplasmin	[65, 66]

Table 10.3 Description of copper-domains.

Description				*Cu-atoms*
T1	Type 1	Copper 1	Blue Copper site	1
T2	Type 2	Copper 2	Normal Copper site	1
T3	Type 3	Copper 3	Coupled Copper site	2

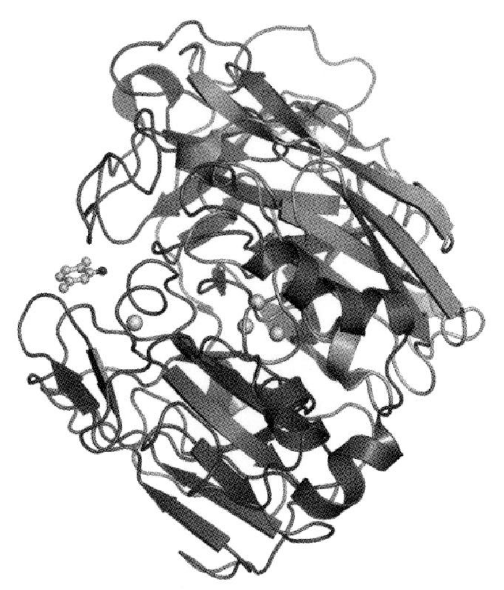

Figure 10.1 Crystal structure of *Trametes versicolor* laccase (ribbon diagram).

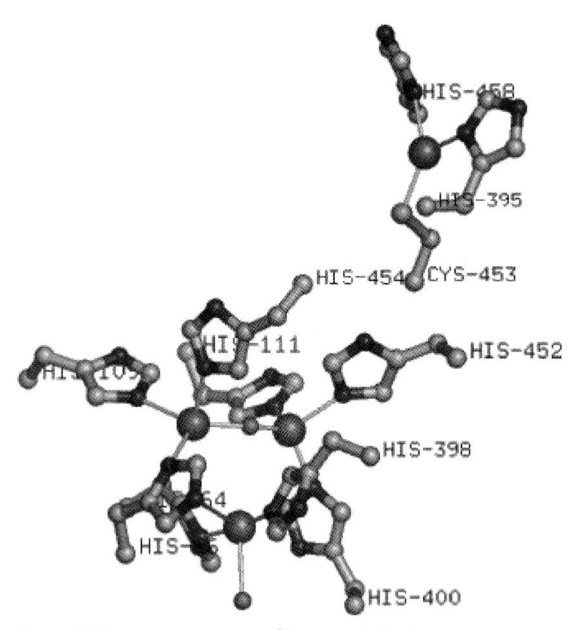

Figure 10.2 Copper centers of laccases [56, 68].

"non-blue" copper site. Due to the fact, that the T3 copper pair is strong anti-ferromagnetic coupled by a hydroxyl bridge the T3 coppers are silent in EPR spectrum [68]. However, a weak UV absorbance at 330 nm can be used for detection [3]. The trinuclear cluster is formed by T2 and T3 copper atoms and is responsible for oxygen binding and its reduction to water. The coordination of T2 copper by two histidines and T3 copper pair by six histidines is evolutionary conserved. The T1 site is responsible for the electron abstraction from reducing substrates (electron donors). Then the electron is transferred to the T2/T3 cluster (Figure 10.2).

The Michaelis constant km and the catalytic efficiency constant (kcat/Km) quantitatively describe the catalytic action of an enzyme. It seems that the catalytic efficiency of laccases for some reducing substrates depend linearly on the redox potential of the T1 copper. It has been shown that the higher the potential of the T1 site, the higher the catalytic efficiency [5, 69]. A high redox potential of the T1 site is of special interest for different applications, for example for efficient bleaching and bioremedition processes [70, 71]. Therefore, the spectrum of the redox potential of the T1 site was determined by using different mediators. The lowest redox potential was found with 430 mV for a laccase from *Rhus vernicitera* However, Trametes, Polyporus and Coriolus laccases usuall have the highest redox potential [5, 69–74]. The Michaelis constant Km and the catalytic efficiency constant have been measured for a large number of laccases (Table 10.4).

Very significant variance has been observed in the catalytic efficiencies of various laccases (Table 10.4). Also, kinetic data of laccases from different sources were reported [83].

Table 10.4 Catalytic efficiency constants of different laccases, adapted from [75].

Substrate	*Laccase*	*Catalytic efficiency* k_{cat}/km $(m^{-1} \times s^{-1})$	*pH*
ABTS	*Pleurotus ostreatus* POXA1 [76]	3900	3
	Trametes pubescens [77, 78]	2960	3
	Pleuratus ostreatus POXC [76]	203	3
	Coprinus cinereus [23]	47	5.5
	Trametes villosa [5]	46.5	5.3
	Trichophyton rubrum [79]	14	5.5
	Bacillus subtiliz CotA [52]	9	4
	Trametes trogii [80]	6.6	3.4
	Myceliophtthora thermophila LCC1 [81]	2.7	6
2,6-DMP	*Pleurotus sajor-caju* Lac4 [82]	483	6
	Trametes pubescens LAP2 [77, 78]	333	3
	Pleurotus ostreatus POXA1 [76]	10	5
Guaiacol	*Trametes pubescens* LAP2 [22]	300	3
	Pleurotus sajor caju Lac4 [82]	103	6.5
	Pleurotus ostreatus POXC [76]	0.125	6
	Trametes trogii POXL3 [80]	0.025	3.4
Syringaldazine	*Trametes pubescens* LAP2 [77, 78]	2800	4.5
	Myceliophthora thermophila Lcc1 [81]	1312	6
	Pleurotus ostreatus POXC [76]	1150	6
	Trametes villosa Lcc1 [5]	769	5.3
	Pleurotus ostreatus POXA1 [76]	215	6
	Pleurotus sajor-caju Lac4 [82]	125	6.5
	Bacillus subtiliz CotA [52]	7.7	6

V_{max} varies with the source of laccase (50–300 M/s), but Km values are similar for the co-substrate oxygen (about 10–5). Depending on the source of enzyme, the substrate and type of reaction the turnover is heterogeneous over a broad range. Also, it is important that the kinetic constants differ in their dependence on pH. The catalytic constant is pH-dependent, while Km is pH-independent for substrate and co-substrate [100]. Of particular importance is the fact that laccases can be very strongly inhibited by many anions. These small anions like cyanide, thiocyanide, azide, and fluoride are able to interact with the copper sites. Also, complexing agents like ethylenediaminetetraacetic acid (EDTA) removing copper from the active site exert a reversible activity inhibition.

Another important property is the temperature stability that varies considerably, dependent on the source. Typically, laccases are stable at 30–50 °C and lose rapidly activity at temperatures above 60 °C [78, 79, 84, 85]. The most thermostable laccases have been isolated from bacteria [48, 52].

10.5 Catalytic Mechanism of Laccases

A number of mechanistic schemes and catalytic cycles for laccases have been proposed. Most of them are consistent with the kinetic and structural data currently available [4]. However, the mechanisms of electron transfer and dioxygen reduction to water are not fully understood for laccases. After binding, the cleft between domain B and C the substrate reduces the T1 site. One full catalytic cycle involves the transfer of four electrons from the T1 site to the T2/T3 cluster. It seems that a conserved His-Cys-His tripeptide is also involved in this transfer [55, 60, 64]. Two possible mechanisms have been described for the reduction of the T2/T3 cluster. The first possibility decribes the reduction of the T3 pair by the T1 and T2 sites (Figure 10.3), and in the second possibility, in the trinuclear cluster each copper is reduced by electron transfer from the T1 site (Figuer 10.3). In this case, the T3 pair does not function as a two-electron acceptor. There is evidence, that the electron transfer to the trinuclear site is too slow to be catalytically relevant [4].

10.6 Laccase-Mediator-System (LMS)

Laccase can oxidize phenols, phenolic fragments and amines directly [6, 86]. Small natural low-molecular weight compounds with high redox potential can be used to expand the substrat spectra of laccases [31]. The combination of laccase, oxygen and these small compounds (mediators) is able to oxidize the other non-phenolic residues in the delignification process [27, 86, 87]. Especially in the field of bioremediation and therefore in the detoxification of organophosphates, the discovery of new and efficient synthetic mediators to extend the laccase catalysis

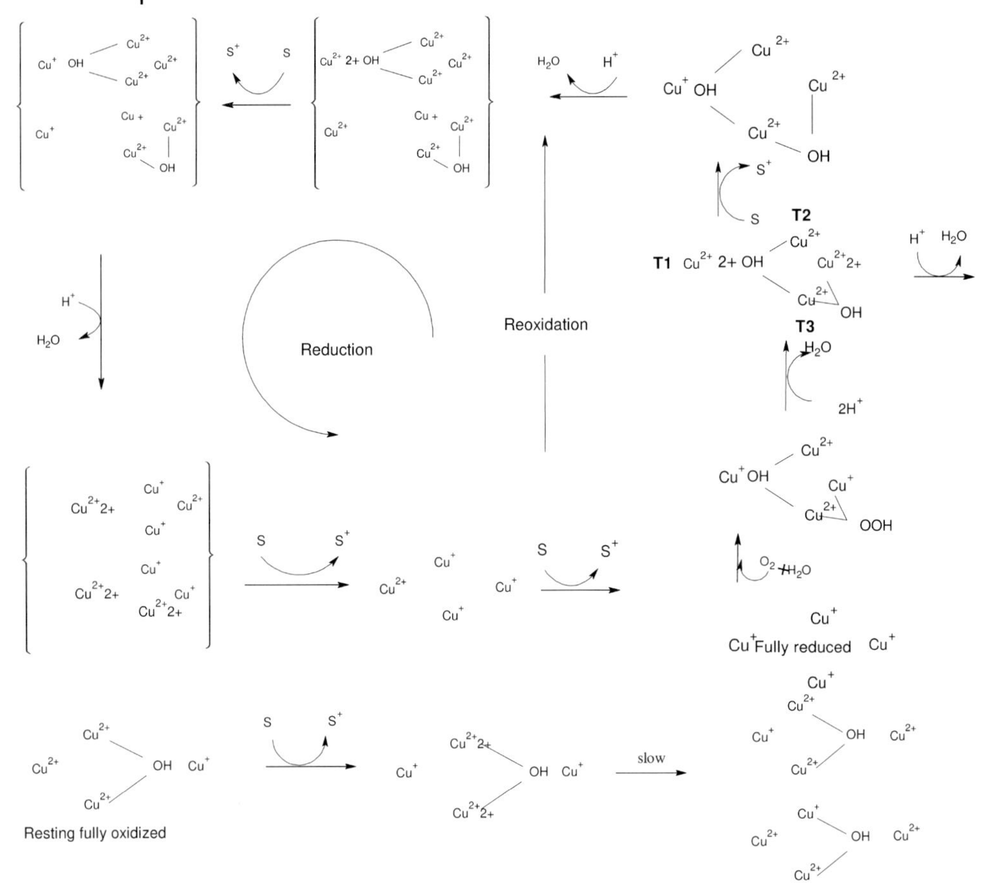

Figure 10.3 Proposed catalytic mechanism of laccases (adapted from [4]).

towards xenobiotic substrates was a breakthrough [27, 31, 86, 88–93]. The oxidation of xenobotic substrates has the results that, on the one hand, the relatively small mediator molecule can oxidize any substrate that, due to its size, cannot directly enter the catalytic site of laccases [94], and on the other hand, with the chemical properties of mediators, it is possible to use new, innovative reaction mechanisms for the enzymatic detoxification process. The generally accepted reaction mechanims can be described as a redox cascade (Figure 10.4) [95].

For a user-friendly application, mediators should fulfill the following properties:

1. Water solubility
2. Low molecular weight
3. Ability to transfer electrons
4. High redox potential
5. Acceptability as a substrate for laccases and no inhibition of laccases

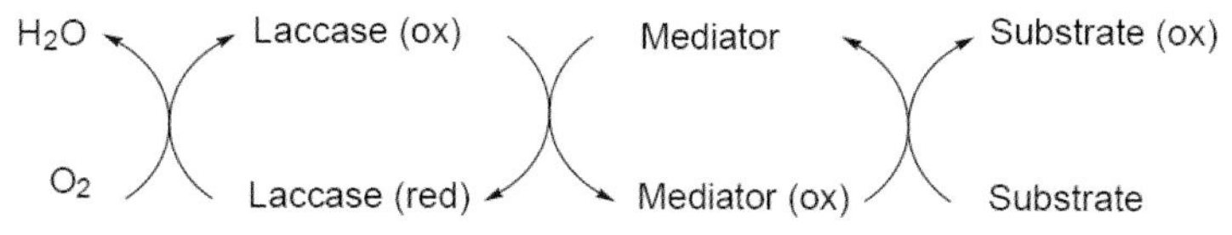

Figure 10.4 Catalytic cycle of a laccase mediator system [94, 95].

6. Biodegradability and environmental compatibility
7. Low prize and merchantable
8. Reversible redoxproperties
9. No toxix by-products and unwanted discoloration [96, 97]

At the moment, there is no known mediator that is able to meet all of these demands. Therefore, the investigation of the nature of the reaction mechanism during the oxidation of non-phenolic substrates like pesticides and chemical warfare agents is of high improtance. Over 100 possible mediator compounds are available at the moment, but the most commonly used are ABTS (azine 2,2′-azino-bis-(3-ethylbenzothiazoline-6- sulfonic acid) and HBT (triazole 1-hydroxybenzotriazole) (Figure 10.5) [88, 93, 98].

Various laccases readily oxidize ABTS to the cation radical ABTS+·. The redox potentials of ABTS+· and ABTS2+ were estimated as 0.680 V and 1.09 V respectively [99]. The concentration of the intensely colored green-blue cation radical can be correlated to the enzyme activity. HBT belongs to the *N*-heterocyclics compounds bearing *N*–OH–groups mediators [100]. By consuming oxygen, laccases are able to convert HBT into the active intermediate, which is oxidized to a reactive radical (R–NO.) [88]. Mediated laccase catalysis has been used in a wide range of applications (Table 10.5).

Regarding the increasing public concerns of the environmental impact of the chlorine-based oxidants currently being used in delignification or bleaching in paper and pulp industry, novel enzymatic bleaching technologies are attracting increasing attention [87, 93, 100, 108].

10.7 Heterologous Expression of Laccases

In their native hosts, laccase concentrations are often found at very low levels. Therefore, to improve laccase production and to achieve a competitive prize for biotechnical applications, laccases have been expressed heterologously in different hosts. For example, fungal laccases have been expressed heterologously in

Saccharomyces cerevisiae [109]
Trichoderma reesei [110]
Pichia pastoris [111]
Aspergillus sojae [112]
Aspergillus niger [113]

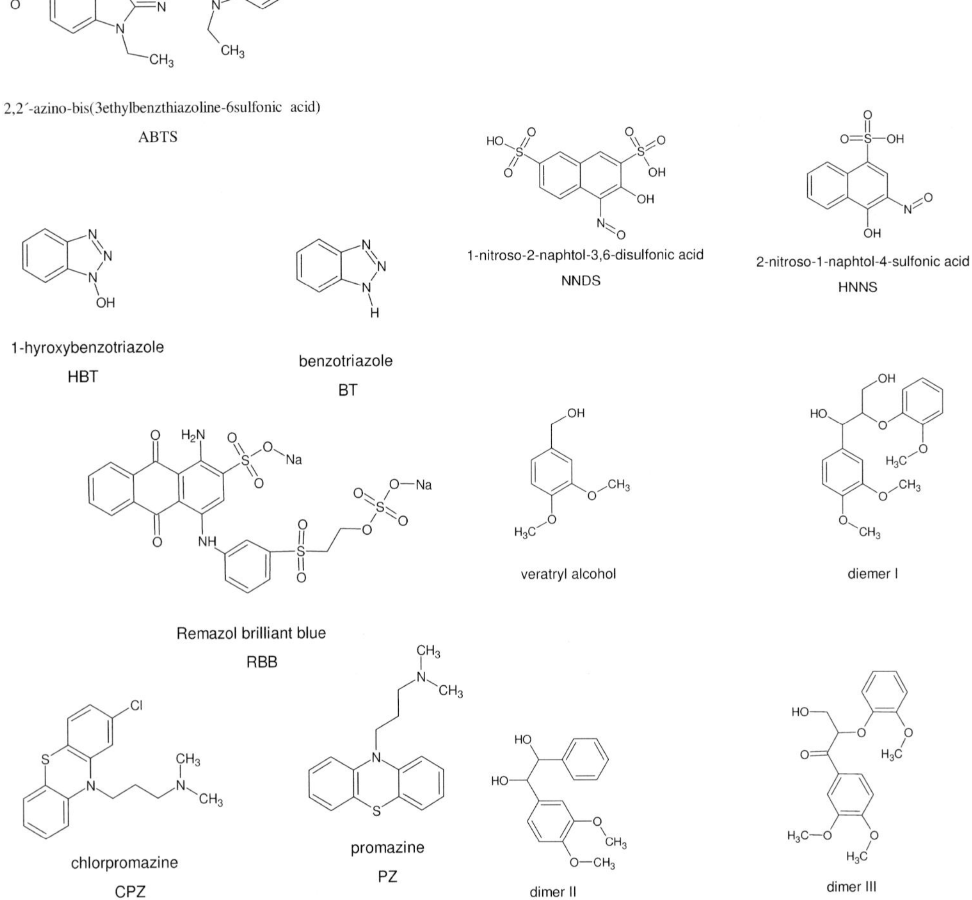

Figure 10.5 Examples for natural and synthetic mediators [adapted from [88]].

Table 10.5 Mediated laccase catalysis.

Application	*References*
Pulp delignification	[87, 91, 100, 101]
Textile dye bleaching	[93, 102]
Polycyclic aromatic hydrocarbon degradation	[103–105]
Toxic compounds degradation	[105, 106]
Organic synthesis	[107]

Aspergillus nidulans [113]
Aspergillus oryzae [108]
tobacco [114] and maize [115].

To facilitate protein engineering of laccases [81] or to improve the resistance of yeast to phenolic growth inhibitors [116], heterologous yeast expression systems have been developed. Although it was shown that bacterial laccases from *Bacillus subtilis* and *Streptomyces lavendulae* could be expressed in *Escherichia coli* successfully, expression of fungal laccases in *E. coli* in sufficient amounts has not been reported yet [48, 52].

For a successful heterologous expression of laccase genes in hosts, the promoter has to be chosen very carefully. It was shown that laccases can be expressed in *Aspergillus* spp. under the control of the strong constitutive TAKA-amylase *(amyA)* promoter [113, 117].

Also, the glyceraldehyde3-phosphate dehydrogenase *(gpdA)* promoter can be used [118]. For another heterologous expression system, *Trichoderma reesei*, the promoter region of the major cellulase gene *cbhl* has been used [110]. An important point for simplifying the purification step of heterologous laccases, the secretion of heterologous laccases has generally been directed by using native laccase signal sequences in the expression constructs by using signal sequences derived from host genes [82, 118]. The highest reported yields have been achieved in filamentous fungi, especially in *Aspergillus* spp. *Aspergillus* spp. is one of the most common heterologous hosts for the production of industrial enzymes. Also, yeast expression systems are very common in the biotechnological industry for the production of enzymes in a technical scale. Therefore, improved laccase production levels were achieved by expression in the yeast *Pichia pastoris* [81, 119]. Homologous production systems seem far more efficient for the production of laccases (Table 10.6). In a shake flask cultivation of *Pycnoporus cinnabarinus*, 1000–1500 mg/L laccase was reported [122] and in a fermentor cultivation of *Trametes pubescens*, 700 mg r1laccase was obtained [78]. However, there are still problems in the production of laccases. An extensive intracellular aggregation of bacterial laccases from *Bacillus subtilis* and *Streptomyces lavendulae* in *E. coli* was reported [48, 52]. The recovery of *Bacillus subtiliz* laccase from the inclusion bodies was not successful [52]. However, the common treatment with urea and 2-mercaptoethanol has led to an active form of the *Streptomyces lavendulae* laccase [48]. Although the laccase production levels have often been improved significantly by expression in heterologous hosts, the reported levels have still been rather low for industrial applications (Table 10.6).

Therefore, there is still need for further improvement in the production of heterologous laccases to reduce the costs for technical applications. Varying the cultivation conditions could lead to better production. For example, better production has been achieved in yeast systems by controlling the pH of the culture, optimizing the medium and by lowering cultivation temperatures [82, 116, 123]. Also, maintaining the pH above 4 in the culture medium seems to be important for stability of secreted laccases and inactivation of acidic proteases [82]. Improved folding of

Table 10.6 Heterologous and homologous expression systems for laccase production, adapted from [75].

Laccase gene	*Production host*	*Laccase production in ($mg\,l^{-1}$) shake flask cultivation*	*Reference*
Ceriporiopsis subvermispora lcs-1	*Aspergillus nidulans*	1.5	[113]
	Aspergillus niger		
Coprinus cinereus lcc1	*Aspergillus oryzae*	135	[117]
Myceliophthora thermophila lcc1	*Aspergillus oryzae*	19	[120]
	Saccharomyces cerevisiae	18	[81]
Pleurotus sajor-caju lac1	*Pichia pastoriz*	4.9	[82]
Pycnoporus cinnabarinus lac1	*Pichia pastoriz*	8	[121]
	Aspergillus niger	70	[118]
	Aspergillus oryzae	80	[108]
Pycnoporus cinnabarinus r1lac	*Pycnoporus cinnabarinus*	1000–1500	[122]
Laccase gene	Production host	Laccase production ($mg\,l^{-1}$) in a laboratory fermenter	Reference
Phlebia radiata lac1	*Trichoderma reesei*	20	[110]
Trametes pubescens r1lac	*Trametes pubescens r1lac*	700	[78]

heterologous proteins could be achieved by lowering cultivation temperatures and could result in better production [116]. In addition, overexpression of Ss02p, a membrane protein involved in the protein secretion machinery [124], has been shown to improve heterologous laccase production in S. *cerevisiae* [119]. The addition of copper into the culture medium has also proved to be important for heterologous laccase production in different host systems like *Pichia pastoris* and *Aspergillus* spp. [113, 123, 125, 126]. However, the achieved levels of recombinant laccase production are still not sufficient and further investigations and research are necessary to obtain higher laccase production. Without easy production of oxidative enzyme systems, results comparable to other decontaminants are difficult to achieve.

10.8 Laccase Applications

Laccasse applications include the detoxification of industrial effluents from the pulp and paper, textile and petrochemical industries. Laccases are used as tools for medical diagnostics and for bioremediation, to clean up herbicides, pesticides, explosives and chemical warfare agents [105, 127, 128]. Different laccase applications are described in more detail in the following sections and these will show the potential of enzyme based solutions for the removal of xenotics and toxic compounds. As a potential replacement for environmental harmful processes as chlorine-based bleaching processes, these applications could lead to more environmentally harmless applications, especially for decontamination purposes.

10.8.1
Dye Degradation

The textile industry consumes large amounts of water and chemicals for wet processing of textiles. Real textile effluents are extremely variable in composition. In addition to dyes, they contain also salts, chelating agents, precursors, by-products and surfactants [129–131]. All of these additional substances might reduce or inhibit enzyme activity and their planned decolorization [132]. Therefore, for an enzymatic decolorization of textile effluents, it is necessary to be careful in the choice of the type of enzyme, as well as of reactor environment [102]. Oxidative enzymes like laccases are able to act on chromophore compounds such as azo, triarylmethane, anthraquinonic and indigoid dyes (Table 10.7).

This activity leads to the suggestion that they could be applied in industrial decolorization processes and that these alternative bio-bleaching systems might lead to a replacement of the commonly used chemical bleach decolorization process [102, 132, 139, 140, 143]. Further studies for phenolic and non-phenolic azo dye degradation showed evidence of several mechanisms for enzymatic degradation [144]. In one proposed model, azo dyes are degraded without direct cleavage of the azo bond which should not limit the number of azo dyes that can be degraded [145, 146]. However, in other laccase reactions some substrate specificity can also be found. Therefore, the further described laccase/mediator systems (LMS) are under investigation to broaden the range of azo dyes and to increase decolorization rates [88, 93, 147, 148]. However, little is known about the efficiency of the LMS and our capacity to evaluate laccase degradation potential remains limited since there is incomplete knowledge of dye decolorization pathways, dye mineralization mechanisms and formation of potentially toxic accumulating intermediates [102]. Despite this, the use of laccases in the textile industry is growing fast [128, 149]. In 1996 a new concept for laccase catalyzed dye bleaching in the finishing of indigo-dyed cotton denim fabric was developed [150, 151].

Two commercial laccase products are available for this type of application, Denilite I and Denilite II, both containing a fungal recombinantly expressed laccase and a redox mediator [150, 151]. By combined action of the laccase and the redox mediator, the blue indigo is oxidized into a virtually colorless compound known

Table 10.7 Laccases for decolourization of dyes [128].

Laccase source	***References***
Aspergillus	[133]
Pleurotus ostreatus	[134]
Pleurotus cinnabarinus	[135, 136]
Trametes hirsute	[132, 137]
Trametes versicolor	[93, 138–140]
Trametes villosa	[141, 142]

as Isatin. The traditional bleaching by chlorine chemistry that is normally applied in the final step in finishing of indigo-dyed denim, is replaced by this more environmentally benign method. Benefits for customers include better processing control, reduced water usage and less release of polluted waste material.

10.8.2 Delignification and Biobleaching

In the industrial preparation of paper, the separation and degradation of lignin in wood pulp is essential. There are growing concerns about the environmental effects of the conventional and polluting chlorine-based delignification and bleaching process [90]. Therefore, enzymatic delignification processes have been industrially introduced in the last years to replace conventional and polluting chlorine-based methods [90]. The pre-treatment of wood pulp with oxidoreductases such as laccase could provide milder and cleaner strategies of delignification [90, 108, 152]. The ability of different laccases to work in a biobleaching process have been investigated (Table 10.8).

Nonphenolic ligno-cellulose can be oxdized a laccase-mediator-system [88, 156]. More recently, the potential of this enzyme for binding fiber-, particle- and paperboards was discovered [157]. Also, it was shown that laccases are able to graft various derivaties onto kraft pulp fibers [158, 159]. However, to get from these preliminary results to a biotechnological product, the different pathways have to be investigated and improvement of the reactions mechanisms have to be done.

10.8.3 Organic Synthesis

Increasing interest has been focused on the application of laccases as new biocatalysts in organic synthesis (Table 10.9) [8, 160]. For this book we give only a short overview. For details of laccase catalysed organic synthesis, the reader is referred to reviews giving detailed lists of laccases and their catalysed reactions [160].

The use of laccases could lead to an environmentally benign process of polymer production in air [121, 167]. The crosslinking reaction of new urushiol analogues for the preparation of polymeric films by laccases has been demonstrated [168]. Also, it was shown that there is potential for crosslinking and functionalizing

Table 10.8 Laccases for biopulping and biobleaching.

Laccase source	*References*
Trametes versicolor	[95, 153]
Coriolus versicolor	[100, 154]
Pleurotus cinnabarinus	[155]

Table 10.9 Laccases for organic synthesis (adapted from [160]).

Laccase source	*References*
Trametes versicolor	[107, 161, 162]
Coriolus hirsute	[163, 164]
Pleurotus cinnabarinus	[165]
Pleurotus oryzae	[149]
Pleurotus coccineus	[166]

lignaceous compounds by using laccases [169]. Moreover, radical polymerization induced by laccase is also possible and this opens the window for new applications [170]. Laccases have also been used for the chemo-enzymatic synthesis of lignin graft-copolymers [157], for polymerization of various amino and phenolic compounds [171–173] and to generate color "in situ" from originally non-colored low-molecular substances [174, 175]. These abilities of laccase can be applied in several natural substrates like cotton, sisal, wool, flax and wood [176].

10.8.4
Wine and Beer Stabilization

Wine stabilization is one of the main applications of laccase in the food industry [177]. Musts and wines are complex mixtures of different chemical compounds, such as ethanol, organic acids (aroma), salts and phenolic compounds (color and taste). The properties of laccases can be used to remove selective polyphenols from wine. This selective removal is necessary to conserve the wine's organoleptic characteristics. Laccase can be used for polyphenol elimination treatment in wines because they have some unique properties such as stability in acid medium and reversible inhibition with sulfite [178–180]. Another wine-related problem is that cork stoppers in bottled wine quite often cause an unpleasant musty taste and flavor in the wine (i.e. the wine has a "corky" taste). Bottles with such a flavor are spoiled. The main cause of this flavor seems to be the presence of phenolic compounds in the cork. In particular, one such compound, trichloroanisole, has an extremely penetrating taste and this compound is suspected to give the wine its musty character [181, 182]. By using a laccase-based enzymatic rinse process, the phenolic compounds may be removed or polymerized in a process either washing out or copolymerizing/entrapping the anisole as well [183]. A commercial laccase product, Suberase from Novozymes, has been developed and marketed for use in such a treatment.

Another interesting application of laccases is the use in the brewing industry. One major problem is haze formation in beers. It has been shown that nucleophilic substitution of phenolic rings by protein sulfydryl groups might be used to lead to a permanent haze. However, storage life of beer can also be improved by laccases, when they were added to the wort [177, 184].

10.8.5 Food Improvement

The ability of laccases to eliminate dissolved oxygen can be used to improve the flavor quality of vegetable oils [185]. Laccases have been used to deoxygenate food items and to improve their flavor and taste. The flavor and taste of cacao and its products were enhanced by laccase containing solutions. Also, the reduction of odors using laccase is well documented in the literature [186]. For further examples the reader is referred to the literature [177, 187–192].

10.8.6 Biosensors

A biosensor is an integrated biological-component probe with an electronic transducer. A biochemical signal is converted into a quantifiable electrical signal. This signal includes information regarding a physiological or biochemical change [193]. Laccases in biosensors have been developed for aromatic amines, phenolic compound and glucose determinations and for immunoassays [194–199]. Further information about this advanced application will be given in the chapter on immobilization.

10.8.7 Bioremediation or Degradation of Xenobiotics Like Pesticides and Chemical Warfare Agents

Industrial wastewater, agricultural biocide treatment and municipal discharges release a large number of hazardous chemicals into the environment. The ability to use laccases to detoxify or degrade such xenobiotic pollutants has been the focus of research for a number of years. Various aromatic compounds, such as the endocrine-disrupting or estrogenic compounds bisphenol A and dichlorophenol, the chlorinated phenols or α-naphthanols (pesticide leftovers), can all be degraded, polymerized or otherwise immobilized by the action of laccases [106, 171, 200–204].

10.8.7.1 Bioremediation of Xenobiotics

Laccases may polymerize the compounds themselves or copolymerize them with other substances present (e.g. humic acid). Laccases have also been studied for degrading sulfur-containing coal substances, for potential applications in coal biodesulfurization/upgrading, bioremediation around coal mines, and emission reduction of acid rain-causing agents from power plants. Organophosphorus compounds and polycyclic aromatic hydrocarbons (PAHs) together with other xenobiotics are of major concern for the environment due to their residence time in soil and their toxicity. They are used as pesticides, insecticides, plasticizers and chemical warfare agents. Contamination of soil from pesticides as a result from their handling in farming leads to worldwide contamination of surface and ground water with high mammalian toxicity. Therefore, it is of growing interest to remove them from the environment by environmentally benign methods. Furthermore, there is still need to destroy about 250 000 metric tons of chemical warfare agents worldwide under the Chemical Weapons Convention. Bioremediation could offer an efficient and alternative method to the commonly used methods of both cleaning and decontamination of polluted environments and destruction of chemical warfare agents [205]. Laccases have been shown to be useful for the removal of these toxic compounds by several mechanisms.

1. Oxidative enzymatic coupling of the contaminants, leading to insoluble complex structures with a lower biological availability [206–210].
2. A variety of persistent environmental pollutants like phenols, chlorinated phenols, pesticides and chemical warfare agents were degraded [105, 109, 127, 146, 211–213] (Table 10.10).

As mentioned before, laccases are able to immobilize soil pollutants by coupling humic substances to soil [206, 207]. Phenolic compounds and anilines such as 3,4-dichloroaniline, 2,4,6-trinitrotoluene or chlorinated phenols can be immobilized in this way [211, 228]. With this technology, it is possible to lower the biological availability of the xenobiotics and thus, their toxicity.

Phenolic compounds are also present in wastes from several industrial processes, such as coal conversion, petroleum refining, production of organic chemicals and olive oil production, among others [229]. Laccase was found to be responsible for the transformation of 2,4,6-trichlorophenol to 2,6-dichloro-1,4-hydroquinol and 2,6-dichloro-1,4-benzoquinone [230, 231]. Laccases from white-rot fungi have also been used to oxidize alkenes, carbazole, N-ethylcarbazole, fluorene and dibenzothiophene in the presence of HBT and ABTS as mediators [220, 232]. Isoxaflutole is a herbicide activated in soils and plants to its diketonitrile derivative, the active form of the herbicide. Laccases are able to convert the diketonitrile into acid. The pesticide Rotenon was degraded by a laccase-mediator-system effectivily [233]. The alteration of chlorinated compounds to non-chlorinated

Table 10.10 Laccases for degradation of xenobiotics [128].

Laccase source	*References*
Coprinus gallica	[213, 214]
Coriolus versicolor	[215, 216]
Pleurotus osteratus	[105, 212, 217, 218]
Rhus vernifecera	[219]
Trametes hirsuta	[220]
Trametes versicolor	[95, 104, 218, 221–224]
Trametes villosa	[106, 225–227]

products leads to less toxic and more biodegradable intermediates [210]. It was shown that the use of laccases and the mediator TEMPO could avoid the formation of polymers [230]. Therefore, the use of TEMPO for the regulation of the degree of the polymerization is under consideration. The degradation of these substances was also shown [225].

Phenolic compounds in olive oil wastewater can be removed by laccases [234–236]. Although some of polycyclic aromatic hydrocarbons (PAHs) can be oxidized by laccases directly, the use of laccase-mediator-systems (LMS) can improve the reactivity rate [210]. For example, the degradation of anthracen can be improved in a combination of laccases with mediators such as HBT and ABTS [104]. Acenaphthene and acenaphthylene can be oxidized by laccase and with laccase and HBT. Also, it was found that the oxidation efficiency of other PAHs can be improved by using the mediator HBT [95, 210].

10.8.7.2 Biodegradation of Chemical Warfare Agents

As for pestizides, there are evidence that laccases can be used for effective decontamination of chemical warfare agents such as VX and mustard. VX is a persistent nerve agent and its decontamination is critical for survival [127]. One ambitious challenge is to find an enzymatic system for the environmentally safe degradation of VX. Nerve agents such as VX can be detoxified rapidly by chemical oxidation of the P-S bond [127]. Therefore, common active chemicals such as hydrogen peroxide [237] and monomagnesium perphthalate [238] can be used for rapid detoxification. The oxidative hydrolysis of phosphonothiolates forms products with moderate toxicity and more environmentally benign degradation than the hydrolysis pathway [105].

A: Hydrolysis leads to an Alkyl Thiol
B: Oxidation leads to an Alkyl Sulfonate
VX: $R = C_2H_5$, $R' = CH_3$, $R'' = CH_2CH_2N(iPr)_2$
RVX: $R = iso\text{-}C_4H_9$, $R' = CH_3$, $R'' = CH_2CH_2N(Et)_2$
DiPrAmiton: $R = C_2H_5$, $R' = OC_2H_5$, $R'' = CH_2CH_2N(iPr)_2$

Although most of the research work has been done by investigating the hydrolysis pathway, there is evidence that the use of oxidoreductases such as laccases for the oxidative breakdown of the P-S bond in VX might lead to more efficient decontamination. Amitai *et al.* were the first to report a laccase-mediator-system for the rapid oxidative degradation of O,O-diethyl S-[N,N-diisopropylaminoethyl] phosphorothiolate (DiPr-Amiton), and the nerve agents O-ethyl S-[N,N diisopropylaminoethyl] methylphosphonothiolate (VX) and O-isobutyl S-[N,N-diethylaminoethyl]-methylphosphonothiolate (RVX) in detail. They used a laccase purified from *Pleurotus ostreatus* and ABTS as a mediator [105]. As shown in Figure 10.6, the oxidative enzymatic degradation of VX, RVX and DiPr-Amiton by purified *Pleurotus ostreatus* laccase in the presence of ABTS as a mediator proceeds to completion. PH 7.4 was found for optimal degradation of VX and RVX. DiPr-Amiton is degraded more rapidly at pH 8. It is important to notice that the optimal molar ratio of ABTS/OP for VX and RVX degradation is 1:20 whereas the

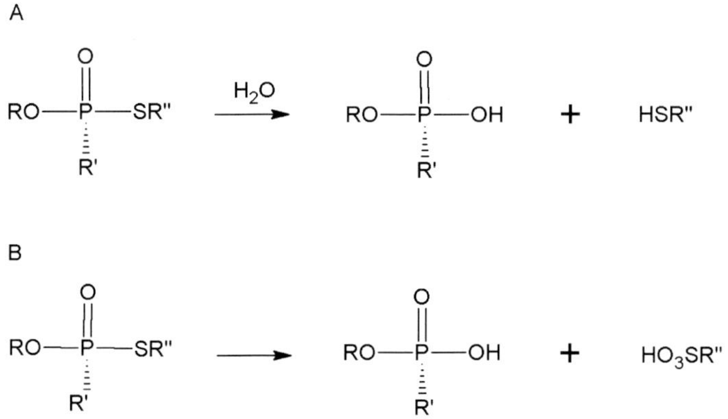

A

$$RO{-}\overset{O}{\overset{\|}{P}}{-}SR'' \; (R') \xrightarrow{H_2O} RO{-}\overset{O}{\overset{\|}{P}}{-}OH \; (R') + HSR''$$

B

$$RO{-}\overset{O}{\overset{\|}{P}}{-}SR'' \; (R') \longrightarrow RO{-}\overset{O}{\overset{\|}{P}}{-}OH \; (R') + HO_3SR''$$

Figure 10.6 Degradation pathways of VX, RVX and DiPr-Amiton by hydrolysis and oxidation (Adapted from Amitai 1998 [105]).

Table 10.11 First order rate constants (k_{obs}, min^{-1}) and specific activity values (k_{sp}, $nmol\,min^{-1}\,mg^{-1}$) of VX, RVX and DiPr-Amiton degradation by Laccase, Adapted from Amitai 1998 [105].

Laccase	*Substrate*	k_{obs} *(min^{-1})*	k_{sp} *($nmol\,min^{-1}\,mg^{-1}$)*
Pleurotus ostreatus	VX	0.066	2200
Pleurotus ostreatus	Amiton	0.055	1833
Pleurotus ostreatus	RVX	0.020	667
Chaetomium thermophilus	VX	0.0013	42

rate of DiPr-Amiton degradation reaches its maximum at a molar ratio of 1:10 [105]. The reaction time for the oxidative degradation of VX, RVX and DiPr-Amiton by the used laccase-mediator-system was calculated t½ for VX degradation not more than 5 seconds (Table 10.11).

However, the exact mechanism of oxidative degradation of phosphorothiolate by laccase/mediator is as yet unknown. It was assumed that the sulfur atom is oxidized followed by cleavage of the P-S bond and that the formation of an N-oxide intermediate may affect the cleavage of the P-S bond [239]. This hypothesis has not been proven yet. Nevertheless, Amitai and his co-workers have shown that a rapid oxidative biodegradation of both optical enantiomers of VX and RVX by a laccases-mediator-system can be achieved. It is important to know that the P(3) enantiomer of VX is 6,4-fold more toxic than its P(+) antipode [240] and therefore, oxidative biodegradation of both optical enantiomers of VX and RVX is necessary. It seems that only an oxidative biodegradation leads to complete detoxification of these toxic nerve agents.

10.9 Inactivation of Spores

Inactivation of spores is of great interest for medical and military purposes. Most of the commonly used methods are based on gaseous and fluid chemicals which are highly toxic during the desinfection process. Also, sometimes there are concerns about the outgasing effect of some of the disinfectants used (e.g. formaldehyde). Therefore, the inactivation of spores by an environmentally more benign method based on oxidative enzymes has been the focus of scientific research over recent years.

10.9.1 Laccase Mediator System (LMS) for Inactivation of Bacillus anthraxis Spores

Enzymes are able to oxidize univalent anion of halogens like Cl-, Br-, iodide. Haloperoxidases that are using hydrogen peroxide are the most common halides utilizing enzymes. But a further method for the oxidation of iodide with the production of elemental iodine is also very interesting, because the microbicidal activity of elemental iodine is well known [241]. Oxidases for this enzymatic production of iodine are possible alternatives and a fungal laccase (*Mycceliophthora thermophila*) was discovered which functions as an iodide oxidase [242]. In aerated solutions, the laccase catalyzes the oxidation of iodide to iodine and the concomitant reduction of dioxygen to water. The produced iodine can then be used for the killing of spores (Figure 10.7).

Mediators are able to significantly enhance the catalysis of the laccase in iodide-oxidation. But also the optimization of pH, temperature, solved oxygen, turnover rate and so on are dependent on the mediator influence the catalysis. Kulys *et al.* (2005) [243] working with laccase-catalyzed iodine-oxidation searched for new mediators. They found that fungal laccases can effectively catalyze potassium iodide (KI) oxidation in presence of methyl-syringate (MS) (Figure 10.8).

The initial rate of triiodide formation was larger in the acid solution. It increased as pH decreased from 6.5 to 4.5. The dependence of iodide oxidation rate on pH is a result of different factor such as the change of enzyme activity and the radical

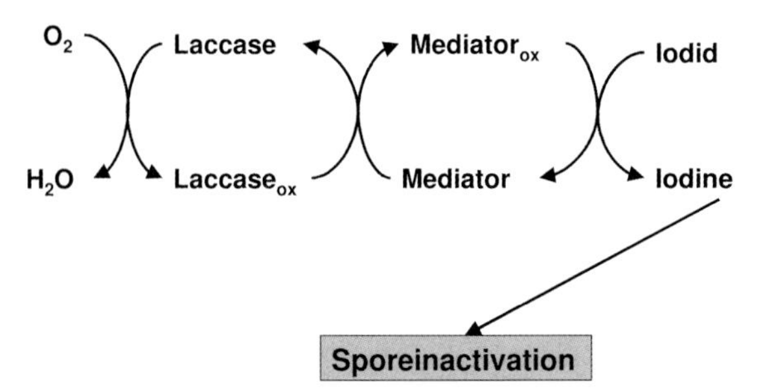

Figure 10.7 Redoxcascade for inactivation of spores.

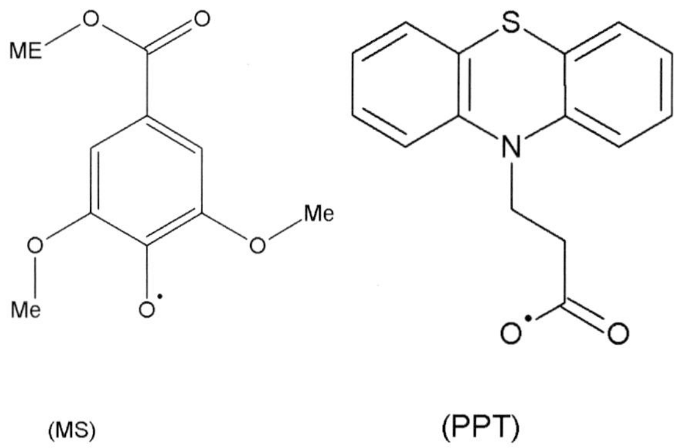

(MS) (PPT)

Figure 10.8 Methylsyringate (MS) and 10-phenothia-zine-propionic acid (PPT) as possible mediators in a laccase-mediator system for inactivation of spores.

cation concentration. The main factor determining the reaction rate increase is the increase of radical cation concentration at acidic pH.

10.9.2 Regeneration of Different Mediators

The regeneration of the used mediator during the redoxcascade is one of the important problems which must be solved for a technical application. The investigation could lead to the following knowledge:

- Deeper understanding of the reaction mechanism
- Usage of the mediator as a catalysator or in stoichiometric amounts
- Estimation of the required amount of the mediator in a technical application.

In screening experiments, the regeneration of different mediators by the laccase used, after oxidizing iodide was tested. The produced iodine can be measured by adding starch. The iodine-starch complex has a deep blue color. Eight different mediators were tested by measuring the iodine-starch complex (Figure 10.9).

Depending on the laccase used, is it necessary to check for each LMS the best mediator. It was found that PPT and MS are two of the best mediators for the investigated laccases.

10.9.3 Oxygen Consumption by Laccase-Mediator-System

As mentioned before, one factor of limitation for the inactivation of spores by a laccase-mediator-system (LMS) could be the amount of the solved oxygen in the solution. To check this hypothesis, the oxygen consumption by the LMS was tested

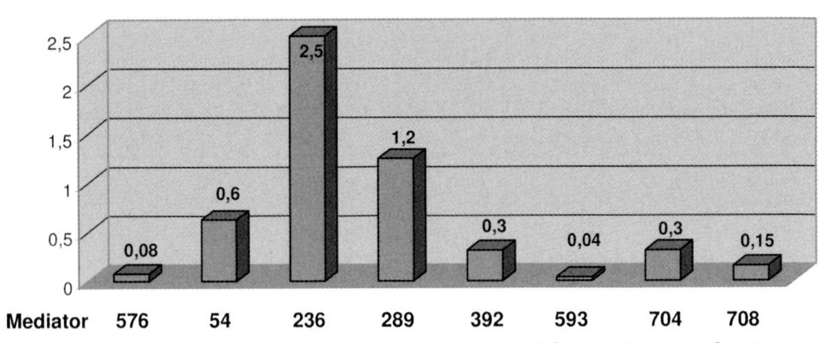

Figure 10.9 Produced iodine by different mediators and for one laccase after 60 minutes.

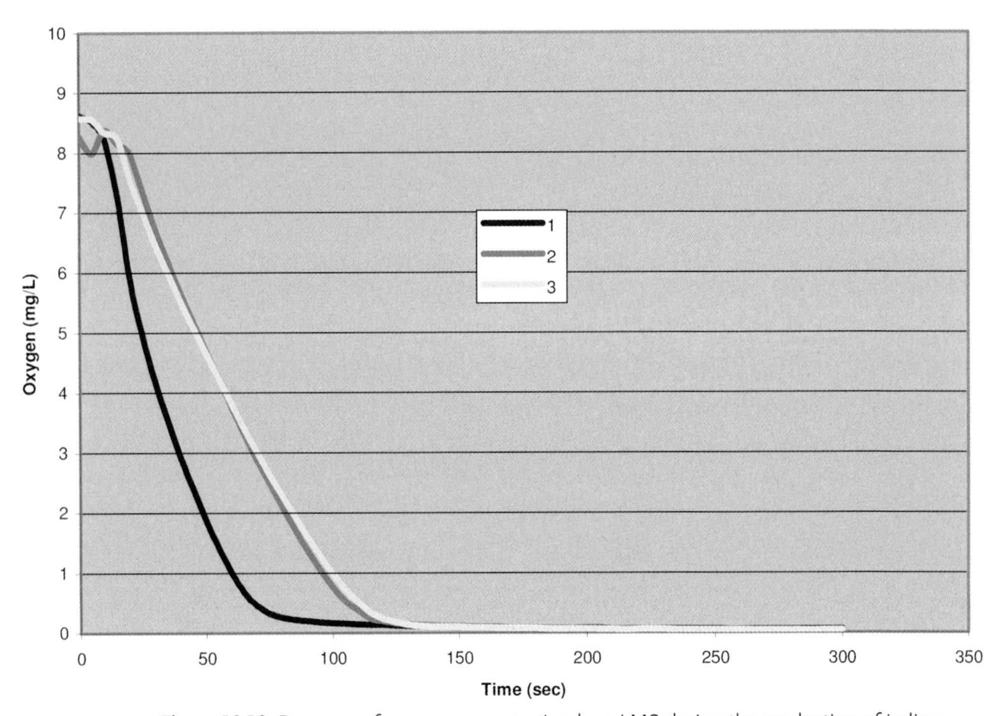

Figure 10.10 Decrease of oxygen concentration by a LMS during the production of iodine.

in different experiments. In independent experiments, it was shown that the oxygen was consumed by the LMS in 1–2 minutes (Figure 10.10).

The iodine produced in the same experiments increased over the time (Figure 10.9). It seems that the decrease of oxygen after 1–2 minutes does not affect the production of iodine.

10.9.4
Inactivation of Spores by Laccase-Mediator-Systems

The effectivity of enzymes for the spore-decontamination was also investigated based on the action of an enzyme laccase based oxidoreductase system that relies on generation of oxidative halide compounds through coupling of enzymatic formation of oxidized organic radicals. A large number of different blue copper proteins or laccases of fungal origin have previously been identified. The main focus was on recombinantly produced laccases from two different filamentous fungi – the thermophylic fungi *Myceliophthora thermophila* (MtL) and the basidiomycete *Polyporus pinsitus* (PpL) (current name for the species is *Trametes villosa*). Their ability using different conditions like temperature, pH and so on was investigated. In this connection a systemically determination and optimization of all parameters for the achievement of efficient sporekilling was carried out with different *Bacillus species*. Tests were started on the recombinant laccase from *Myceliophthora thermophila* (MtL). The influence of the ingredients on the reation mix using methyl-syringate as the mediator was investigated. Then the testing was performed with variations of reaction time, concentration of laccase and spores of *Bacillus thuringiensis* DSM 2046, NZ). On this basis, a systematic determination and optimization of the laccase concentration and incubation time, using an adapted method for working with pathogenic spores, was carried out. The effectivity of the laccase for spores of *Bacillus thuringiensis* (ATCC 10792), *Bacillus subtilis* (DSM 347), *Bacillus cereus* (ATCC 12826) and *Bacillus anthracis* (stars) was investigated in suspension and on surfaces. The reaction mix consisted of potassium-iodide, methyl-syringate as the mediator and MtL. The results demonstrated that using final concentrations of 10^6 spores/ml and at least 0.1 g MtL/L the number of viable spores of *Bacillus thuringiensis* (NZ), *Bacillus cereus* and *Bacillus anthracis* in suspension was reduced by at least 4–5 log-units. The spores of Bacillus subtiliz seemed to be more resistant, so that higher concentrations for a longer time-period would be necessary for killing the spores. Afterwards, investigations for the sporicidal effect of the Pp-Laccase (PpL) and new mediator in suspension were started. The reaction mix contained potassium-iodide, the mediator PPT- and PpL. Using final concentrations this reaction mix and at least 0.05 g Pp-Laccase/L the number of viable spores of *Bacillus thuringiensis* (NZ), *Bacillus cereus* and *Bacillus anthracis* in suspension was reduced by at least 4–5 log-units.

The spores of two tested *Bacillus subtilis*-strains seem to be more resistant, so that higher concentrations for a longer time-period are necessary for killing the spores.

10.9.4.1 Temperature Dependency for Spore Inactivation by Laccase-Mediator-Systems

The temperature dependency for inactivation of spores with an enzyme based system is one of the crucial factors which has to be noted during the development of an enzyme based decontamination system. Therefore, different experiments began to check the efficiency of the investigated LMS in suspension (Figure 10.11).

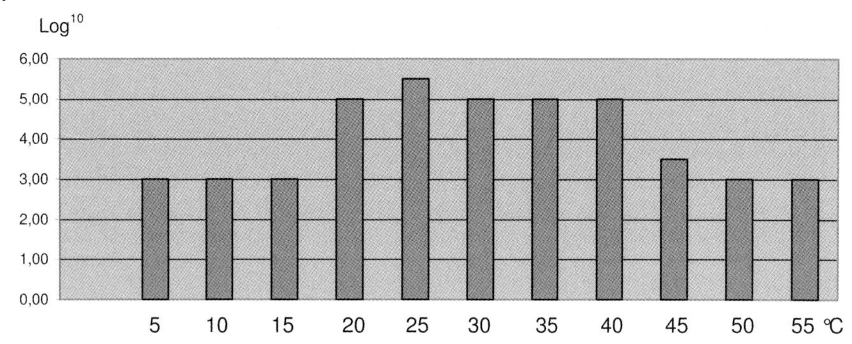

Figure 10.11 Temperature dependency inactivation of DSM350 spores by a laccase mediator system.

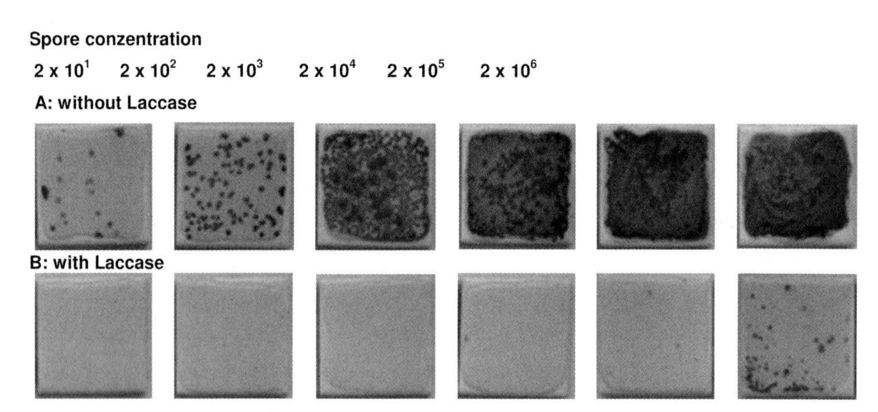

Figure 10.12 Inactivation of spores on ceramic plates by a laccase-mediator-system (Figure: Novozymes, Denmark).

In lab trials it was found that the laccase mediator system works very well in solution between +20 °C and +55 °C. However, even at lower temperatures, significant log reduction of the spores were found. These results show clearly that under defined conditions, laccase mediator systems could be used for the inactivation of spores.

10.9.4.2 Inactivation of Spores on Surfaces

Another crucial point for a technical enzyme system for the inactivation of spores is the efficient decontamination of different surfaces. Therefore, the efficiency of the laccase mediator systems was investigated on ceramic plates (Figure 10.12).

It was found that the investigated laccase-mediator-system was able to inactivate spores on ceramic plates under defined conditions. Tests with the Mt-Laccase system on surfaces show that with a final concentration of above 0.1 g MtL/L for 120 minutes on the glazed ceramic plates, the number of viable *Bacillus thuringiensis* spores was reduced by about 3–4 log-units. Using 0.5–2.0 g MtL/L the number of viable *Bacillus cereus*–spores dried on the ceramic surface was reduced by 2–3 log-units. The ceramic surface seems not to have provided the optimum condition determined for killing Bacillus species spores in suspension. It was possible to reduce the number of viable spores by 2–4 log-units depending of the test organism. The surface tests with Pp-Laccase and spores of Bacillus cereus and Bacillus thuringiensis gave very different results. Using 2 g PpL/L, reduction of the *Bacillus cereus* spore concentration by about 5 log-steps was possible within 150 minutes.

10.10
Conclusions

Oxidative enzymes like laccases are well known for detoxification of xenotics and have been used for different biotechnical application. Investigations have shown that the oxidative principle can be used for decontamination of toxic compounds like pesticides and chemical warfare agents like VX. Also, initial results give strong evidence that oxidoreductaces can be used for the killing of spores. However, to achieve the ambitious stipulations for a technical application based on oxidoreductases, the existing enzyme systems have to improved in terms of production of enzymes, new improved mediators, efficieny in the killing of spores, reaction velocity and specifity. Therefore, there are different oxidative enzyme systems under investigations worldwide to find the best formulation for technical decontamination applications based on a oxidative biocatalyst.

References

1 Reinhammar, B. (1984) Laccase, in *Copper Proteins And Copper Enzymes*, Vol. **3** (ed. Lontie, R.), CRC Press, Boca Raton, pp. 1–35.

2 Messerschmidt, A. (1997) *Copper Metalloenzymes*, in Comprehensive biological catalysis, Vol. **3** (ed. M. Sinnet), Academic Press Limited, London, pp. 401–26

3 Solomon, E.I., Sundaram, U.M. and Machonkin, T.E. (1996) Multicopper oxidases and oxygenases. *Chemical Reviews*, **96**, 2563–606.

4 Shleev, S., Tkac, J., Christenson, A., Ruzgas, T., Yaropolov, A.I., Whittaker, J.W. and Gorton, L. (2005) Direct electron transfer between copper-containing proteins and electrodes. *Biosensors and Bioelectronics*, **20**, 2517–54.

5 Xu, F. (1996) Oxidation of phenols, anilines, and benzenethiols by fungal laccases: correlation between activity and redox potentials as well as halide inhibition. *Biochemistry*, **35**, 7608–14.

6 Kersten, P.J., Kalyanaraman, B., Hammel, K.E., Reinhammar, B. and Kirk, T.K. (1990) Comparison of lignin peroxidase, horseradish peroxidase and laccase in the oxidation of methoxybenzenes. *The Biochemical Journal*, **268**, 475–80.

7 Thurston, C. (1994) The structure and function of fungal laccases. *Microbiology*, **140**, 19–26.

8 Mayer, A.M. and Staples, R.C. (2002) Laccase: new functions for an old enzyme. *Phytochemistry*, **60**, 551–65.

9 Claus, H. (2003) Laccases and their occurrence in prokaryotes. *Archives of Microbiology*, **179**, 145–50.

10 Yoshida, H. (1883) Chemistry of Lacquer (Urishi) part 1. *Journal of the Chemical Society (Tokyo)*, **43**, 472–86.

11 Bertrand, G. (1894) Sur le latex de l'arbre à laque. *Comptes Rendus hebdomadaires des Séances de L'Académie des Sciences (Paris)*, **118**, 1215–8.

12 Lehmann, E., Harel, E. and Mayer, A.M. (1974) Copper content and other characteristics of purified peach laccase. *Phytochemistry*, **13**, 1713–7.

13 Bligny, R. and Douce, R. (1983) Excretion of laccase by sycamore (*Acer pseudoplatanus* L.) cells. Purification and properties of the enzyme. *The Biochemical Journal*, **209**, 489–96.

14 De Marco, A. and Roubelakis-Angelakis, K.A. (1997) Laccase activity could contribute to cell-wall reconstitution in regenerating protoplasts. *Phytochemistry*, **46**, 421–5.

15 Ranocha, P., McDougall, G., Hawkins, S., Sterjiades, R., Borderies, G., Stewart, D., Cabanes-Macheteau, M., Boudet, A.M. and Goffner, D. (1999) Biochemical characterization, molecular cloning and expression of laccases–a divergent gene family in poplar. *European Journal of Biochemistry*, **259**, 485–95.

16 Bao, W., O'Malley, D.M., Whetten, R. and Sederoff, R.R. (1993) A laccase associated with lifgnification in Loblolly pine xylem. *Science*, **260**, 672–4.

17 Sterjiades, R., Dean, J.F.D. and Eriksson, K.E. (1992) Laccase from sycamore maple (*Acer pseudoplatanus*) polymerizes monolignols. *Plant Physiology*, **99**, 1162–8.

18 Liu, L., Dean, J.F.D., Friedman, W.E. and Eriksson, K.E. (1994) A laccase-like phenoloxidase is correlated with lignin biosynthesis in *Zinnia elegans* stem tissue. *The Plant Journal*, **6**, 213–24.

19 Richardson, A., Duncan, J. and McDougall, G.J. (2000) Oxidase activity in lignifying xylem of taxonomically diverse range of trees: identification of a conifer laccase. *Tree Physiology*, **20**, 1039–47.

20 Johnson, D.L., Thompson, J.L., Brinkmann, S.M., Schuller, K.A. and Martin, L.L. (2003) Electrochemical characterization of purified *Rhus vernicifera* laccase: voltammetric evidence for a sequential four-electron transfer. *Biochemistry*, **42**, 10229–37.

21 Marbach, I., Harel, E. and Mayer, A.M. (1984) Molecular properties of extracellular *Botrytis cinerea* laccase. *Phytochemistry*, **23**, 2713–7.

22 Chefetz, B., Chen, Y. and Hadar, Y. (1998) Purification and characterization of laccase from *Chaetomium thermophilium* and its role in humification. *Applied and Environmental Microbiology*, **64**, 3175–9.

23 Schneider, P., Caspersen, M.B., Mondorf, K., Halkier, T., Skov, L.K., Østergaard, P.R., Brown, K.M., Brown, S.H. and Xu, F. (1999) Characterization of a *Coprinus cinereus* laccase. *Enzyme and Microbial Technology*, **25**, 502–8.

24 Froehner, S.C. and Eriksson, K.-E. (1974) Purification and properties of *Neurospora crassa* laccase. *Journal of Bacteriology*, **120**, 458–65.

25 Niku-Paavola, M.L., Karhunen, E., Salola, P. and Raunio, V. (1988) ligniolytic enzymes of the white-rot fungus *Phlebia radiata*. *The Biochemical Journal*, **254**, 877–84.

26 Bezalel, L., Hadar, Y. and Cerniglia, C.E. (1996) Mineralization of polycyclic aromatic Hydrocarbons by the white rot fungus *Pleurotus ostreatus*. *Applied and Environmental Microbiology*, **62**, 292–5.

27 Eggert, C., Temp, U., Dean, J.F.D. and Eriksson, K.E.L. (1996) A fungal metabolite mediates degradation of non-phenolic lignin structures and synthetic lignin by laccase. *FEBS Letters*, **391**, 144–8.

28 Leonowicz, A., Cho, N.S., Luterek, J., Wilkolazka, A. *et al.* (2001) Review: fungal laccase: properties and activity on lignin. *Journal of Basic Microbiology*, **41**, 185–227.

29 Hatakka, A. (2001) Biodegradation of lignin, in *Biopolymers. Biology, Chemistry, Biotechnology, Applications, Lignin, Humic Substances and Coal*, Vol. **1** (eds M. Hofrichter and A. Steinbüchel) Wiley-VCH Verlag GmbH, Weinheim, Germany, pp. 129–80.

30 Van Etten, H.D., Mansfield, J.W., Bailey, J.A. and Farmer, E.E. (1994) Two classes of plant antibiotics: phytoalexins versus "phytoanticipins". *The Plant Cell*, **6**, 1191–2.

31 Bourbonnais, R. and Paice, M.G. (1990) Oxidation of non-phenolic substrates An expanded role for laccase in lignin biodegradation. *FEBS Letters*, **267**, 99–102.

32 Pezet, R., Pont, V. and Hoang-Van, K. (1992) Enzymatic detoxication of stilbenes by *Botrytis cinerea* and inhibition by grape berries proanthrocyanidins, in *Recent Advances in Botrytis Research* (eds K. Verhoeff, N.E. Malathrakis and B. Williamson), Pudoc Scientific, Wageningen, pp. 87–92.

33 Pipe, N.D., Brasier, C.M. and Buck, K.W. (2000) Evolutionary relationships of the Dutch elm disease fungus *Ophiostoma novo-ulmi* to other *Ophiostoma* species investigated by restriction fragment length polymorphism analysis of the rDNA region. *Journal of Phytopathology*, **148**, 533–9.

34 Brasier, C.M. and Kirk, S.A. (2001) Designation of the EAN and NAN races of *Ophiostoma novo-ulmi* as subspecies. *Mycological Research*, **105**, 547–54.

35 Schouten, A., Wagemakers, C.A.M., Stefanato, F., van der Kaaij, R.M. and Van Kan, J.A.L. (2002) Resveratrol acts as a natural profungicide and induces self-intoxication by a specific laccase. *Molecular Microbiology*, **43**, 883–94.

36 Gil-ad, N.L., Bar-Nun, N. and Mayer, A.M. (2001) The possible function of the glucan sheath of *Botrytis cinerea*: effects on the distribution of enzyme activities. *FEMS Microbiology Letters*, **199**, 109–13.

37 Van Etten, H., Temporini, E. and Wasmann, C. (2001) Phytoalexin (and phytoanticipin) tolerance as a virulence trait: why is it not required by all pathogens. *Physiological and Molecular Plant Pathology*, **59**, 83–93.

38 Bar-Nun Tal, N., Lev, A., Harel, E. and Mayer, A.M. (1998) Repression of laccase formation in *Botrytis cinerea* and its possible relation to phytopathogenicity. *Phytochemistry*, **27**, 2505–9.

39 Alexandre, G. and Zhulin, I.B. (2000) Laccases are widespread in bacteria. *Trends in Biotechnology*, **18**, 41–2.

40 Givaudan, A., Effosse, A., Faure, D., Potier, P., Bouillant, M.L. and Bally, R. (1993) Polyphenol oxidase in *Azospirillum lipoferum* isolated from rice rhizosphere: evidence for laccase activity in non-motile strains of *Azospirillum lipoferum*. *FEMS Microbiology Letters*, **108**, 205–10.

41 Faure, D., Bouillant, M.L. and Bally, R. (1994) Isolation of *Azospirillum lipoferum* 4T Tn5 mutants affected in melanization and laccase activity. *Applied and Environmental Microbiology*, **60**, 3413–5.

42 Freeman, J.C., Nayar, P.G., Begley, T.P. and Villafranca, J.J. (1993) Stoichiometry and spectroscopic identity of copper centers in phenoxazinone synthase: a new addition to the blue copper oxidase family. *Biochemistry*, **32**, 4826–30.

43 Claus, H. and Filip, Z. (1997) The evidence of a laccase-like activity in a *Bacillus sphaericus* strain. *Microbiological Research*, **152**, 209–15.

44 Sanchez-Amat, A. and Solano, F. (1997) A pluripotent polyphenol oxidase from the melanogenic marine Alteromonas sp. shares catalytic capabilities of tyrosinases and laccases. *Biochemical and Biophysical Research Communications*, **240**, 787–92.

45 Diamantidis, G., Effosse, A., Potier, P. and Bally, R. (2000) Purification and characterization of the first bacterial laccase in the rhizospheric bacterium *Azospirillum lipoferum*. *Soil Biology and Biochemistry*, **32**, 919–27.

46 Sanchez-Amat, A., Lucas-Elio, P., Ferandez, E., Garcia-Borron, J.C. and Solano, F. (2001) Molecular cloning and functional characterization of a unique multipotent polyphenol oxidase from Marinomonas mediterranea. *Biochimica et Biophysica Acta*, **1547**, 104–16.

47 Endo, K., Hosono, K., Beppu, T. and Ueda, K. (2002) A novel extracytoplasmatic phenol oxidase of Streptomyces: its possible involvement in the onset of morphogenesis. *Microbiology*, **148**, 1767–76.

48 Suzuki, T., Endo, K., Ito, M., Tsujibo, H., Miyamoto, K. and Inamori, Y. (2003) A thermostable laccase from *Streptomyces lavendulae* REN-7: purification, characterization, nucleotide sequence, and expression. *Bioscience, Biotechnology, and Biochemistry*, **67**, 2167–75.

49 Arias, M.E., Arenas, M., Rodríguez, J., Soliveri, J., Ball, A.S. and Hernández, M. (2003) Kraft pulp biobleaching and mediated oxidation of a nonphenolic substrate by laccase from *Streptomyces cyaneus* CECT 3335. *Applied and Environmental Microbiology*, **69**, 1953–8.

50 Hullo, M.F., Moszer, I., Danchin, A. and Martin-Verstraete, I. (2001) CotA of *Bacillus substilis* is a copper-dependent laccase. *Journal of Bacteriology*, **183**, 5426–30.

51 Hirose, J., Nasu, M. and Yokoi, H. (2003) Reaction of substituted phenols with thermostable laccase bound to Bacillus subtilis spores. *Biotechnology Letters*, **25**, 1609–12.

52 Martins, L.O., Soares, C.M., Pereira, M.M., Teixera, M., Costa, T., Jones, G.H. and Henriques, A.O. (2002) Molecular and biochemical characterization of a highly stable bacterial laccase that occurs as a structural component of the *Bacillus subtilis* endospore coat. *The Journal of Biological Chemistry*, **277**, 18849–59.

53 Antorini, M., Herpoel-Gimbert, I., Choinowski, T., Sigoillot, J.C., Asther, M., Winterhalter, K. and Piontek, K. (2001) Purification, crystallization and X-ray diffraction study of fully functional laccases from two ligninolytic fungi. *Biochimica et Biophysica Acta*, **1594**, 109–14.

54 Hakulinen, N., Kiiskinen, L.L., Kruus, K., Saloheimo, M., Paananen, A., Koivula, A. and Rouvinen, J. (2002) Crystal structure of a laccase from *Melanocarpus albomyces* with an intact trinuclear copper site. *Nature Structural Biology*, **9**, 601–5.

55 Piontek, K., Antorini, M. and Choinowski, T. (2002) Crystal structure of a laccase from the fungus *Trametes versicolor* at 1.90 Å resolution containing a full complement of coppers. *The Journal of Biological Chemistry*, **277**, 37663–9.

56 Enguita, F.J., Martins, L.O., Henriques, A.O. and Carrondo, M.A. (2003) Crystal structure of a bacterial endospore coat component: a laccase with enhanced thermostability properties. *The Journal of Biological Chemistry*, **278**, 19416–25.

57 Christenson, A., Dimcheva, N., Ferapontova, E.E., Gorton, L., Ruzgas, T., Stoica, L., Shleev, S., Yaropolov, A.I., Haltrich, D., Thorneley, R.N.F. and Aust, S.D. (2004) Direct electron transfer between ligninolytic redox enzymes and electrodes. *Electroanalysis*, **16**, 1074–92.

58 Palmer, A.E., Szilagyi, R.K., Cherry, J.R., Jones, A., Xu, F. and Solomon, E.I. (2003) Spectroscopic characterization of the Leu513His variant of fungal laccase: effect of increased axial ligand interaction on the geometric and electronic structure of the type 1 Cu site. *Inorganic Chemistry*, **42**, 4006–17.

59 Ducros, V., Brzozowski, A.M., Wilson, K.S., Brown, S.H., Ostergaard, P., Schneider, P., Yaver, D.S., Pedersen, A.H. and Davies, G.J. (1998) Crystal structure of the type-2 Cu depleted laccase from *Coprinus cinereus* at 2.2 Å resolution. *Nature Structural Biology*, **5**, 310–6.

60 Bertrand, T., Jolivalt, C., Briozzo, P., Caminade, E., Joly, N., Madzak, C. and Mougin, C. (2002) Crystal structure of a four-copper laccase complexed with an arylamine: insights into substrate recognition and correlation with kinetics. *Biochemistry*, **41**, 7325–33.

61 Guss, J.M. and Freeman, H.C. (1983) Structure of oxidized poplar plastocyanin at 1.6 Å resolution. *Journal of Molecular Biology*, **169**, 521–63.

62 Inoue, T., Gotowda, M., Sugawara, H., Kohzuma, T., Yoshizaki, F., Sugimura, Y. and Kai, Y. (1999) Structure comparison between oxidized and reduced plastocyanin from a fern, *Dryopteris crassirhizoma*. *Biochemistry*, **38**, 13853–61.

63 Norris, G.E., anderson, B.F. and Baker, E.N. (1983) Structure of azurin from *Alcaligenes denitrificans* at 2.5 Å resolution. *Journal of Molecular Biology*, **165**, 501–21.

64 Messerschmidt, A., Ladenstein, R., Huber, R., Bolognesi, M., Avigliano, L., Petruzzelli, R., Rossi, A. and Finazzi-Agro, A. (1992) Refined crystal structure of ascorbate oxidase at 1.9 Å resolution. *Journal of Molecular Biology*, **224**, 179–205.

65 Zaitseva, I., Zaitsev, V., Card, G., Moshkov, K., Bax, B., Ralph, A. and Lindley, P. (1996) The X-ray structure of human serum ceruloplasmin at 3.1 Å: nature of the copper centres. *Journal of Biological Inorganic*, **1**, 15–23.

66 Murphy, M.E.P., Lindley, P.F. and Adman, E.T. (1997) Structural Comparison of Cupredoxin domains: domain recycling for construct proteins with novel functions. *Protein Science : A Publication of the Protein Society*, **6**, 761–70.

67 Durán, N., Rosa, M.A., D'Annibale, A. and Gianfreda, L. (2002) Applications of laccases and tyrosinases (phenoloxidases) immobilized on different supports: a review. *Enz. Microbial Technol.* **31**, 907–31.

68 Claus, H. (2004) Laccases: structure, reactions, distribution, *Micronesica*, **35**, 93–6.

69 Xu, F., Kulys, J.J., Duke, K., Li, K., Krikstopaitis, K., Deussen, H.J., Abbate, E., Galinyte, V. and Schneider, P. (2000) Redox chemistry in laccase-catalyzed oxidation of N-hydroxy compounds. *Applied and Environmental Microbiology*, **66**, 2052–6.

70 Reinhammar, B. and Vanngard, T.I. (1971) Electron-accepting sites in *Rhus vernicifera* laccase as studied by anaerobic oxidation-reduction titrations. *European Journal of Biochemistry*, **18**, 463–8.

71 Reinhammar, B.R.M. (1972) Oxidation-reduction potentials of the electron acceptors in laccases and stellacyanin. *Biochimica et Biophysica Acta*, **275**, 245–59.

72 Xu, F., Palmer, A.E., Yaver, D.S., Berka, R.M., Gambetta, G.A., Brown, S.H. and Solomon, E.I. (1999) Targeted mutations in a Trametes villosa laccase. Axial perturbations of the T1 copper. *The Journal of Biological Chemistry*, **274**, 12372–5.

73 Koroleva, O.V., Yavmetdinov, I.S., Shleev, S.V., Stepanova, E.V. and Gavrilova, V.P. (2001) Isolation and study of some properties of laccase from the Basidiomycetes *Cerrena maxima*. *Biochemistry (Moscow)*, **66**, 618–22.

74 Klonowska, A., Gaudin, C., Fournel, A., Asso, M., Le Petit, J., Giorgi, M. and Tron, T. (2002) Characterization of a low redox potential laccase from the basidiomycete C30. *European Journal of Biochemistry*, **269**, 6119–25.

75 Kiiskinen, L.L., (2005) Characterization and Heterologous Production of a Novel Laccase from Melanocarpus albomyces, *VTT publications*, dissertation.

76 Palmieri, G., Giardina, P., Bianco, C., Scaloni, P., Capasso, A. and Sannia, G. (1997) A novel white laccase from *Pleurotus ostreatas*. *The Journal of Biological Chemistry*, **272**, 31301–7.

77 Galhaup, C., Goller, S., Peterbauer, C.K., Strauss, J. and Haltrich, D. (2002) Characterization of the major laccase isoenzyme from *Trametes pubescens* and regulation of its synthesis by metal ions. *Microbiology*, **148**, 2159–69.

78 Galhaup, C., Wagner, H., Hinterstoisser, B. and Haltrich, D. (2002) Increased production of laccase by the wood-degrading basidiomycete Trametes pubescens. *Enzyme and Microbial Technology*, **30**, 529–36.

79 Jung, H., Xu, F. and Li, K. (2002) Purification and characterization of laccase from wood-degrading fungus *Trichophyton rubrum* LKY-7. *Enzyme and Microbial Technology*, **30**, 161–8.

80 Garzillo, A.M., Colao, M.C., Caruso, C., Caporale, C., Celletti, D. and Buonocore, V. (1998) Laccase from the white-rot fungus *Trametes trogii*. *Applied Microbiology and Biotechnology*, **49**, 545–51.

81 Bulter, T., Alcalde, M., Sieber, V., Meinhold, P., Schlachtbauer, C. and Arnold, F.H. (2003) Functional expression of a fungal laccase in *Saccharomyces cerevisiae* by directed evolution. *Applied and Environmental Microbiology*, **69**, 987–95.

82 Soden, D.M., Callaghan, O.J. and Dobson, A.D.W. (2002) Molecular cloning of a laccase isozyme gene from *Pleurotus sajor-caju* and expression in the heterologous *Pichia pastoris* host. *Microbiology*, **148**, 4003–14.

83 Yaropolov, A.I., Skorobogat'ko, O.V., Vartanov, S.S. and Varfolomeyev, S.D. (1994) Laccase. Properties, catalytic mechanism and applicability. *Applied Biochemistry and Biotechnology*, **49**, 257–80.

84 Wood, D.A. (1980) Production, purification and properties of extracellular Laccase of *Agaricus bisporus*. *Journal of General Microbiology*, **117**, 327–38.

85 Palonen, H., Saloheimo, M., Viikari, L. and Kruus, K. (2003) Purification, characterization and sequence analysis of a laccase from the ascomycete *Mauginiella* sp. *Enzyme and Microbial Technology*, **33**, 854–62.

86 Bourbonnais, R. and Paice, M.G. (1992) Demethylation and delignification of kraft pulp by *Trametes versicolor* laccase in the presence of 2,2′-azinobis-(3 ethylbenzthiazoline-6-sulphonate). *Applied Microbiology and Biotechnology*, **36**, 823–7.

87 Bourbonnais, R. and Paice, M.G. (1996) Enzymatic delignification of kraft pulp using laccase and a mediator. *Tappi Journal*, **79**, 199–204.

88 Bourbonnais, R., Paice, M.G., Freiermuth, B., Bodie, E. and Borneman, S. (1997) Reactivities of various mediators and laccases with kraft pulp and lignin model compounds. *Applied and Environmental Microbiology*, **12**, 4627–32.

89 Hammel, K.E. (1996) Extracellular free radical biochemistry of ligninolytic fungi. *New Journal of Chemistry*, **20**, 195–8.

90 Kuhad, R.C., Singh, A. and Eriksson, K.E.L. (1997) Biotechnology in the Pulp and Paper Industry, in *Advances in Biochemical Engineering Biotechnology*, Vol. **57** (ed. K.-E.L. Eriksson), Springer, Berlin, Germany.

91 Crestini, C.L. and Argyropoulos, D.S. (1998) The early oxidative biodegradation steps of residual kraft lignin models with laccase. *Bioorganic and Medicinal Chemistry*, **6**, 2161–9.

92 Van Aken, B. and Agathos, S.N. (2001) Biodegradation of nitro-substituted explosives by ligninolytic white-rot fungi: a mechanistic approach. *Advances in Applied Microbiology*, **48**, 1–77.

93 Camarero, S., Ibarra, D., Martinez, M.J. and Martinez, A.T. (2005) Lignin-derived compounds as efficient laccase mediators for decolorization of different types of recalcitrant dyes. *Applied and Environmental Microbiology*, **71**, 1775–84.

94 Banci, L., Ciofi-Baffoni, S. and Tien, M. (1999) Lignin and Mn peroxidase-catalyzed oxidation of phenolic lignin oligomers. *Biochemistry*, **38**, 3205–10.

95 Majcherczyk, A., Johannes, C. and Hüttermann, A. (1998) Oxidation of polycyclic aromatic hydrocarbons (PAH) by laccase of *Trametes versicolor*. *Enzyme and Microbial Technology*, **22**, 335–41.

96 Amann, M. (1997) *The lignozym(r) process coming closer to the mill*. Proc. 9th Int. Symp. On Wood and Pulping Chem. (Montreal): F4-1–F4-5

97 Potthast, A., Rosenau, T. and Fischer, K. (2001) Oxidation of benzyl alcohols by the laccase-mediator system (LMS)–a comprehensive kinetic description. *Holzforschung*, **55**, 47–56.

98 Bourbonnais, R., Paice, M., Reid, I., Lanthier, P. and Yaguchi, M. (1995) Lignin oxidation by laccase isozymes from *Trametes versicolor* and role of the mediator 2,2-azinobis(3-ethylbenzthiazoline-6-sulfonate) in kraft lignin depolymerization. *Applied and Environmental Microbiology*, **61**, 1876–80.

99 Scott, S.L., Chen, W.J., Bakac, A. and Espenson, J.H. (1993) Spectroscopic parameters electrode potentials, acid ionization constants, and electron exchange rates of the 2,2-azinobis(3-ethylbenzothi- azolone-6-sulfonate) radicals and ions. *The Journal of Physical Chemistry*, **1993**, 6710–4.

100 Call, H.P. and Mucke, I. (1997) History, overview and applications of mediated lignolytic systems, especially laccase-mediator-systems. *Journal of Biotechnology*, **53**, 163–202.

101 Schneider, P. and Pedersen, A.H. (1995) Enhancement of laccase reactions. PCT world patent WO95/01426.

102 Wesenberg, D., Kyriakides, I. and Agathos, S.N. (2003) White-rot fungi and their enzymes for the treatment of industrial dye effluents. *Biotechnology Advances*, **22**, 161–87.

103 Paice, M., Bourbonnais, R., Archibald, F.S., Reid, I.D., Renaud, S. and Rochefort, D. (1999) *Delignification mechanisms for the bleaching of Kraft pulps* *with the enzymes laccase and manganese peroxidase*. Proc. 10th Biennal Int. Symp. on Wood and Pulping Chem. (Yokohama): Vol. 1, 578–82.

104 Johannes, C., Majcherczyk, A. and Hüttermann, A. (1996) Degradation of anthracene by laccase of *Trametes versicolor* in the presence of different mediator compounds. *Applied Microbiology and Biotechnology*, **46**, 313–7.

105 Amitai, G., Adani, R., Sod-Moriah, G., Rabinovitz, I., Vincze, A., Leader, H., Chefetz, B., Leibovitz-Persky, L., Friesem, D. and Hadar, Y. (1998) Oxidative biodegradation of phosphorothiolates by laccase. *FEBS Letters*, **438**, 195–200.

106 Kang, K.H., Dec, J., Park, H. and Bollag, J.M. (2002) Transformation of the fungicide cyprodinil by a laccase of Trametes villosa in the presence of phenolic mediators and humic acid. *Water Research*, **36**, 4907–15.

107 Fritz-Langhals, E. and Kunath, B. (1998) Synthesis of aromatic aldehydes by laccase-mediator assisted oxidation. *Tetrahedron Letters*, **39**, 5955–6.

108 Sigoillot, C., Camarero, S., Vidal, T., Record, E., Asther, M., Pérez-Boada, M., Martínez, M.J., Sigoillot, J.C., Asther, M., Colom, J.F. and Martínez, A.T. (2005) Comparison of different fungal enzymes for bleaching highquality paper pulps. *Journal of Biotechnology*, **115**, 333–43.

109 Ryan, D., Leukes, W. and Burton, S. (2007) Improving the bioremediation of phenolic wastewaters by *Trametes versicolor*. *Bioresource Technology*, **98**, 579–87.

110 Saloheimo, M. and Niku-Paavola, M.-L. (1991) Heterologous production of a ligninolytic enzyme: expression of the *Phlebia radiata* laccase gene in Trichoderma reesei. *Biotechnology*, **9**, 987–90.

111 Hong, F., Meiander, N.Q. and Jönsson, L.J. (2002) Fermentation strategies for improved heterologous expression of laccase in Pichia pastoris. *Biotechnology and Bioengineering*, **79**, 438–449.

112 Hatamoto, O., Sekine, H., Nakano, E. and Abe, K. (1999) Cloning and expression of a cDNA encoding the laccase from Schizophyllum commune. *Bioscience, Biotechnology, and Biochemistry*, **63**, 58–64.

113 Larrondo, L.F., Avila, M., Salas, L., Cullen, D. and Vicuña, R. (2003) Heterologous expression of laccase cDNA from *Ceriporiopsis subvermispora* yields copper-activated apoprotein and complex isoform patterns. *Microbiology*, **149**, 1177–82.

114 LaFayette, P.R., Eriksson, K.-E. and Dean, J.F. (1999) Characterization and heterologous expression of laccase cDNAs from xylem tissues of yellow-poplar (*Liriodendron tulipifera*). *Plant Molecular Biology*, **40**, 23–35.

115 Bailey, M.R., Woodard, S.L., Callaway, E., Beifuss, K., Magallanes-Lundback, M., Lane, J.R., Horn, M.E., Mallubhotla, H., Delaney, D.D., Ward, M., Van Gastel, F., Howard, J.A. and Hood, E.E. (2004) Improved recovery of active recombinant laccase from maize seed. *Applied Microbiology and Biotechnology*, **63**, 390–7.

116 Cassland, P. and Jönsson, L.J. (1999) Characterization of a gene encoding *Trametes versicolor* laccase A and improved heterologous expression in Saccharomyces cerevisiae by decreased cultivation temperature. *Applied Microbiology and Biotechnology*, **52**, 393–400.

117 Yaver, D.S., Overjero, M.J., Xu, F., Nelson, B.A., Brown, K.M., Halkier, T., Bernauer, S., Brown, S.H. and Kauppinen, S. (1999) Molecular characterization of laccase genes from the basidiomycete Coprinus cinereus and heterologous expression of the laccase lcc1. *Applied and Environmental Microbiology*, **65**, 4943–8.

118 Record, E., Punt, P.J., Chamkha, M., Labat, M., van Den Hondel, C.A. and Asther, M. (2002) Expression of the Pycnoporus cinnabarinus laccase gene in Aspergillus niger and characterization of the recombinant enzyme. *European Journal of Biochemistry*, **269**, 602–9.

119 Larsson, S., Cassland, P. and Jönsson, L.J. (2001) Development of a Saccharomyces cerevisiae strain with enhanced resistance to phenolic fermentation inhibitors in lignocellulose hydrolysates by heterologous expression of laccase. *Applied and Environmental Microbiology*, **67**, 1163–70.

120 Berka, R.M., Schneider, P., Golightly, E.J., Brown, S.H., Madden, M., Brown, K.M., Halkier, T., Mondorf, K. and Xu, F. (1997) Characterization of the gene encoding an extracellular laccase of Myceliophtora thermophila and analysis of the recombinant enzyme expressed in *Aspergillus oryzae*. *Applied and Environmental Microbiology*, **63**, 3151–7.

121 Otterbein, L., Record, E., Longhi, S., Asther, M. and Moukha, S. (2000) Molecular cloning of the cDNA encoding laccase from *Pycnoporus cinnabarinus* I-937 and expression in Pichia pastoris. *European Journal of Biochemistry*, **267**, 1619–25.

122 Lomascolo, A., Record, E., Herpoel-Gimbert, I., Delattre, M., Robert, J.L., Georis, J., Dauvrin, T., Sigoillot, J.C. and Asther, M. (2003) Overproduction of laccase by a monokaryotic strain of Pycnoporus cinnabarinus using ethanol as inducer. *Journal of Applied Microbiology*, **94**, 618–24.

123 Liu, W., Chao, Y., Liu, S., Bao, H. and Qian, S. (2003) Molecular cloning and characterization of a laccase gene from the basidiomycete Fome lignosus and expression in Pichia pastoris. *Applied Microbiology and Biotechnology*, **63**, 174–81.

124 Aalto, M.K., Rönne, H. and Keränen, S. (1993) Yeast syntaxins Sso1p and Sso2p belong to a family of related membrane proteins that function in vesicular transport. *The EMBO Journal*, **12**, 4095–104.

125 O'Callaghan, J., O'Brien, M.M., McClean, K. and Dobson, A.D.W. (2002) Optimisation of the expression of a *Trametes versicolor* laccase gene in Pichia pastoris. *Journal of Industrial Microbiology and Biotechnology*, **29**, 55–9.

126 Uldschmid, A., Dombi, R. and Marbach, K. (2003) Identification and functional expression of ctaA, a P-type ATPase gene involved in copper trafficking in *Trametes versicolor*. *Microbiology*, **149**, 2039–48.

127 Amitai, G., Adani, R., Hershkovitz, M., Bel, P., Rabinovitz, I. and Meshulam, H. (2003) Degradation of VX and sulfur mustard by enzymatic haloperoxidation. *Journal of Applied Toxicology*, **23**, 225–33.

128 Couto, S.R. and Herrera, J.L.T. (2006) Industrial and biotechnological applications of laccases: a review. *Biotechnology Advances*, **24**, 500–13.

129 Mishra, A. and Bajpai, M. (2005) Flocculation behaviour of model textile wastewater treated with a food grade polysaccharide. *Journal of Hazardous Materials*, **118**, 213–7.

130 Banat, I.M., Nigam, P., Singh, D. and Marchant, R. (1996) Microbial decolorization of textile-dye-containing effluents: a review. *Bioresource Technology*, **58**, 217–27.

131 Juang, R.S., Tseng, R.L., Wu, F.C. and Lin, S.J. (1996) Use of chitin and chitosan in lobster shell wastes for colour removal from aqueous solutions. *Journal of Environmental Science and Health. Part A, Toxic/hazardous Substances and Environmental Engineering*, **31**, 325–38.

132 Abadulla, E., Tzanov, T., Costa, S., Robra, K.-H., Cavaco-Paulo, A. and Gübitz, G.M. (2000) Decolorization and detoxification of textile dyes with a laccase from *Trametes hirsuta*. *Applied and Environmental Microbiology*, **66**, 3357–62.

133 Soares, G.M.B., Pessoa Amorim, M.T., Hrdina, R. and Costa-Ferreira, M. (2002) Studies on the biotransformation of novel disazo dyes by laccase. *Process biochem*, **37**, 581–7.

134 Palmieri, G., Cennamo, G. and Sannia, G. (2005) Remazol brilliant blue R decolourisation by the fungus Pleuroteus osreatus and its oxidative enzymatic system. *Enzyme and Microbial Technology*, **36**, 17–24.

135 Mccarthy, J.T., Levy, V.C., Lonergan, G.T. and Fecondo, J.V. (1999) Development of optimal conditions for the decolourisation of a range of industrial dyes using Pyenoporus cinnabarinus laccase. *Hazardous and Industrial Wastes*, **31**, 489–98.

136 Hou, H., Zhou, J., Wang, J., Du, C. and Yan, B. (2004) Enhancement of laccase production by Pleurotus ostrearus and its use for the decolorization of anthraquinone dye. *Process Biochemistry*, **39**, 1415–9.

137 Rodriguez Couto, S. and Sanroman, M.A. (2005) Coconut flesh: a novel raw material for laccase production by *Trametes hirsuta* under solidstate conditions. Application to Lissamine Green B decolourization. *Journal of Food Engineering*, **71**, 208–13.

138 Tavares, A.P.M., Gamelas, J.A.F., Gaspar, A.R., Evtuguin, D.V. and Xavier, A.M.R.B. (2004) A novel approach for the oxidative catalysis employing polyoxometalate-laccase system: application to the oxygen bleaching of kraft pulp. *Catalysis Communications*, **5**, 485–9.

139 Blanquez, P., Casas, N., Font, X., Gabarrell, M. and Sarra, M., *et al.* (2004) Mechanism of textile metal dye biotransformation by *Trametes versicolor*. *Water Research*, **38**, 2166–72.

140 Peralta-Zamora, P., Pereira, C.M., Tiburtius, E.R.L., Moraes, S.G. and Rosa, M.A., *et al.* (2003) Decolorization of reactive dyes by immobilized laccase. *Applied Catalysis B: Environmental*, **40**, 131–44.

141 Zille, A., Tzanov, T., Guebitz, G.M. and Cavaco-Paulo, A. (2004) Immobilized laccase for decolourization of Reactive Black 5 dyeing effluent. *Biotechnology Letters*, **25**, 1473–7.

142 Knutson, K. and Ragauskas, A. (2004) Laccase-mediator biobleaching applied to a direct yellow dyed paper. *Biotechnology Progress*, **20**, 1893–6.

143 Damsus, T., Kirk, O., Pedersen, G. and Venegas, M.G. (1991) Novo Nordisk A/S, The Procter & Gamble Company, Patent WO9105839.

144 Chivukula, M. and Renganathan, V. (1995) Phenolic Azo Dye Oxidation by Laccase from *Pyricularia oryzae*. *Applied and Environmental Microbiology*, **61**, 4374–7.

145 Wong, Y. and Yu, J. (1999) Laccase-catalyzed decolorization of synthetic dyes. *Water Research*, **33**, 3512–20.

146 Gianfreda, L., Xu, F. and Bollag, J.M. (1999) Laccases: a useful group of oxidoreductive enzymes. *Bioremediation Journal*, **3**, 1–25.

147 Rodriguez Couto, S., Sanroman, M.A. and Gübitz, G.M. (2005) Influence of redox mediators and metal ions on synthetic acid dye decolourization by crude laccase from *Trametes hirsuta*. *Chemosphere*, **58**, 417–22.

148 Fabbrini, M., Galli, C. and Gentili, P. (2002) Comparing the catalytic efficiency of some mediators of laccase. *Journal of Molecular Catalysis B: Enzymatic*, **16**, 231–40.

149 Setti, L., Giuliani, S., Spinozzi, G. and Pifferi, P.G. (1999) Laccase catalyzed-oxidative coupling of 3-methyl 2-benzothiazolinone hydrazone and methoxyphenols. *Enzyme and Microbial Technology*, **25**, 285–9.

150 Campos, R., Kandelbauer, A., Robra, K.H., Cavaco-Paulo, A. and Cübitz, G.M. (2001) Indigo degradation with purified laccases from *Trametes hirsuta* and Scelerotium rolfsii. *Journal of Biotechnology*, **89**, 131–9.

151 Kierulff, J.V. and Pedersen, A.H. (1996) International patent WO96/12845.

152 Gamelas, J.A.F., Tavares, A.P.M., Evtuguin, D.V. and Xavier, A.M.B. (2005) Oxygen bleaching of kraft pulp with polyoxometalates and laccase applying a novel multi-stage process. *Journal of Molecular Catalysis B: Enzymatic*, **33**, 57–64.

153 Archibald, F.S., Bourbonnais, R., Jurasek, L., Paice, M.G. and Reid, I.D. (1997) Kraft pulp bleaching and delignification by *Trametes versicolor*. *Journal of Biotechnology*, **53**, 215–36.

154 Call, H.-P. and Mücke, I. (1997) History, overview and applications of mediated lignolytic systems, especially laccase-mediator-systems (Lignozym(r)-process). *Journal of Biotechnology*, **53**, 163–202.

155 Georis, J., Lomascolo, A., Camarero, S., Dorgeo, V., Herpoel, I. and Asther, M. (2003) Pycnoporus cinnabarinus laccases: an interesting tool for food or non-food applications. *Mededelingen Faculteit Landbouwkundige en Toegepaste Biologische Wetenschapen*, **68**, 263–6.

156 Camarero, S., Garcia, O., Vidal, T., Colom, J., del Rio, J.C., *et al.* (2004) Efficient bleaching of non-wood high-quality paper pulp using laccase-mediator system. *Enzyme and Microbial Technology*, **35**, 113–20.

157 Gübitz, G.M. and Cavaco Paulo, A. (2003) New substrates for reliable enzymes: enzymatic modification of polymers. *Current Opinion in Biotechnology*, **14**, 577–82.

158 Lund, M. and Ragauskas, A.J. (2001) Enzymatic modification of kraft lignin through oxidative coupling with water-soluble phenols. *Applied Microbiology and Biotechnology*, **55**, 699–703.

159 Chandra, R.P. and Ragauskas, A.J. (2002) Evaluating laccase-facilitated coupling of phenolic acids to high-yield kraft pulps. *Enzyme and Microbial Technology*, **30**, 855–61.

160 Burton, S.G. (2003) Laccases and phenol oxidases in organic synthesis – a review. *Current Organic Chemistry*, **7**, 1317–31.

161 Akta, N., Cicek, H., Tapinar, Ü.A., Kibarer, G., Kolankaya, N. and Tanyolac, A. (2001) Reaction kinetics for laccase-catalyzed polymerization of 1-naphthol. *Bioresource Technology*, **80**, 29–36.

162 Akta, N. and Tanyolac, A. (2003) Reaction conditions for laccase catalyzed polymerization of catechol. *Bioresource Technology*, **87**, 209–14.

163 Baker, W.L., Sabapathy, K., Vibat, M. and Lonergan, G. (1996) Lactase catalyzes formation of an indamine dye between 3-methyl-2-benzothiazolinone, hydrazone and 3-dimethylaminobenzoic acid. *Enzyme and Microbial Technology*, **18**, 90–4.

164 Karamyshev, A.V., Shleev, S.V., Koroleva, O.V., Yaropolov, A.I. and Sakharov, I.Y. (2003) Laccase-catalyzed synthesis of conducting polyaniline. *Enzyme and Microbial Technology*, **33**, 556–64.

165 Mikolasch, A., Hammer, E., Jonas, U., Popowski, K., Stielow, A. and Schauer, F. (2002) Synthesis of 3-(3,4-dihydroxyphenyl)-propionic acid derivatives by N-coupling of amines using laccase. *Tetrahedron*, **58**, 7589–93.

166 Uyama, H. and Kobayashi, S. (2002) Enzyme-catalyzed polymerization to functional polymers. *Journal of Molecular Catalysis B: Enzymatic*, **20**, 117–27.

167 Mita, N., Tawaki, S.I., Hiroshi, U. and Kobayashi, S. (2003) Laccase-catalyzed oxidative polymerization of phenols. *Macromolecular Bioscience*, **3**, 253–7.

168 Ikeda, R., Tanaka, H., Oyabu, H., Uyama, H. and Kobayashi, S. (2001) Preparation of artificial urushi via an environmentally benign process. *Bulletin of the Chemical Society of Japan*, **74**, 1067–73.

169 Grönqvist, S., Buchert, J., Rantanen, K., Viikari, L. and Suurnäkki, A. (2003) Activity of laccase on unbleached and bleached thermomechanical pulp. *Enzyme and Microbial Technology*, **32**, 439–45.

170 Ikeda, R., Tanaka, H., Uyama, H. and Kobayashi, S. (1998) Laccase-catalyzed polymerization of acrylamide. *Macromolecular Rapid Communications*, **19**, 423–5.

171 Aktas, N., Kibarer, G. and Tanyolaç, A. (2000) Effects of reaction conditions on laccase-catalysed 1-naphthol polymerisation. *Journal of Chemical Technology and Biotechnology*, **75**, 840–6.

172 Aktas, N. and Tanyolaç, A. (2003) Kinetics of laccase-catalyzed oxidative polymerization of catechol. *Journal of Molecular Catalysis B: Enzymatic*, **22**, 61–9.

173 Güreir, M., Akta, N. and Tanyolaç, A. (2005) Influence of reaction conditions on the rate of enzymic polymerization of pyrogallol using laccase. *Process Biochemistry*, **40**, 1175–82.

174 Barfoed, M., Kirk, O. and Salmonm, S. (2001), Novozymes A/S, Patent US2001037532.

175 Pilz, R., Hammer, E., Schauer, F. and Kragl, U. (2003) Laccase-catalysed synthesis of coupling products of phenolic substrates in different reactors. *Applied Microbiology and Biotechnology*, **60**, 708–12.

176 Tzanov, T., Basto, C., Guebitz, G.M. and Cavaco-Paulo, A. (2003) Laccases to improve the whiteness in a conventional bleaching of cotton. *Macromolecular Materials and Engineering*, **288**, 807–10.

177 Minussi, S.C., Pastore, G.M. and and Durán, N. (2002) Potential applications of Laccase in the food industry. *Trends in Food Science and Technology*, **13**, 205–16.

178 Plank, P.F.H. and Zent, J.B. (1993) Use of enzymes in wine making and grape processing. Technological advances, in *Analysis, Characterization, and Technological Advances*, (eds B.H. Gump and D.J. Pruett), *ACS Symposium Series, An American Chemical Society Publication*, Vol. **536**, pp. 181–96.

179 Servili, M., De Stefano, G., Piacquadio, P. and and Sciancalepore, V. (2000) A novel method for removing phenols from grape must. *American Journal of Enology and Viticulture*, **51**, 357–61.

180 Tanriöven, D. and Eksi, A. (2005) Phenolic compounds in pear juice from different cultivars. *Food Chemistry*, **93**, 89–93.

181 Sponholz, W.R. and Muno, H. (1994) Der Korkton – ein mikrobiologisches Problem 2. *Die Wein-Wissenschaft*, **49**, 17–22.

182 Sponholz, W.R., Grossmann, M.K., Muno, H. and Hoffmann, A. (1997) The distribution of chlorophenols and chloroanisoles in cork and a microbiological method to prevent their formation. *Industrie delle Bevande*, **26**, 602–7.

183 Conrad, L.S., Sponholz, W.R. and Berker, O., (2000) US patent US6152966.

184 Mathiasen, T.E. (1995) Laccase and Beer Storage. PCT Int. Appl. WO95/21240 A2.

185 Petersen, B.R. and Mathiasen, T.E. (1996) Deoxygenation of a Food Item Using a Laccase. PCT Int. Appl. WO96/31133 A1.

186 Zille, A. (2005) *Laccase Reactions for Textile Applications*, Universidade do Minho, Escola de Engenharia.

187 Bouwens, E.M., Trivedi, K., Van Vliet, C. and Winkel, C. (1999) Method of Enhancing Color in a Tea Based Foodstuff. US patent, US5879730 A.

188 Micard, V. and Thibault, J.F. (1999) Oxidative gelation of sugar-beet pectins: use of laccases and hydration properties of the cross-linked pectins. *Carbohydrate Polymers*, **39**, 265–73.

189 Piacquadio, P., De Stefano, G., Sammartino, M. and Sciancalepore, V. (1998) Apple juice stabilization by laccase (EC 1.10.3.2) immobilized on metalchelate regenerable carries. *Industrie delle Bevande*, **27**, 378–83.

190 Alper, N. and Acar, J. (2004) Removal of phenolic compounds in pomegranate juices using ultrafiltration and laccase-ultrafiltration combinations. *Die Nahrung*, **48**, 184–7.

191 Figueroa-Espinoza, M.C., Morel, M.H., Surget, A., Asther, M., Moukha, S., Sigoillot, J.C. and Rouau, X. (1999) Attempt to cross-link feruloylated arabinoxylans and proteins with a fungal laccase. *Food Hydrocolloids*, **13**, 65–71.

192 Labat, E., Morel, M.H. and Rouau, X. (2001) Effect of laccase and manganese peroxidase on wheat gluten and pentosans during mixing. *Food Hydrocolloids*, **15**, 47–52.

193 D'Souza, S.F. (2001) Microbial biosensors. *Biosensors and Bioelectronics*, **16**, 337–53.

194 Simkus, R.A., Laurinavicius, V., Boguslavsky, L., Skotheim, T., Tanenbaum, S., Nakas, J.P. and Slomczynski, D.J. (1996) Laccase containing sol-gel based optical biosensors. *Analytical Letters*, **29**, 1907–19.

195 Bauer, C.G., Kuhn, A., Gajovic, N., Skorobogatko, O., Holt, P.J., Bruce, N.C., Makower, A., Lowe, C.R. and Scheller, F.W. (1999) New enzyme sensors for morphine and codeine based on morphine dehydrogenase and laccase. *Fresenius' Journal of Analytical Chemistry*, **364**, 179–83.

196 Huang, T., Warsinke, A., Koroljova-Skorobogat'ko, O.V., Makower, A., Kuwana, T. and Scheller, F.W. (1999) A bienzyme carbon paste electrode for the sensitive detection of NADPH and the measurement of glucose-6-phosphate dehydrogenase. *Electroanalysis*, **11**, 295–300.

197 Ghindilis, A. (2000) Direct electron transfer catalysed by enzymes: application for biosensor development. *Biochemical Society Transactions*, **28**, 84–9.

198 Freire, R.S., Durán, N., Wang, J. and Kubota, L.T. (2002) Laccase-based screen printed electrode for amperometric detection of phenolic compounds. *Analytical Letters*, **35**, 29–38.

199 Gomes, S.A.S., Nogueira, J.M.F. and Rebelo, M.J.F. (2004) An amperometric biosensor for polyphenolic compounds in red wine. *Biosensors and Bioelectronics*, **20**, 1211–6.

200 Jaouani, A., Guillen, F., Penninckx, M.J., Martinez, A.T. and Martinez, M.J. (2005) *Enzyme and Microbial Technology*, **36**, 478–86.

201 Fukuda, T., Uchida, H., Suzuki, M., Miyamoto, H., Morinaga, H., Nawata, H. and Uwajima, T. (2004) *Journal of Chemical Technology and Biotechnology*, **79**, 1212–8.

202 Nakamura, Y. and Mtui, G. (2003) Anaerobic fermentation of woody biomass treated by various methods. *Biotechnology and Bioprocess Engineering*, **8**, 179–182.

203 Schultz, A., Jonas, U., Hammer, E. and Schauer, F. (2001) Dehalogenation of Chlorinated Hydroxybiphenyls by Fungal Laccase. *Applied and Environmental Microbiology*, **67**, 4377–81.

204 Ullah, M.A., Bedford, C.T. and Evans, C.S. (2000) *Reactions of penta chlorophenol with laccase from Coriolus versicolor. Applied Microbiology and Biotechnology*, **53**, 230–4.

205 Singh, B.K. and Walker, A. (2006) Microbial degradation of organophosphorus compounds. *FEMS Microbiology Reviews*, **30**, 428–41.

206 Bollag, J.M. (1992) Enzymes catalyzing oxidative coupling reactions of pollutants. *Metal Ions in Biological Systems*, **28**, 205–17.

207 Bollag, J.M. and Myers, C. (1992) Detoxification of aquatic and terrestrial sites through binding of pollutants to humic substances. *The Science of the Total Environment*, **117** (118), 357–66.

208 Dawel, G., KÄStner, M., Michels, J., Poppitz, W., GÄnther, W. and Fritsche, W. (1997) Structure of a Laccase-Mediated Product of Coupling of 2,4-Diamino-6 Nitrotoluene to Guaiacol, a Model for Coupling of 2,4,6-Trinitrotoluene Metabolites to a Humic Organic Soil Matrix. *Applied and Environmental Microbiology*, **63**, 2560–5.

209 Wang, C.J., Thiele, S. and Bollag, J.M. (2002) Interaction of 2,4,6-Trinitrotoluene(TNT) and 4-Amino-2,6-Dinitrotoluene with Humic Monomers in the Presence of Oxidative Enzymes. *Archives of Environmental Contamination and Toxicology*, **42**, 1–8.

210 Hager, A. (2003) *Zum Effekt von Laccasen beim Altpapier-Deinking. Dissertation an der Universität Hamburg*, Fachbereich Biologie.

211 Ahn, M.Y., Dec, J., Kim, J.E. and Bollag, J.M. (2002) Treatment of 2,4-dichlorophenol polluted soil with free and immobilized laccase. *Journal of Environmental Quality*, **31**, 1509–15.

212 Hublik, G. and Schinner, F. (2000) Characterization and immobilization of the laccase from *Pleurotus ostreatus* and its use for the continuous elimination of phenolic pollutants. *Enzyme and Microbial Technology*, **27**, 330–6.

213 Ehlers, G.A. and Rose, P.D. (2005) Immobilized white-rot fungal biodegradation of phenol and chlorinated phenol in trickling packed-bed reactors by employing sequencing batch operation. *Bioresource Technology*, **96**, 1264–75.

214 Pickard, M.A., Roman, R., Tinoco, R. and Vazquez-Duhalt, R. (1999) Polycyclic aromatic hydrocarbon metabolism by white rot fungi and oxidation by *Coriolopsis gallica* UAMH 8260 laccase. *Applied and Environmental Microbiology*, **65**, 3805–909.

215 Itoh, K., Fujita, M., Kumano, K., Suyama, K. and Yamamoto, H. (2000) Phenolic acids affect transformations of chlorophenols by a Coriolus versicolor laccase. *Soil Biology and Biochemistry*, **32**, 85–91.

216 Okazaki, S., Michizoe, J., Goto, M., Furusaki, S., Wariishi, H. and Tanaka, H. (2000) Oxidation of bisphenol A catalysed by Laccase hosted in reversed micelles in organic media. *Enzyme and Microbial Technology*, **31**, 227–32.

217 Eggen, T. (1999) Application of fungal substrate from commercial mushroom production *Pleuorotus ostreatus* for bioremediation of creosote contaminated soil. *International Biodeterioration and Biodegradation*, **44**, 117–26.

218 Keum, Y.S. and Li, Q.X. (2004) Fungal laccase-catalyzed degradation of hydroxy polychlorinated biphenyls. *Chemosphere*, **56**, 23–30.

219 Moeder, M., Martin, C. and Koeller, G. (2004) Degradation of hydrpxylated compounds using laccase and horseradish peroxidase immobilized on microporous polypropylene hollow fiber membranes. *Journal of Membrane Science*, **245**, 183–90.

220 Niku, P.M.L. and Viikari, L. (2000) Enzymatic oxidation of alkenes. *Journal of Molecular Catalysis B: Enzymatic*, **10**, 435–44.

221 Collins, P.J., Kotterman, M.J.J., Field, J.A. and Dobson, A.D.W. (1996) Oxidation of anthracene and benzo[a]pyrene by laccase from *Trametes versicolor. Applied and Environmental Microbiology*, **62**, 4563–7.

222 Johannes, C., Majcherezyk, A. and Huttermann, A. (1998) Oxidation of acenaphthene and acenaphthylene by laccase of *trametes versicolor* in a laccase-mediator system. *Journal of Biotechnology*, **61**, 151–6.

223 Castro, A., Evtuguin, D. and Xavier, A.M.B. (2003) Degradation of biphenyl lignin model compounds by laccase of *Trametes versicolor* in the presence of 1-hydroxybeazotriazole and heteropolyanion [SiW11VO40]5. *Journal of Molecular Catalysis B: Enzymatic*, **22**, 13–20.

224 Dodor, D.E., Hwang, H.M. and Ekunwe, S.N. (2004) Oxidation of anthracene and benzo[a]pyrene by immobilized laccase from *Trametes versicolor. Enzyme and Microbial Technology*, **35**, 210–7.

225 Fukuda, T., Uchida, H., Takashima, Y. and Uwajima, T. (2001) Degradation of bisphenol A by purified laccase from Trametes villosa. *Biochemical and Biophysical Research Communications*, **284**, 704–6.

226 Fabbrini, M., Galli, C., Gentili, P. and Macchitella, D. (2001) An oxidation of alcohols by oxygen with the enzmye laccase and mediation by TEMPO. *Tetrahedron Letters*, **42**, 7551–3.

227 Cantarella, G., Galli, C. and Gentili, P. (2003) Free radical versus electron transfer routes of oxidation of hydrocarbons by laccase/mediator systems. Catalytic or stoichiometric procedures. *Journal of Molecular Catalysis B: Enzymatic*, **22**, 135–44.

228 Dec, J. and Bollag, J.M. (1990) Detoxification of substituted phenols by oxidoreductive enzymes through polymerization reactions. *Archives of Environmental Contamination and Toxicology*, **19**, 543–50.

229 Aggelis, G., Iconomou, D., Christou, M., Bokas, D., Kotzailias, S., Christou, G., Tsagou, V. and Papanikolaou, S. (2003) Phenolic removal in a model olive oil mill wastewater using *Pleurotus ostreatus* in bioreactor cultures and biological evaluation of the process. *Water Research*, **37**, 3897–904.

230 Leontievsky, A.A., Myasoedova, N.M., Baskunov, B.P., Evans, C.S. and Golovleva, L.A. (2000) Transformation of 2,4,6-trichlorophenol by the white rot fungi Panus tigrinus and Coriolus versicolor. *Biodegradation*, **11**, 331–40.

231 Pointing, S.B. (2001) Feasibility of bioremediation by white-rot fungi. *Applied Microbiology and Biotechnology*, **57**, 20–33.

232 Bressler, D.C., Fedorak, P.M. and Pickard, M.A. (2000) Oxidation of carbazole, Nethylcarbazole, fluorene, and dibenzothiophene by the laccase of *Coriolopsis gallica. Biotechnology Letters*, **22**, 1119–25.

233 Sariaslani, F.S., Beale, J.M. and Rosazza, P. (1984) Oxidation of rotenone by Polyporus anceps laccase. *Journal of Natural Products*, **47**, 692–7.

234 Manzanares, P., Fajardo, S. and Martín, C. (1995) Production of ligninolytic activities when treating paper pulp effluents by *Trametes versicolor. Journal of Biotechnology*, **43**, 125–32.

235 D'Annibale, A., Stazi, S.R., Vinciguerra, V.G., DI Mattia, E. and Giovannozzi Sermanni, G. (1999) Characterization of immobilized laccase from *Lentinula edodes* and its use in olive-mill wastewater treatment. *Process Biochemistry*, **34** (6–7), 697–706.

236 D'Annibale, A., Stazi, S.R., Vinciguerra, V., Sermanni, G.G. (2000) Oxirane-immobilized Lentinula edodes laccase: stability and phenolics removal efficiency in olive mill wastewater. *Journal of Biotechnology*, **77**, 265–73.

237 Yang, Y.C., Baker, J.A. and Ward, J.R. (1992) Decontamination of chemical warfare agents. *Chemical Reviews*, 1729–43.

238 Magnaud, G., Lion, C., Delmas, G. and Reynaud, A. (1998). *Thorough decontamination system.* In Proc. 6th CBW Protection Symposium, Stockholm, Sweden, 309–12.

239 Yang, Y., Szafraniac, L., Beaudry, W.T. and Bohrbaugh, D.K. (1990) Oxidative biodegradation of phosphorothiolates by fungal laccase. *Journal of the American Chemical Society*, **112**, 6621–4.

240 Hall, C.R., Inch, T.D., Inns, R.H., Muir, A.W., Sellers, D.J. and Smith, A.P. (1977) Differences between some biological properties of enantiomers of alkyl S-alkyl methylphosphophonothiolates. *The Journal of Pharmacy and Pharmacology*, **29**, 574–6.

241 Reimer, K., Schreier, H., Erdos, G., König, B., König, W. and Fleischer, W. (1998) Molekulare Effekte eines mikrobiziden Wirkstoffes auf relevante Mikroorganismen: Elektronenmikroskopische und biochemische Untersuchungen zu Polyvidon-Iod. *Zentralblatt für Hygiene und Umweltmedizin*, **200**, 423–34.

242 XU, F. (1996) Catalysis of Novel Enzymatic iodide Oxidation by Fungal Laccase. *Applied Biochemistry and Biotechnology*, **59**, 221–30.

243 Kulys, J., Bratkovskaja, I. and Vidziunaite, R. (2005) Laccase-catalysed Iodide Oxidation in Presence of Metyl Syringate. *Biotechnology and Bioengineering*, **92**, 124–8.

11
Medical Aspects of Chemical Warfare Agents

Kai Kehe, Franz Worek and Horst Thiermann

11.1
Introduction

Chemical warfare agents (CWA) are toxic chemicals designed to cause death or other harm through their toxic properties which are produced for military purposes. CWAs can induce incapacitation, serious injury or even death in humans and animals.

A chemical weapon consists of a CWA and a delivery system, for example, bombs, mines, rockets or grenades. Any agricultural sprayer mounted on a vehicle or aircraft may be used as a delivery system by terrorists, for example.

The classical CWA are usually characterized and classified according to their toxicological action into five categories: nerve agents, vesicants, blood agents (cyanides), pulmonary agents, and incapacitants (Table 11.1). This classification is not scientifically sound. Regarding "blood" agents, it should be mentioned that they do not interact with any blood constituents.

Nerve agents and vesicants are regarded as the most important chemical warfare agents. Cyanides and pulmonary agents appear to be of minor importance as huge amounts are necessary to produce highly toxic concentrations in the field. Incapacitating agents produce various (also unwanted) effects and have limited relevance in warfare. Thus, the medical aspects of nerve agents and vesicants are described in detail.

11.2
Decontamination

The harming effects of highly toxic chemical increase with the absorbed dose. To prevent, minimize or reduce the toxic effects of CWA it is of utmost importance, to remove them from skin as fast as possible. In military context, soldiers are usually trained to perform self-decontamination. This is indispensable to prevent further absorption as well as prevent spreading of the agent. It should be mentioned that effective amounts of vapour may be trapped in clothing and be released

Decontamination of Warefare Agents.
Edited by André Richardt and Marc-Michael Blum

ISBN: 978-3-527-31756-1

Table 11.1 Classification of chemical warfare agents (US army code).

Nerve agents	Tabun (GA), sarin (GB), soman (GD), cyclosarin (GF), VX
Vesicants	Sulfur mustard (HD), nitrogen mustard (HN-3), lewisite (L)
Cyanides	Hydrogen cyanide (AC), cyanogen chloride (CK)
Pulmonary agents	Phosgene (CG), diphosgen (DP), chlorpikrin (PS)
Incapacitating agents	3-Quinuclidinyl benzilate (BZ)

at undressing. This was dramatically shown in the Tokyo Sarin theatre, when 20–30% of hospital personnel were affected by Asarin vapour that was released from the patients clothes [1]. Accordingly, effective decontamination is necessary before admission of patients in hospitals or medical facilities. The harming effects of highly toxic chemicals increase with the absorbed dose. Removal of the agent stops this process and should be done as early as possible. Decontamination can be done by physical, chemical or enzymatic means.

It is not necessary or even possible to decontaminate vaporous CWAs, but it is crucial for liquid agents. The medical personnel at the decontamination point will be confronted especially with these poisonings and their inherent risks. Thus, they should be highly trained concerning the management of CWA intoxicated patients. For wounded patients the deconfamination procedure is much more difficult and need much more resources. Specific effective and mild decontamination for wounds are not available. It is necessary to develop new decontamination procedures which may be useful for wound decontamination. Hereby, use of systems based on enzymatic detoxification of nerve agents may be advantageous because most chemical procedures are irritating to skin and mucous membranes.

11.3 Nerve Agents

11.3.1 Physico-Chemical Properties

Highly toxic organophosphorus (OP) compound-type chemical warfare agents (nerve agents) are potent inhibitors of cholinesterases. At present, the nerve agents soman, sarin, tabun, cyclosarin, VX and VR are considered as most important threat agents. In addition, OP pesticides (e.g. parathion, malathion, chlorpyrifos) may be used as chemical agents by terrorists.

Nerve agents are liquids at ambient temperature and evaporate at different rates. Pure nerve agents are colourless and without odor. They are of different persistency in the field, ranging from 1 hour to several weeks. Nerve agents can easily cross the skin or other barriers of the human body due to their lipophilic properties. Therefore, a protective suit is necessary to enter an area where contamination is possible.

11.3.2
Mechanism of Action

Nerve agents are able to bind covalently to serine esterases, mainly to acetylcholinesterase (AChE), butyrylcholinesterase (BuChE) throughout the whole body. Here, the active site serine is phosphylated (denotes both phosphorylation and phosphonylation) leading to a dramatic loss of enzyme activity (Figure 11.1). AChE and catalyzes the hydrolysis of the neurotransmiter acetylcholine, thereby terminating the signal transmission by acetylcholine. AChE BuChE inhibition results in a reduced hydrolysis of the neurotransmitter acetylcholine (ACh) leading to accumulation of ACh at synapses. Inhibited AChE and BuChE can be regenerated by spontaneous dephosphylation (spontaneous reactivation) of the enzyme. The velocity and extent of this process depends on the OP structure, is negligible in case of soman, sarin, tabun and cyclosarin, but may lead to a partial restoration of enzyme activity in case of VX, VR and different pesticides. In addition, the phosphyl-enzyme complex may undergo post-inhibitory transformation by loss of an alkyl group (Figure 11.2). This process, called "aging", results in a stabilization of the phosphylated enzyme preventing nucleophilic attack by water or oximes. Again, aging half-times are depending on the OP structure and are extremely rapid with soman ($t_{1/2}$ AChE ~ 2 minutes) and much slower with sarin (3 hours) or VX (40 hours).

Clinical effects of nerve agent exposure are consequences of ACh accumulation and subsequent over-stimulation of central and peripheral nicotinic and muscarinic receptors.

Additional targets of organophosphates are other serine esterases, for example, carboxylesterase (CaE), neuropathy target esterase (NTE), trypsin and chymotrypsin [2–5]. Even a tyrosine residue of human serum albumin has been shown being attacked by organophosphates [6]. The clinical importance to these additional targets is of minor importance in the acute toxicity.

Nerve agents may directly alter the release of neurotransmitters such as γ-aminobutyric acid (GABA). G-Agents have also been linked to activate the kinase-mediated protein Ca^{++}/calmodulin kinase II (Ca++/CaM kinase II) which may be responsible for axonal degeneration [7]. However, these non-cholinergic effects are mostly described in *in vitro* models and not in humans. The physiological relevance has to be shown for the complex situation of a whole body.

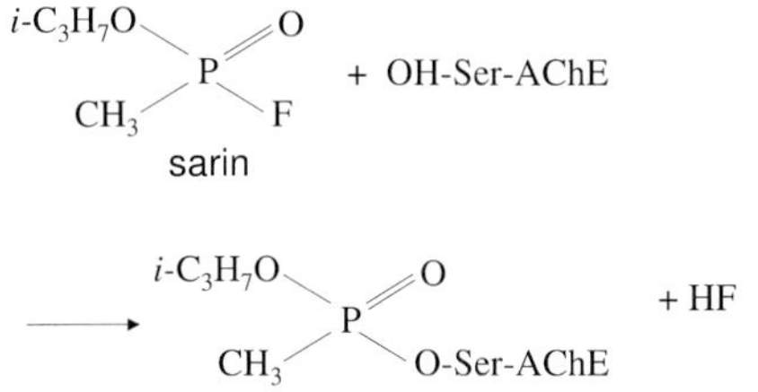

i-C_3H_7O(P=O)(F)CH_3 (sarin) + OH-Ser-AChE → i-C_3H_7O(P=O)(O-Ser-AChE)CH_3 + HF

Figure 11.1 Reaction of sarin with acetylcholinesterase produces inhibited enzyme.

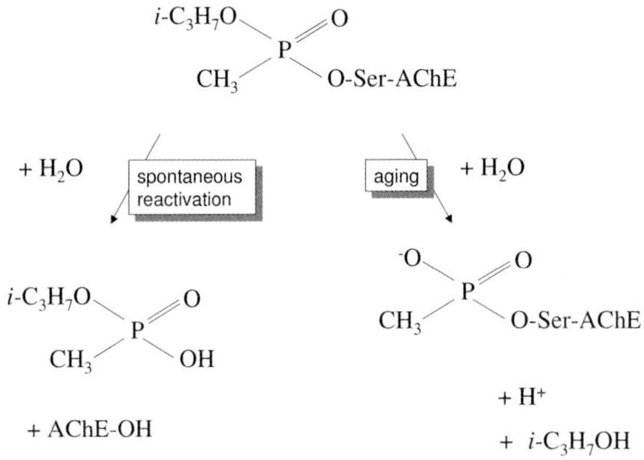

$i\text{-}C_3H_7O$ — P(=O) — CH_3, O-Ser-AChE

$+ H_2O$ spontaneous reactivation

aging $+ H_2O$

$i\text{-}C_3H_7O$ — P(=O) — CH_3, OH

$+$ AChE-OH

^-O — P(=O) — CH_3, O-Ser-AChE

$+ H^+$

$+ i\text{-}C_3H_7OH$

Figure 11.2 Aging and spontaneous hydrolysis of sarin-inhibited acetylcholinesterase. Aging results in an irreversibly inhibited enzyme, spontaneous hydrolysis in releasing of less toxic metabolites. Spontaneous hydrolysis is of no relevance in sarin poisoning.

11.3.3 Symptoms of Poisoning

Nerve agents may be absorbed through skin and mucous membranes, for example, the eyes or the respiratory or gastrointestinal tract. Local effects appear at the site of absorption followed by systemic effects if a toxic dose is absorbed. The effects of nerve agent poisoning are summarized in Table 11.2.

11.3.3.1 Local Effects

Local ocular effects start within seconds or minutes after vapor exposure. The earliest sign is miosis followed by conjunctival hyperaemia and ocular pain. Dim or blurred vision may occur. After vapour exposure ocular signs and symptoms are the most frequent in the early phase [1]. Local respiratory effects start with nasal congestion followed by a tight chest and increased bronchial secretion. A liquid drop of nerve agent on skin can induce localized sweating and muscular fasciculation. Following ingestion of (tasteless) nerve agent, abdominal cramps, nausea, vomiting and diarrhea are the first gastrointestinal symptoms.

11.3.3.2 Systemic Effects

The most life-threatening effects are at the respiratory system, which is affected by central respiratory depression and muscular weakness of the diaphragm. This situation is worsened by excessive upper and lower airway secretion leading to severe dyspnea and cyanosis. Only rapid and appropriate therapeutic intervention will be life-saving. Other symptoms are miosis, ocular secretions, sweating, intestinal hypermotility, bradycardia, muscle fasciculations, twitching, weakness, paralysis, loss of consciousness, convulsions. Minimal effects observed at lower doses include miosis, tightness of the chest, rhinorrhoea, and dyspnea.

Table 11.2 Signs and symptoms of nerve agent poisoning.

Organ	*Muscarinic (m) nicotinic (n)*	*Signs/Symptoms*
Eye	m	Miosis, blurred vision, frontal headache, dim vision, tears
Nose, pharyngeal	m	Rhinorrhoea, hypersalivation
Respiratory system	m	Tightness in chest, wheezing, bronchorrhoea, dyspnea, cyanosis
Gastrointestinal tract	m	Nausea, vomiting, diarrhea, abdominal cramps, involuntary defecation
Skin	m	Increased sweating
Bladder	m	Involuntary micturition, urgency
Heart	m	Bradycardia (after initial tachycardia), arrhythmia
Skeletal muscle	n	Weakness, fasciculation, twitching, cramps
Sympathetic ganglia	n	Transitory elevated blood pressure, hypotension
CNS		
– respiration	m + n	Respiratory depression
– activity	m + n	Restlessness, weakness, tremor, ataxia, convulsions
– behaviour	m + n	Sleeplessness, nightmares, memory and learning impairments

11.3.4 Diagnosis

Besides clinical diagnosis by typical signs and symptoms of cholinergic crisis, measurement of decreased AChE activity in blood is, at present, the most rapid method for diagnosis of organophosphate poisoning. Therefore, it is essential that medical personnel are trained to recognize the cholinergic toxidrome and to run AChE field or bedside tests (e.g. Testmate).

The analysis of intact nerve agent or its metabolites or adducts in blood and urine is possible, but only available in specialized laboratories.

11.3.5 Medical Management

11.3.5.1 Prophylactic Pharmacological Measures

Carbamoylation of AChE prevents a complete inhibition of the enzyme in case of OP poisoning. In addition, carbamoylated AChE is prone to spontaneous decarbamoylation resulting in a gradual increase of active AChE sufficient to sustain vital functions. However, the administration of atropine and an oxime is still needed. Carbamate pre-treatment is of special importance in case of soman and tabun since, with these agents, the therapeutic efficacy of atropine and oximes is limited. Experimental animal studies showed that a partial inhibition of AChE by carbamates (pyridostigmine, physostigmine) in advance of exposure to soman or tabun may enhance the therapeutic effects of atropine and oximes. On the other hand, the antidotal efficacy is adequate with sarin, cyclosarin and VX and carbamate pre-treatment has only a limited effect. Presently, pyridostigmine is used in different countries, for example, USA, UK, Germany. However, pyridostigmine is a quaternary amine and can only poorly cross the blood-brain barrier. This has the disadvantage that brain AChE is not protected. Therefore, physostigmine, a non-polar tertiary amine, is presently under investigation to replace pyridostigmine.

11.3.5.2 Enzymatic Prophylaxis and Therapeutics

Alternative prophylactic countermeasures may be the administration of stochiometric or catalytic scavengers. It has been demonstrated in various animal models that BuChE can protect against the toxic effects of organophosphates (up to $5 \times LD_{50}$), for example, volatile soman [8]. Thus, great effort has been made to enhance the production of recombinant BuChE. Human BuChE has recently been cloned into goats and is secreted via the milk of the animals (Protexia). Additionally, recombinant and pegylated AChE has been shown to protect against soman intoxication [9]. AChE and BuChE are stochiometric scavengers, that is, are only able to bind nerve agents in a molar ratio of 1:1. Hence, different OP hydrolyzing enzymes, for example, human paraoxonase (PON1), diisopropylfluorophosphatase (DFPase), organophosphorus hydrolase (OPH) and organophosphorous acid anhydrolase (OPAA), are under investigation as catalytic scavengers. Besides an inadequate activity of these enzymes towards different OP, the use of enzymes from non-human origin has limitations because of possible immunologic reactions and potential adverse effects. Several attempts were made to mask such proteins by carrier red blood cells or liposomes. Using sterically stabilized liposomes, OPAA showed some protection against diisopropylfluorophosphate (DFP) [10].

11.3.5.3 Pharmacotherapy

As noted above, both muscarinic and nicotinic effects of nerve agents may be life-threatening. Thus, for the treatment of OP poisoning muscarinic antagonists (e.g. atropine), oximes to reactivate inhibited AChE (e.g. obidoxime) and drugs for neuroprotection (e.g. diazepam) are needed (Table 11.3).

Atropine antagonizes the effects of ACh at muscarinic receptors. The initial dose of atropine, 2–5 mg, may be doubled every 10 minutes until therapeutic effects like dry mouth, mydriasis and an increase in heart rate are observed. Atropine may produce an impaired performance in soldiers. Therefore, autoinjectors contain 2 mg atropine sulphate. The side effects of three auto-injectors are tolerable to healthy adults. Atropine should be applied immediately after exposure towards nerve agents by self- or buddy-aid. However, atropine in therapeutic tele-

Table 11.3 Medical treatment of nerve agent poisoning.

Symptoms	*Therapy*
Self- and Buddy-Aid	
Miosis, salivation, impaired breathing, blurred vision	2 mg atropine i.m. + oxime (e.g. 250 mg obidoxime) i.m.
After 8–10 min persistent or worsening symptoms	2 mg atropine i.m.
After 8–10 min persistent or worsening symptoms	2 mg atropine i.m.
Medical Treatment (Physician)	
Symptoms of cholinergic crisis: for example, miosis, salivation, impaired breathing, blurred vision	2 mg atropine i.v. + oxime (e.g. 220 mg obidoxime) i.v.
Atropine therapy: dose is doubled every 10 min until therapeutic effects occur (dry mouth, reversal of miosis, increase of heart rate)	2-4-8-16-32 mg atropine i.v.
Persistent symptoms of cholinergic crisis and/or no recovery of RBC-AChE	Oxime i.v. (e.g. 750 mg/24 h obidoxime infusion) (maximum 1000 mg/d)
Convulsions	Diazepam (bolus dose of 10 mg i.v. slowly until cessation)
Intensive care	Ventilatory support, substitution of fluids, and electrolytes

vant doses has no clinical relevant effect at nicotinic receptors, for example, at the neuromuscular junction. Therefore, cholinesterase reactivators (oximes, e.g. obidoxime) are used to casually restore AChE activity. Early treatment with oximes is crucial to improve neuromuscular transmission and to prevent aging of inhibited AChE. Beneficial effects on CNS have been shown in numerous animal studies when administered early [11]. Accordingly, diazepam should be given as early as possible, even before the occurrence of convulsions. Alternatively, the anticonvulsant midazolam may be used.

The therapeutic effect of oximes may be monitored by repeated determination of RBC-AChE activity measurements.

11.3.5.4 Supportive Therapeutic Measures

In termination of exposure by removal of the agent from skin and hair, mucous membranes have to be flushed with water in excess. Gastric lavage may be necessary if the agent was swallowed within 30 min. Airway management is crucial. Artificial ventilation may be life-saving. Blood pressure should be monitored and cardiovascular depression should be treated adequately, e.g. by catecholamines.

11.3.6 Long-term Effects

Organophosphate induced delayed neuropathy (OPIN) has been related to organophosphate insecticides, although it has not been observed in survivors of nerve agent poisoning.

11.3.7
Special Toxicology

There is no evidence about mutagenic, carcinogenic or teratogenic effects after sarin, tabun or VX poisoning.

11.4
Alkylating Agents

11.4.1
Introduction/Definition

Sulfur mustard (bis(2-chloroethyl)sulfide); CASRN: 505-60-2) was first synthesised in 1822 by Despretz and rediscovered by Niemann and Guthrie 1860. The typical vesicant properties of the agent were rapidly noticed. Meyer improved synthesis and produced sulfur mustard with higher purity. In the distilled form, sulfur mustard has being designated "HD" by the US military. Other synonyms are mustard gas (typical odor and volatility), Yperite (first use during the battle at Ypres), Lost (acronym of the German chemists Lommel and Steinkopf who investigated the mass production), pyro (British code) and yellow cross (German shells were marked with a yellow cross). Agent T (Bis-(2-(2-chloroethylthio)ethyl)ether; CASRN: 63918-89-8) is a similar alkylating agent. The mixture of 40% agent T and 60% sulfur mustard (HD) is designated HT. Nitrogen analogues are ethylbis(2-chloroethyl)amine (HN-1; CASRN: 538-07-8), bis(2-chloroethyl)methylamine (mechlorethamine, HN-2; CASRN: 51–75-2) and tris(2-chloroethyl)amine (trichlormethine, HN-3; CASRN: 555-77-1).

Until the present, sulfur mustard has been the most produced and weaponized vesicant. It was used in the Iran-Iraq war. More than 100 000 Iranian soldiers were injured, of which 45 000 are still suffering from lasting effects today [12]. Poisoning by sulfur mustard produces ocular and dermal injury, respiratory tract damage, reproductive and developmental toxicity, gastrointestinal effects, haematological effects, and cancer.

Extensive research during World War II also resulted in medical applications of alkylating agents. Their strong cytotoxic effect on dividing cell population was used in the therapy of cancer [13]. Several nitrogen mustards (R-N-bis-(2-chloroethylchlorides)) are in use as cytostatic drugs until today: mechlorethamine (HN-2), cyclophosphamide, melphalan, chlorambucil, or ifosfamide [12]. The strong immunosuppressive effect of sulfur mustard has promoted the study of chemical immunosuppression and, finally, smoothed the way for organ transplantation. The nitrogen mustard cyclophosphamide is still used as an immunosuppressant. The cytostatic effect of sulfur mustard has been used to treat psoriasis [14].

11.4.2
Mechanism of Action

Sulfur mustard can easily penetrate body surfaces within minutes. In an aqueous environment, each 2-chloroethyl sidechain undergoes first order (SN1) intramolecular cyclization which is accompanied by the release of chloride ions. The resulting ethylene sulfonium cation intermediate opens to form the highly reactive carbonium ion. In consequence, sulfur mustard reacts immediately with all cell constituents like DNA, RNA, proteins and other molecules. It has been shown that the presence of a nucleus is required for the toxicity of sulfur mustard [15]. Therefore, it can be concluded that the most important alkylations affect the DNA. Sulfur mustard forms several DNA adducts, for example, 7-(2-hydroxyethylthioethyl) guanine (7-HETE-G) which accounted for 61% of total DNA alkylation [16]. Less likely alkylations are on the 3 position of adenine (16%) and 0.1% on the O6 position of guanine [17]. Despite the low occurence of O6-(2-ethylthioethyl) guanine, it has been assumed that this DNA lesion accounts for the strong carcinogenicity [18]. Sulfur mustard can also form intra- and interstrand cross-links. Approximately, 17% of alkylations involve two guanines (G-alkyl-G) [19]. It has previously been described that after alkylation of the DNA, damaged cells are arrested at certain cell cycle checkpoints. Cells are arrested at cell cycle checkpoints depending on sulfur mustard concentration: human keratinocytes exposed to vesicating doses (>50 μM) show a G1 block and at concentrations 10-fold below a G2 block occurs [20]. This cell cycle arrest is, in part, mediated by autophosphorylation of ataxia-teleangiectasia mutated (ATM) protein and its downstream targets γ-H2Ax, p53 and p21. During the cell cycle arrest, DNA repair takes place and the cells recover or, in case of persistent DNA damage, die. Keratinocyte cell death after sulfur mustard exposure can be from either apoptosis, necrosis or terminal differentiation [21, 22]. Activation of the nuclear enzyme poly(ADP-ribose) polymerase type 1 (PARP-1) can be observed within one hour after exposure. Severe DNA damage from sulfur mustard (>500 μM) highly activates PARP-1 which causes a depletion of its substrate nicotine adenine dinucleotide (NAD^+) and, in consequence, glycolysis inhibition. Therefore, the cellular energy supply is heavily disturbed. This negative energy balance is worsened by NAD^+-resynthesis which consumes ATP [23]. In consequence, intracellular ATP loss determines necrotic cell death [24]. NAD^+ levels may be maintained by PARP inhibitors, for example, nicotinamide, but this can only prevent cell survival in the first hours after injury. PARP inhibition also blocks DNA repair and potentially elevates the mutation frequency in surviving cells [25, 26]. Mild genotoxic stress of sulfur mustard exposure activates PARP-1 to a lesser extent thereby stimulating DNA repair [25]. However, if DNA damage can not be repaired, apoptosis will be initiated [27–31]. Apoptotic stimuli cause an early burst of (ADP-ribose) polymer formation with consumption of NAD and ATP leading to energy depletion of the cell [32]. Cleavage of PARP-1 through caspase 3 during apoptosis eleminates enzyme activity and stops energy consumption [33]. However, overstimulation of PARP-1 in the initial phase favours necrotic cell death. It has been shown that the p38 MAP kinase (MAPK14) signaling pathway is activated in response to sulfur mustard induced cellular stress and is involved in the upregulation of IL-8, IL-6, and TNF-α after sulfur mustard exposure [34]. Furthermore, activation of the FAS/FAS-Ligand-system has been reported to be responsible for apoptotic cell death in sulfur mustard treated keratinocytes [35]. In addition, the apoptotic response can be attenuated by Ca^{++} chelators [22]. This is further supported by the finding that calmodulin has a crucial role in sulfur mustard mediated toxicity [36].

In conclusion of the discussed pathways, the most investigated pharmacological compounds for treating sulfur mustard injuries are intracellular scavengers (e.g. N-acetyl-cysteine, amifostine), cell cycle inhibitors (e.g. mimosine), PARP inhibitors (niacinamide), calcium modulators (e.g. BAPTA), protease inhibitors (e.g. sulfonyl fluorides, ilomastat) and anti-inflammatory compounds (e.g. indomethacin) [37]. However, early decontamination will prevent sulfur mustard injury.

11.4.3 Toxicokinetic

Mustards are lipophilic compounds and toxicologically relevant concentrations can therefore penetrate epithelial tissues rapidly. The eyes, respiratory tract and skin of unprotected persons will be most likely damaged after exposure to sulfur mustard. In the eyes and especially in the respiratory tract, the epithelium is thinner than in skin and these tissues lack a stratum corneum which acts as a little barrier. At body sites with thicker stratum corneum (palms, soles) higher concentrations of sulfur mustard are needed for blistering. Moistening of the skin enhances absorption. Hair follicles provide a more direct entry route into the dermis. Much of local applied sulfur mustard (80%) evaporates and 20% penetrates the skin. 20% from the penetrated dose is fixed to macromolecules in skin. The remaining 80% are rapidly distributed by circulation. Thus, only 4% of the local applied dose is fixed in skin [38–40].

Besides of their pronounced local damaging capacity sulfar mustard can cause severe intoxication. Unhydrolyzed sulfur mustard can accumulate in the brain and fat deposits even days after exposure. Biotransformation of sulfur mustard in man has not been extensively studied. In a study in terminally ill cancer patients, 80–90% radioactivity of the injected ^{14}C-labeled sulfur mustard disappeared from the blood after several minutes and was excreted mainly in urine within 24 hours [41]. More recent investigations demonstrated that 60% of the dose is excreted in urine in 24 hours. Analysis revealed that most metabolites result from reactions with glutathione, hydrolysis, and direct oxidation of the sulfur. Thiodiglycol sulfoxide, 1,1′-sulfonylbis [2-S(N-acetylcysteinyl)ethane], and 1,1′-sulfonylbis [2-(methylsulfinyl)-ethane] or 1-methylsulfinyl-2-[2-(methylthio)ethylsulfonyl]ethane are the most prevalent metabolites which were presumed to be produced by β-lyase and assumed to be diagnostic and forensic indicators of sulfur mustard poisoning in man [42]. It has been shown in rats that conjugation with glutathione is more important than hydrolysis [41, 43].

11.4.4 Symptoms of Poisoning

11.4.4.1 General

The first contact with mustard-like vesicants is mostly painless. In the case of sulfur mustard, a garlic or mustard odor can be experienced which results from technical impurities. After exposure, a symptom-less interval lasts several hours.

Table 11.4 Signs and symptoms of sulfur mustard poisoning.

Organ system	Symptoms
Eye	Chronic conjunctivitis with impaired vision
Respiratory system	Polyps in the paranasal sinus, chronic bronchitis, chest pain, bronchial carcinoma
Gastrointestinal tract	Anorexia, vomiting, weight loss
Nervous system	Insomnia, irritability
Urinary tract	Bladder carcinoma
Hematological system	Leukemia

The absorbed sulfur mustard dose correlates inversely with duration of this interval. The eyes, the pulmonary tract and the skin are the major sites of entry for mustards (Table 11.4). As noted above, exposure to large doses of sulfur mustard can cause systemic poisoning with impact on the hemopoetic and immune systems.

11.4.4.2 Eyes

After sulfur mustard exposure, short latent periods (30–60 minutes) were reported. This latent period is much shorter than that found in skin or in the respiratory tract exposure. The high susceptibility of the eye is due to the thin epithelial cells covered by tear fluid which favours the formation of reactive sulfur mustard intermediates. Additionally, corneal cells show a high cell turn over and intense energy metabolism. Firstly, irritation, conjunctivitis, grittiness and tearing are noted at the threshold dose (vapour: 50–100 mg min/m^3). 3–12 hours after exposure to higher doses (vapour: >200 mg min/m^3, ICt_{50}) corneal odema, vision impairment, eyelid odema, photophobia and severe blepharospasm occur [44, 45]. Blepharospasm usually accounts for temporary blindness, but this improves in more than 80% of exposed people after 1–2 weeks. When the exposure is higher than some 400 mg min/m^3 severe full thickness corneal damage with ulcerations and occlusion of conjunctival blood vessels due to endothelial damage may take place [46].

11.4.4.3 Respiratory Tract

Although sulfur mustard is regarded as a “persistent” chemical agent, most of the casualties in former wars were caused by vapor. Sulfur mustard dose-dependently damages the mucosa of the respiratory tract. Sulfur mustard is lipophilic and thus produces more damage in upper airways than in lower airways. The inhaled dose is dependent on the respiration rate. Therefore, it is difficult to correlate environmental sulfur mustard concentrations with the inhaled dose. However, usually the concentration (C) time (t) product (Ct) has been widely used to describe an exposure to sulfur mustard vapor. Inhaled sulfur mustard dose correlates inversely to the onset of signs and symptoms. Thus, an onset of symptoms two hours after exposure has to be regarded as a consequence inhalation of a high dose with infaust prognosis.

After a period of no or only limited symptoms, single exposure casualties might notice only discomfort or irritation of nasopharynx or sinuses. Hoarseness with husky voice, lacrimation, rhinorrhoea, loss of smell and taste, discharge of mucus from nose and throat occurs later. This bronchitis during the initial phase after exposure is due to toxic mechanisms. Thus, antibiotics should be only given after confirmation of a bacterial superinfection. Exposures to higher doses result in damage of medium-sized airways presenting as tracheobronchitis, painful and hacking coughing, and diphteric like pseudomembrane formation. The epithelial sloughing may loosen from the bronchiotracheal wall and may obstruct lower airways. Odema in upper and lower airways with necrotic epithelium are consequences of severe exposure. In most severe poisoned patients an adult respiratory distress syndrome (ARDS) with pulmonary odema may develop.

11.4.4.4 **Skin**

Keratinocytes of the basal layer are the most proliferating keratinocyte subpopulation in skin. They are the most susceptible cells of the skin after exposure for the DNA damaging action of sulfur mustard. Thus, signs of heavy cell damage (apoptosis, pyknosis, necrosis, acantholysis) are characteristically found in the epidermal basal layer. Epidermis and dermis start to separate by forming blisters [47]. The vesicles have a thin wall and the yellowish blister fluid is clear. The blister fluid is not toxic as a consequence of the high reactivity of sulfur mustard with tissue. Skin susceptibility to sulfur mustard mostly depends on three factors: skin temperature, moistness and anatomical location [48, 49]. Thin skin at moist body areas (axilla, scrotum, anal region) has a low dermal barrier function (30 times lower than forearm). Thus, these regions are highly sensitive. Sulfur mustard vapour may easily penetrate clothes resulting in severe burns of the genito-anal region, the chest, back and axilla. As mentioned above, the onset of symptoms depends on mustard vapour exposure concentration and duration of exposure. A higher concentration of mustard vapour will result in a shorter delay until symptoms occur. A Ct of about 200 $mg\,min/m^3$ is considered to be a vesicant threshold. After low dose single exposure (vapor: 100–300 $mg\,min/m^3$, liquid: 10–20 $\mu g/cm^2$) dermal symptoms occur within 4–8 hours after contact. Erythema and itching are the first symptoms (Figure 11.3). After exposure to higher doses (vapor: 1000–2000 $mg\,min/m^3$, liquid: 40–100 $\mu g/cm^2$) epidermal-dermal separation (blister) begins with development of small vesicles which may flow together to form large bullae [45]. The Nikolsky sign is positive which means that mechanical stress results in blister formation. Secondary blisters can occur during or after several weeks after exposure without any further exposure to sulfur mustard. At present, an explanation for this phenomenon is still lacking. Skin re-epithelialization after exposure is often observed from the skin stem cell reservoir, for example, the outer root shell keratinocytes of hair follicles.

Another important phenomenon is hyper- or hypopigmentation of the skin after exposure. Poikiloderma as a combination of hyper- and hypopigmentation may occur as a late effect dermal characteristic after sulfur mustard injury.

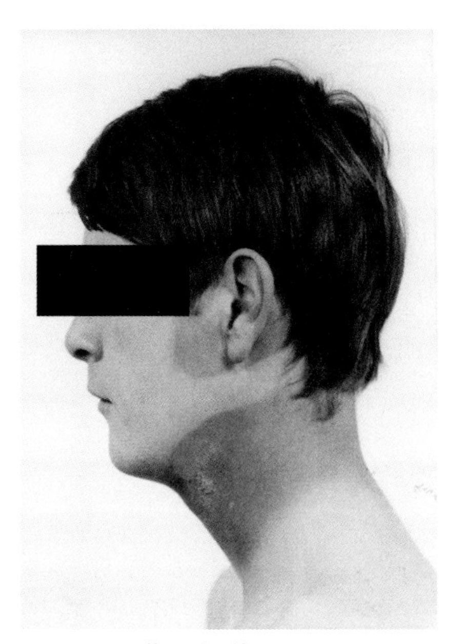

Figure 11.3 Effect of sulfur mustard on unprotected skin. Patient was wearing a protective mask when he was accidentally exposed. He noticed first symptoms after 2 hours. The picture was taken 7 days after exposure.

11.4.4.5 Systemic Toxicity

Signs of systemic sulfur mustard intoxication are similar to tumor chemotherapy with alkylating drugs (e.g. cyclophosphamide): headache, nausea, vomiting, and loss of appetite. High dose exposure may affect the gastrointestinal tract and the bone marrow resulting in immune suppression, leucopenia, diarrhea, fever, cachexia and in very severe cases excitation of the central nervous system with convulsions [12].

11.4.5 Diagnosis

The first diagnosis of poisoning by S- or N-mustard is regularly based on clinical assessment of delayed onset of characteristic blister formation. When patients report of garlic of mustard like smell sulphur mustard exposure may be anticipated, whereas in N-mustard poisoning, a more fishy odour is common. The delayed onset of characteristic blistering is typical for the use of mustard. Special diagnostic methods for the detection of sulfur mustard in tissues are only available in a limited number of laboratories around the world. Sulfur mustard is very labile in serum plasma. Therefore, a direct determination of free agent will practically not be possible in blood of victims. Alternatively, other methods are currently available in specialized laboratories:

- determination of urinary metabolites
- determination of protein adducts
- determination of DNA adducts.

Free sulfur mustard is hard to detect in urine because of its rapid hydrolysis in aqueous solutions [50, 51]. Thus, the modern approach is to detect urine or plasma metabolites conjugated to cysteine or glutathione by mass spectrometric analysis [52–54]. The sulfur mustard reaction product with water, thiodiglycol, can be detected in urine up to two weeks after exposure [55]. However, traces of thiodiglycol can be also detected in urine of unexposed persons. Sulfur mustard exposure detection mostly relies on mass spectrometric or immunochemical analysis of DNA or protein adducts.

11.4.6 Medical Management

Sulfur mustard is a dangerous poison because the patient may not recognize exposure and could feel safe without realizing that medical care is necessary. Unfortunately, no antidote against sulfur mustard poisoning is available. First responders and medical personnel should wear protective clothing, mask and gloves. Protecting clothes (overgarments, gears) may prevent penetration of mustard vapours to a certain extent (dependent on vapour concentration and time of exposure). Penetration of liquid mustard can only be prevented by wearing of impermeable suits.

Exposure should be terminated by rapid evacuation from the contaminated environment, and decontamination. Contaminated normal clothing has to be quickly removed because sulfur mustard penetrates to the skin within several minutes. Military forces are usually equipped with special skin decontaminants, for example, M291 (US army), RSDL (Canada), or Fuller's earth (United Kingdom). If these decontaminants were not available e.g. after a terrorist attack against a civilian population, it would be possible to use talcum powder, or flour [56]. The use of small amounts of water is not recommended because it reduces the skin barrier and after spreading the agent it enhances skin absorption. If possible, only *prolonged* washing with copious amounts of water could be a successful decontamination procedure.

Eye injury has to be treated as soon as possible. Contact lenses have to be removed and flushing of the eyes with water or 0.9% saline is mandatory. Upper respiratory symptoms like sore throat or hoarseness can be relieved with moistened air. Bronchoconstriction may be treated with β_2-sympathomimetic drugs. Repeated bronchial suctioning may be necessary to remove pseudomembranes. Severely intoxicated patients may need tracheotomia. Skin lesions should be treated like burn wounds. Large blisters should not be opened and denuded skin needs a proper wound dressing.

11.4.6.1 Enzymatic Therapeutics

Bacterial dehalogenases have been described as having some reactivity towards sulfur mustard. So far, no attempts have been made to use this group as a therapeutic option against sulfur mustard poisoning.

11.4.7
"Low Level" Exposure

The eye is recognized as the most sensitive organ. Only limited ocular symptoms have been observed after acute low vapor exposure to sulfur mustard. These symptoms including photophobia, ocular irritation, and slight conjunctivitis are transient and disappear after cessation of exposure. Acute low level exposure to non-vesicating doses of sulfur mustard seem not to be related to a higher cancer risk.

Workers employed in mustard gas factories were repeatedly exposed to low concentrations of sulfur mustard vapor for longer periods. Several health problems were related to this type of exposure [57–59].

11.4.8
Long-term Effects

During the Iran-Iraq-war more than 100 000 people were exposed to sulfur mustard [60, 61]. Thus, most of these patients are still living today and are subject of a lot of epidemiological studies mostly performed in Iran [62]. Khateri *et al.* [61] examined 34 000 Iranians, 13–20 years after exposure to sulfur mustard. The most common late effects were found in the

- respiratory tract (42.5%)
- eyes (39%)
- skin (24.5%).

11.4.8.1 Respiratory Tract

Late respiratory effects due to sulfur mustard exposure have to be regarded as severely disabling complications. The patients regularly report of a long history with chronic bronchitis (59%), asthmatoid bronchitis (11%), pulmonary fibrosis (12%), or bronchial stenosis (10%). Modern investigation techniques, for example, high resolution chest computer tomography, revealed persistent pulmonary damage like air trapping (76%), bronchiectases (74%), mosaic parenchymal attenuation (72%), irregular and dilated major airways (66%), bronchial wall thickening (90%), and interlobular septal wall thickening (26%). Bronchoalveolar lavage showed increased level of transforming growth factor β (TGF-β) which has been related to several fibrogenic disorders. However, it is a matter of debate, whether this finding justifics classification of pulmonary fibrosis as late effect of sulfur mustard exposure. Sulfur mustard affects the upper part of the respiratory tract to a greater extent. Severe damage of the bronchio-tracheal epithelium which is frequently observed lead to persistent mucociliary malfunction. The resulting chronic inflammation may be responsible for chronic obstructive pulmonary disease and stenosis of the major airways [63].

11.4.8.2 Eye

Corneal stem cells in the pericorneal conjunctiva are apparently very sensitive. Pleyer *et al.* reported that nearly 90% of eye injuries due to sulfur mustard exposure experienced later on some kind of ocular symptoms. However, only 0.5% of exposed soldiers developed a delayed type of ulcerative keratitis up to 40 years after severe sulfur mustard eye injury [46]. The most striking observation is that this keratopathy shows a long asymptomatic interval but then results in opacification of the entire cornea. The onset of this keratopathy is characterized by chronic conjunctivitis.

11.4.8.3 Skin

A recent study on 40 Iranian veterans by Balali-Mood *et al.* [62] on heavily exposed soldiers to the sulfur mustard 16 to 20 years ago revealed the most common cutaneous lesions:

- hyperpigmentation (55%) and hypopigmentation (25%)
- erythematous papular rash (42.5%)
- dry skin (40%)
- multiple cherry angiomas (37.5%)
- atrophy (27.5%).

These distribution of the lesions were located on

- genital areas (48%)
- the back (48%)
- the front thorax and abdomen (44%)
- lower extremities (mainly inguinal) (44%)
- upper extremities (mainly axillary) (41%)
- head and neck (15%).

This shows that most of the soldiers were exposed to sulfur mustard vapor. Interestingly, abnormal pigmentation (hyper- and hypopigmentation) can be seen in formerly exposed areas. Hypopigmentation at areas that were exposed to high concentrations and hyperpigmentation at areas exposed to lower ones. It is thought that the altered pigmentation of the skin depends on the survival of melanocytes after sulfur mustard exposure. Local destruction of melanocytes results in depigmentation, otherwise (postinflammatory) hyperpigmentation predominates.

11.4.9 Special Toxicology

Sulfur mustard is a known human carcinogen, rated by the IARC. It has been identified as a risk factor for occupational lung cancer [64, 65].

Sulfur mustard is a highly genotoxic agent because of its bifunctional reactions with DNA resulting in interstrand crosslinks. Although most cells possess effective DNA repair mechanisms, interstrand crosslinks are difficult to repair. Additionally, alkylation of O6-guanine by sulfur mustard seems to be highly critical. The DNA repair enzyme O6-alkylguanine-DNA alkyltransferase has only limited efficacy to repair the sulfur mustard guanine adduct O6-ethylthioethylguanine [17].

Thus, this O6-lesion is believed to be the most important mutagenic lesion resulting in base exchange after replication. However, only limited data are available on the mutation spectra after sulfur mustard exposure. Notably, mutations in the p53 tumor suppressor gene ("guardian of the genome") have been detected in Japanese mustard gas workers [66]. Mutations in this tumor suppressor have been described 50% of all known human cancers.

Bebe *et al.* reported that American World War I veterans with a single (high dose) battlefield exposure showed an elevated but not significant increase in lung cancer in comparison with the non-gas wounded control [67].

Nishimoto investigated former Japanese gas workers and described that larynx, pharynx and lung cancer occurred 30 years after exposure. [64]. Former employees of a British sulfur mustard producing facility showed a high relative risk of 9.3 for laryngeal cancer as well as significant excess for cancers of the pharynx, the buccal cavity and other upper respiratory sites [68]. These observations were backed up by a follow-up study that confirmed these data and also added a significant excess of lung cancers compared with national rates for cancer [69].

Former workers hired for disbandment of the "Heeresmunitionsanstalt St. Georgen" in Germany after WWII were equipped by the Allies with inadequate protective measures. Multiple skin tumors such as basal cell carcinoma, Bowen's disease, Bowen's carcinoma and carcinoma spinocellulare were found later [70]. Furthermore, a higher rate of chronic myeloic leukemia (CML) has also been reported in Iranian mustard gas victims as well as abnormal lymphocytes [63, 71, 72]. It has to be noted that the soldiers were also exposed to several other highly toxic chemicals and that it is difficult to correlate the observed higher cancer risk to the single factor "sulfur mustard exposure."

11.5
Outlook

Chemical degradation of CWAs is the mostly used approach for decontamination. This may be critical in the medical context, as it is not possible to use bleach or other highly oxidizing chemicals in wounds, eyes or other muceous membranes without damage to the organ system. These limitations are potentiated when CWA are released in combination with explosives resulting in a high amount of injured and contaminated individuals. Thus, the use of classical decontamination products is limited. It is important to develop new decontaminants which are aggressive against CWAs but are tolerated by the human body. Enzyme systems may be one approach to reach this goal. As mentioned above, several specific enzymes have been identified with activity against CWAs. A combination of these enzyme systems could be beneficial. Also non-specific approaches were developed to oxidize for example, VX and sulfur mustard with enzymatic systems [73]. Such enzyme systems might also be used as internal drugs in order to facilitate degradation of already absorbed CWAs. It will be a big challenge to overcome the immunologic problems with foreign proteins.

References

1 Okumura, T., Takasa, N., Jshimatsu, S. (1996) Report on 640 victims of the Tokyo subway sarin attack. *Ann. Emerg. Med.*, **28**, 129–135.

2 Ecobichon, D.J. and Comeau, A.M. (1973) Hepatic aliesterase sensitivity to dichlorvos and diisopropylfluorophosphate. *Toxicology and Applied Pharmacology*, **26**, 260–3.

3 Fonnum, F., Sterri, S.H., Aas, P. and Johnsen, H. (1985) Carboxylesterase, importance for detoxification of organophosphorus anticholinesterases and trichothecenes. *Fundamental and Applied Toxicology*, **5**, S29-S38.

4 Johnson, M.K. and Glynn, P. (1995) Neuropathy target esterase (NTE) and organophosphorus-induced delayed polyneuropathy (OPIDP): recent advances. *Toxicology Letters*, **82/83**, 459–63.

5 Ooms, A.J.J. and van Dijk, C. (1966) The reaction of organophosphorus compounds with hydrolytic enzymes–III The inhibition of chymotrypsin and trypsin. *Biochemical Pharmacology*, **15**, 1361–77.

6 Black, R.M., Harrison, J.M. and Read, R.W. (1999) The interaction of sarin and soman with plasma proteins: the identification of a novel phosphonylation site. *Archives of Toxicology*, **73**, 123–6.

7 Abou-Donia, M.B. and Lapadula, D.M. (1990) Mechanisms of organophosphorus ester-induced delayed neurotoxicity: type I and type II. *Annual Review of Pharmacology and Toxicology*, **30**, 405–40.

8 Allon, N., Raveh, L., Gilat, E., Cohen, E., Grunwald, J. and Ashani, Y. (1998) Prophylaxis against soman inhalation toxicity in guinea pigs by pretreatment alone with human serum butyrylcholinesterase. *Toxicological Sciences*, **43**, 121–8.

9 Kronman, C., Cohen, O., Raveh, L., Mazor, O., Ordentlich, A. and Shafferman, A. (2007) Polyethylene-glycol conjugated recombinant human acetylcholinesterase serves as an efficacious bioscavenger against soman intoxication. *Toxicology*, **233**, 40–6.

10 Petrikovics, I., Cheng, T.C., Papahadjopoulos, D., Hong, K., Yin, R., DeFrank, J.J., Jaing, J., Song, Z.H., McGuinn, W.D., Sylvester, D., Pei, L., Madec, J., Tamulinas, C., Jaszberenyi, J.C., Barcza, T. and Way, J.L. (2000) Long circulating liposomes encapsulating organophosphorus acid anhydrolase in diisopropylfluorophosphate antagonism. *Toxicological Sciences*, **57**, 16–21.

11 Marrs, T.C. (2003) Diazepam in the treatment of organophosphorus ester pesticide poisoning. *Toxical. Rev.*, **22** (*2*), 75–81.

12 Dacre, J.C. and Goldman, M. (1996) Toxicology and pharmacology of the chemical warfare agent sulfur mustard. *Pharmacological Reviews*, **48**, 289–326.

13 Goodman, L.S., Wintrobe, M.M., Dameshek, W., Goodman, M.J., Gilman, A. and McLennan, M.T. (1984) Landmark article Sept. 21, 1946: Nitrogen mustard therapy. Use of methyl-bis(beta-chloroethyl)amine hydrochloride and tris(beta-chloroethyl)amine hydrochloride for Hodgkin's disease, lymphosarcoma, leukemia and certain allied and miscellaneous disorders. *The Journal of the American Medical Association*, **251**, 2255–61.

14 Illig, L. (1977) Management of psoriasis vulgaris using an external sulphur mustard compound with special reference to its possible carcinogenic hazard (1st continuation and conclusion). Carcinogenesis of sulphur mustard compound in animal experiment and in man] Die Behandlung der Psoriasis vulgaris mit Schwefel-Lost extern unter besonderer Berucksichtigung ihres moglichen Carcinogenese-Risikos (1. Fortsetzung und Schluss). Zur Cancerogenitat von Schwefel-Lost im Tier-Versuch und beim. *Zeitschrift für Hautkrankheiten*, **52**, 1035–44.

15 Lodhi, I.J., Sweeney, J.F., Clift, R.E. and Hinshaw, D.B. (2001) Nuclear dependence of sulphur mustard-mediated cell death. *Toxicology and Applied Pharmacology*, **170**, 69–77.

16 Ludlum, D.B., Austin-Ritchie, P., Hagopian, M., Niu, T.Q. and Yu, D. (1994) Detection of sulphur mustard-induced DNA modifications. *Chemico-Biological Interactions*, **91**, 39–49.

17 Ludlum, D.B., Kent, S. and Mehta, J.R. (1986) Formation of O6-ethylthioethylguanine in DNA by reaction with the sulphur mustard, chloroethyl sulfide, and its apparent lack of repair by O6-alkylguanine-DNA alkyltransferase. *Carcinogenesis*, **7**, 1203–6.

18 Ludlum, D.B., Tong, W.P., Mehta, J.R., Kirk, M.C. and Papirmeister, B. (1984) Formation of O6-ethylthioethyldeoxyguanosine from the reaction of chloroethyl ethyl sulfide with deoxyguanosine. *Cancer Research*, **44**, 5698–701.

19 Lawley, P.D., Lethbridge, J.H., Edwards, P.A. and Shooter, K.V. (1969) Inactivation of bacteriophage t7 by mono- and difunctional sulphur mustards in relation to cross-linking and depurination of bacteriophage dna. *Journal of Molecular Biology*, **39**, 181–98.

20 Smith, W.J., Sanders, K.M., Ruddle, S.E. and Gross, C.L. (1993) Cytometric analysis of DNA changes induced by sulphur mustard. *Journal of Toxicology. Cutaneous and Ocular Toxicology*, **12**, 337–47.

21 Kehe, K. and Szinicz, L. (1999) Features of apoptotic cell death in SCL II cells after exposure to 2,2-dichlorodiethylsulfide, in *NBC Risks* (eds T. Sohns and V.A. Voicu), Kluwer Academic Publishers, the Netherlands, pp. 155–9.

22 Rosenthal, D.S., Simbulan-Rosenthal, C.M., Iyer, S., Spoonde, A., Smith, W., Ray, R. and Smulson, M.E. (1998) Sulphur mustard induces markers of terminal differentiation and apoptosis in keratinocytes via a Ca2+-calmodulin and caspase-dependent pathway. *The Journal of Investigative Dermatology*, **111**, 64–71.

23 Burkle, A. (2001) Physiology and pathophysiology of poly(ADP-ribosyl)ation. *Bioessays*, **23**, 795–806.

24 Berger, S.J., Sudar, D.C. and Berger, N.A. (1986) Metabolic consequences of DNA damage: DNA damage induces alterations in glucose metabolism by activation of poly (ADP-ribose) polymerase. *Biochemical and Biophysical Research Communications*, **134**, 227–32.

25 Bhat, K.R., Benton, B.J., Rosenthal, D.S., Smulson, M.E. and Ray, R. (2000) Role of poly(ADP-ribose) polymerase (PARP) in DNA repair in sulphur mustard-exposed normal human epidermal keratinocytes (NHEK). *Journal of Applied Toxicology*, **20**, (Suppl. 1), S13–S17.

26 Chatterjee, S., Berger, S.J. and Berger, N.A. (1999) Poly(ADP-ribose) polymerase: a guardian of the genome that facilitates DNA repair by protecting against DNA recombination. *Molecular and Cellular Biochemistry*, **193**, 23–30.

27 Dabrowska, M.I., Becks, L.L., Lelli J, Levee, M.G. and Hinshaw, D.B. (1996) Sulphur mustard induces apoptosis and necrosis in endothelial cells. *Toxicology and Applied Pharmacology*, **141**, 568–83.

28 Hinshaw, D.B., Lodhi, I.J., Hurley, K.B. and Dabrowska, M.I. (1999) Activation of poly [ADP ribose] polymerase in endothelial cells and keratinocytes: role in an *in vitro* model of sulphur mustard-mediated vesication. *Toxicology and Applied Pharmacology*, **156**, 17–29.

29 Kehe, K., Reisinger, H. and Szinicz, L. (2000) Sulphur mustard induces apoptosis and necrosis in SCL II cells in vitro. *Journal of Applied Toxicology*, **20**, S81-S86.

30 Rosenthal, D.S., Simbulan-Rosenthal, C.M., Liu, W.F., Velena, A., Anderson, D., Benton, B., Wang, Z.Q., Smith, W., Ray, R. and Smulson, M.E. (2001) PARP determines the mode of cell death in skin fibroblasts, but not keratinocytes, exposed to sulphur mustard. *The Journal of Investigative Dermatology*, **117**, 1566–73.

31 Stöppler, H., Stöppler, M.C., Johnson, E., Smulson, M.E., Iyer, S., Rosenthal, D.S. and Schlegel, R. (1998) The E7 protein of human papillomavirus type 16 sensitizes primary human keratinocytes to apoptosis. *Oncogene*, **17**, 1207–14.

32 Simbulan-Rosenthal, C.M., Aaronson, S.A., Rosenthal, D.S., Iyer, S., Boulares, H. and Smulson, M.E. (1999) Involvement of PARP and poly(ADP-ribosyl)ation in the early stages of apoptosis and DNA replication. *Molecular and Cellular Biochemistry*, **193**, 137–48.

33 Soldani, C. and Scovassi, A.I. (2002) Poly(ADP-ribose) polymerase-1 cleavage during apoptosis: an update. *Apoptosis*, **7**, 321–8.

34 Dillman, J.F., III, McGary, K.L. and Schlager, J.J. (2004) An inhibitor of p38 MAP kinase downregulates cytokine release induced by sulphur mustard exposure in human epidermal keratinocytes. *Toxicology in Vitro*, **18**, 593–9.

35 Rosenthal, D.S., Velena, A., Chou, F.P., Schlegel, R., Ray, R., Benton, B., Anderson, D., Smith, W.J. and Simbulan-Rosenthal, C.M. (2003) Expression of dominant-negative Fas-associated death domain blocks human keratinocyte apoptosis and vesication induced by sulphur mustard. *The Journal of Biological Chemistry*, **278**, 8531–40.

36 Simbulan-Rosenthal, C.M., Ray, R., Benton, B., Soeda, E., Daher, A., Anderson, D., Smith, W.J. and Rosenthal, D.S. (2006) Calmodulin mediates sulphur mustard toxicity in human keratinocytes. *Toxicology*, **227**, 21–35.

37 Smith, W.J. (2002) Vesicant agents and antivesicant medical countermeasures: clinical toxicology and psychological implications. *Military Psychology*, **14**, 145–57.

38 Chilcott, R.P., Jenner, J., Carrick, W., Hotchkiss, S.A. and Rice, P. (2000) Human skin absorption of Bis-2-(chloroethyl)sulphide (sulphur mustard) in vitro. *Journal of Applied Toxicology*, **20**, 349–55.

39 Chilcott, R.P., Jenner, J., Hotchkiss, S.A. and Rice, P. (2001) In vitro skin absorption and decontamination of sulphur mustard: comparison of human and pig-ear skin. *Journal of Applied Toxicology*, **21**, 279–83.

40 Cullumbine, H. (1947) Medical aspects of mustard gas poisoning. *Nature*, **159**, 151–3.

41 Davison, C., Rozman, R.S. and Smith, P.K. (1961) Metabolism of bis-beta-chloroethyl sulfide (sulfur mustard gas). *Biochemical Pharmacology*, **7**, 65–74.

42 Black, R.M. and Read, R.W. (1995) Improved methodology for the detection and quantitation of urinary metabolites of sulphur mustard using gas chromatography-tandem mass spectrometry. *Journal of Chromatography. B, Biomedical Applications*, **665**, 97–105.

43 Roberts, J.J. and Warwick, G.P. (1963) Studies of the mode of action of alkylating agents – VI The metabolism of bis-2-chlorethylsulphide (mustard gas) and related compounds. *Biochemical Pharmacology*, **12**, 1329–34.

44 Mandel, M. and Gibson, W.S. (1917) Clinical manifestations and treatment of gas poisoning. *The Journal of the American Medical Association*, **69**, 1970–1.

45 Willems, J.L. (1989) Clinical management of mustard gas casualties. *Annales Mediciniae Militaris*, **3**, 1–60.

46 Solberg, Y., Alcalay, M. and Belkin, M. (1997) Ocular injury by mustard gas. *Survey of Ophthalmology*, **41**, 461–6.

47 Petrali, J.P., Oglesby, S.B. and Justus, T.A. (1991) Morphologic effects of sulfur mustard on a human skin equivalent. *Journal of Toxicology. Cutaneous and Ocular Toxicology*, **10**, 315–24.

48 Renshaw, B. (1946) Mechanism in production of cutaneous injuries by sulfur and nitrogen mustards, in *Summary Technical Report of Dinsion 9, NDRC Volume 1: Chemical Warfare Agents and Related Problems* (ed. N.D.C. US Office of Scientific Research and Development), Washington DC, pp. 479–518.

49 Renshaw, B. (1947) Observation on the role of water in the susceptibility of human skin to injury by vesicant vapors. *The Journal of Investigative Dermatology*, **9**, 75–85.

50 Vycudilik, W. (1985) Detection of mustard gas bis(2-chloroethyl)-sulfide in urine. *Forensic Science International*, **28**, 131–6.

51 Vycudilik, W. (1987) Detection of bis(2-chlorethyl)-sulfide (Yperite) in urine by high resolution gas chromatography-mass spectrometry. *Forensic Science International*, **35**, 67–71.

52 Black, R.M. and Read, R.W. (1995) Biological fate of sulphur mustard, 1,1′-thiobis(2-chloroethane): identification of beta-lyase metabolites and hydrolysis products in human urine. *Xenobiotica*, **25**, 167–73.

53 Noort, D., Benschop, H.P. and Black, R.M. (2002) Biomonitoring of exposure to chemical warfare agents: a review.

Toxicology and Applied Pharmacology, **184**, 116–26.

54 Noort, D., Fidder, A., Benschop, H.P., De Jong, L.P. and Smith, J.R. (2004) Procedure for monitoring exposure to sulfur mustard based on modified edman degradation of globin. *Journal of Analytical Toxicology*, **28**, 311–5.

55 Wils, E.R., Hulst, A.G. and van Laar, J. (1988) Analysis of thiodiglycol in urine of victims of an alleged attack with mustard gas, Part II. *Journal of Analytical Toxicology*, **12**, 15–9.

56 Borak, J. and Sidell, F.R. (1992) Agents of chemical warfare: sulfur mustard. *Annals of Emergency Medicine*, **21**, 303–8.

57 Morgenstern, P., Koss, F.R. and Alexander, W.W. (1947) Residual mustard gas bronchiti: effects of prolonged exposure to low concentrations of mustard gas. *Annals of Internal Medicine*, **26**, 27–40.

58 Weiss, A. and Weiss, B. (1975) Carcinogenesis due to mustard gas exposure in man, important sign for therapy with alkylating agents. *Deutsche Medizinische Wochenschrift*, **100**, 919–23.

59 Yanagida, J., Hozawa, S., Ishioka, S., Maeda, H., Takahashi, K., Oyama, T., Takaishi, M., Hakoda, M., Akiyama, M. and Yamakido, M. (1988) Somatic mutation in peripheral lymphocytes of former workers at the Okunojima poison gas factory. *Japanese Journal of Cancer Research*, **79**, 1276–83.

60 Aghanouri, R., Ghanei, M., Aslani, J., Keivani-Amine, H., Rastegar, F. and Karkhane, A. (2004) Fibrogenic cytokine levels in bronchoalveolar lavage aspirates 15 years after exposure to sulphur mustard. *American Journal of Physiology. Lung Cellular and Molecular Physiology*, **287**, L1160–4.

61 Khateri, S., Ghanei, M., Keshavarz, S., Soroush, M. and Haines, D. (2003) Incidence of lung, eye, and skin lesions as late complications in 34 000 Iranians with wartime exposure to mustard agent. *Journal of Occupational and Environmental Medicine*, **45**, 1136–43.

62 Balali-Mood, M., Hefazi, M., Mahmoudi, M., Jalali, E., Attaran, D., Maleki, M., Razavi, M.E., Zare, G., Tabatabaee, A. and Jaafari, M.R. (2005) Long-term complications of sulphur mustard poisoning in severely intoxicated Iranian veterans. *Fundamental and Clinical Pharmacology*, **19**, 713–21.

63 Ghanei, M. and Harandi, A.A. (2007) Long term consequences from exposure to sulfur mustard: a review. *Inhalation Toxicology*, **19**, 451–6.

64 Nishimoto, Y., Yamakido, M., Ishioka, S., Shigenobu, T. and Yukutake, M. (1987) Epidemiological studies of lung cancer in Japanese mustard gas workers. *Princess Takamatsu Symposia*, **18**, 95–101.

65 Wada, S., Miyanishi, M., Nishimoto, Y., Kambe, S. and Miller, R.W. (1968) Mustard gas as a cause of respiratory neoplasia in man. *Lancet*, **1**, 1161–3.

66 Takeshima, Y., Inai, K., Bennett, W.P., Metcalf, R.A., Welsh, J.A., Yonehara, S., Hayashi, Y., Fujihara, M., Yamakido, M. and Akiyama, M. (1994) p53 mutations in lung cancers from Japanese mustard gas workers. *Carcinogenesis*, **15**, 2075–9.

67 Beebe, G.W. (1960) Lung cancer in world war i veterans: possible relation to mustard-gas injury and 1918 influenza epidemic. *Journal of the National Cancer Institute*, **25**, 1231–52.

68 Manning, K.P., Skegg, D.C., Stell, P.M. and Doll, R. (1981) Cancer of the larynx and other occupational hazards of mustard gas workers. *Clinical Otolaryngology*, **6**, 165–70.

69 Easton, D.F., Peto, J. and Doll, R. (1988) Cancers of the respiratory tract in mustard gas workers. *British Journal of Industrial Medicine*, **45**, 652–9.

70 Klehr, N.W. (1984) Late manifestations in former mustard gas workers with special reference to cutaneous findings. *Zeitschrift fur Hautkrankheiten*, **59**, 1161–70.

71 Ghanei, M. and Vosoghi, A.A. (2002) An epidemiologic study to screen for chronic myelocytic leukemia in war victims exposed to mustard gas. *Environmental Health Perspectives*, **110**, 519–21.

72 Ghanei, M. (2004) Delayed haematological complications of mustard gas. *Journal of Applied Toxicology*, **24**, 493–5.

73 Amitai, G., Adani, R., Hershkovitz, M., Bel, P., Rabinovitz, I. and Meshulam, H. (2003) Degradation of VX and sulphur mustard by enzymatic haloperoxidation. *Journal of Applied Toxicology*, **23**, 225–33.

12
Microemulsions: A Versatile Carrier for Decontamination Agents

Thomas Hellweg, Stefan Wellert, S.J. Mitchell and André Richardt

12.1
Introduction

Chemical warfare agents present a significant hazard to personnel who come into contact with contaminated vehicle platforms and equipment. Of particular concern are the liquid organophosphorus nerve agents of the G-type series (soman, sarin, tabun etc.), the low-volatility organophosphorus nerve agent VX and the liquid vesicant agent sulfur mustard (HD). The persistence of these liquid agents can also be increased significantly by the use of polymeric thickeners. In addition, penetration of the agents into capillary features present on a contaminated surface poses a particular decontamination challenge. There is therefore a clear requirement to develop and deploy effective decontaminant formulations which are capable of removing and neutralizing this hazard.

12.2
Requirements for Decontaminants

Development of such formulations is not a trivial exercise. Potential decontaminants must be both highly effective and militarily practical. A summary of the requirements of a practical military decontaminant formulation are given below, and annotated with reference to the use of microemulsions as decontaminants:

12.2.1
Rapid and Complete Solubilization of the Agent from the Contaminated Surface

No liquid-borne decontamination chemistry can act against a chemical agent that is not effectively solubilized by the decontaminant. Thickened HD in particular is highly challenging to solubilize as it is a viscous, highly hydrophobic liquid.

Decontamination of Warefare Agents.
Edited by André Richardt and Marc-Michael Blum

ISBN: 978-3-527-31756-1

12.2.2 Penetration into Complex and Porous Surfaces to Extract Entrapped Chemical Agent

Decontamination media with high surface tension with respect to the solid surface, which has to be cleaned cannot be applied. The decontamination liquid has to wet the surface. The ultra-low surface tensions inherent to microemulsion formulations have the potential to provide excellent surface penetration. In addition, it is possible to choose a microemulsion which wets the surface concerned.

12.2.3 Fast and Complete Decomposition of the Agent

After the solubilization (physical removal) the rate of the decomposition of the agent (chemical destruction) by the active component must be complete in a short period of time. Therefore, the microemsulsion, as the basic medium, has to compromise between physical requirements for the solubilization of different agents and maintenance of the activity of chemicals, reactive nano-particles and enzymes to achieve complete cleaning of surfaces and complete destruction of agents.

12.2.4 Minimal Logistical Burden

The logistics of transporting bulk decontaminants are substantial. In the case of microemulsion decontaminants, the logistics burden can be minimised through the use of oil-in-water (O/W) microemulsion formulations with a high aqueous volume fraction. The formulation can then conceivably be transported to the point of use in a compact, concentrated form, to be heavily diluted immediately before use with local water supplies. Local water supplies can include sea-water, so a formulation must be stable to the resultant increase in ionic-strength.

12.2.5 Stability to Variation in Temperature

Operational scenarios require formulation stability to the conditions prevalent in environments from the arctic to the desert. Such a range of temperature stability is a particularly challenging requirement for microemulsion decontaminant formulation.

12.2.6 High Solubilization Capacity

The formulation must be capable of solubilizing a militarily realistic agent challenge (i.e. 2–3% w/w agent).

12.2.7 Stability to the Incorporation of Decontamination Reagents and Chemical Agent Degradation By-Products

Incorporation of decontaminant chemistry into a microemulsion in most of the cases results in an increase in ionic strength. Sufficient ionic strength capacity must be present to support a quantity of decontaminant reagents sufficient to completely neutralise a realistic agent challenge (see above). The use of catalytic decontamination chemistries in microemulsions is therefore highly desirable, as catalytic turnover provides a route to reducing the quantity of reagents necessary. In this context, enzyme based decontaminants are of great interest. An additional burden is presented by the need to incorporate high buffer capacities for the management of acidic by-products produced by nerve agent hydrolysis. The last important feature of a carrier for decontaminants, which should be mentioned here, is its flammability. The ideal system should not burn at all.

12.2.8 Environmental Compatibility

Historically, the environmental impact of decontamination formulations has received little consideration. However, with increasing emphasis on duty of care to the environment, an emerging requirement is that formulation components must as far as possible be environmentally benign and bio-degradable. Significant factors are the high toxicity of some surfactants to aquatic organisms and the avoidance of environmentally damaging hydrocarbon oil-phases (particularly aromatic hydrocarbons).

Besides the active component a crucial part of a decontamination system is the carrier medium. This medium has to solubilize the active component as well as the respective warfare agent. This is not an easy task, since often one is hydrophobic and the other one hydrophilic. A versatile medium for this purpose are so-called microemulsions. These systems might even allow for the solubilization of decontaminating enzymes. Based on the requirements and arguments listed above, microemulsions seem to be the ideal carrier medium for the design of new decontamination systems.

Microemulsions are thermodynamically stable homogeneous mixtures of oil and water, stabilised by surfactants and in some cases additionally by co-surfactants. In 1943, Schulman and Hoar were first to report a transparent homogeneous system formed by water, oil, and surfactant [1]. In 1959, it was again Schulman and co-workers who introduced the name microemulsion for this type of systems [2].

During the last decades, numerous works dealing with microemulsions have been published. This is due the importance of solubilization problems in industrial production processes (reaction media), oil recovery, and cleaning applications. Moreover, cosmetic and pharmaceutical products are also often based on microemulsions.

The present contribution will give a comprehensive introduction to the subject with a focus on the physical basis and the important principles for the study of microemulsions. Publications dealing with microemulsions have already been the subject of several review articles [3–9].

Beside more fundamental work concerning the microstructure of microemulsions, there is still a lot of interest in using microemulsion phases as reaction media [10], especially with respect to preparing nanostructured inorganic materials like for example, silica [11] or metal nanoparticles [12–14], but also for polymerization reactions [15–18].

This chapter is organised as follows. The next section will give a short introduction with respect to the parameters controlling phase behavior and examples for phase diagrams will be discussed. In the fourth section, methods for the physical characterization of microemulsion will be introduced. This subsection will be followed by a review of the recent literature on water continuous systems (L_1-phases), oil-continuous systems (L_2) and describes results on "sponge"-like structures (L_3-phases). L_3-phases might be of great interest with respect to the developement of decontamination media. Due to their bi-continuous character they are able to wet hydrophobic and hydrophilic surfaces equally well. Also, some publications dealing with lyotropic mesophases (not "strictly" microemulsions) will be discussed briefly. Finally, some results obtained for microemulsion designed for the decontamination of chemical warfare agents will be shown.

12.3 Basic Principles

12.3.1 Theory

In addition to the normal mixing contribution to the free energy the most important quantity, which controls all structuring in microemulsions, is the curvature free energy, sometimes called bending elastic energy. This curvature free energy was introduced by Helfrich and can be written as [19]

$$F = \int_A \frac{1}{2}\kappa(C_1 + C_2 - 2C_0)^2 + \bar{\kappa} C_1 C_2 dA. \tag{12.1}$$

Here, κ and $\bar{\kappa}$ are the mean (splay) and the Gaussian (saddle-splay) bending elastic constant. Knowing the elastic constants for a specific microemulsions system allows us to fully describe the behavior of the respective mixture. Therefore, it is still of great interest to determine theses quantities together with the curvatures C_1, C_2 and the natural curvature C_0 of the formed structure. A is the internal surface in the microemulsion.

The natural curvature of surfactant films in binary mixtures of surfactant and solvent but also in microemulsions is mainly determined by the geometry of the surfactant molecules. Israelachvili introduced a method to summarise these influences.

In surfactant solutions the shape of the formed self-assembled structures can be predicted introducing the packing parameter p [20].

$$p = \frac{v}{al} \tag{12.2}$$

Here, a is the head group area, l the chain length, and v the chain volume of the surfactant. For $p < \frac{1}{3}$ spherical structures are formed. p values between $\frac{1}{3}$ and $\frac{1}{2}$ lead to cylindrical structures. Finally $p > \frac{1}{2}$ favours flat layered structures (for example lamellar phases L_α). Oil continuous microemulsion are favored for $p \geq 1$. The different possible surfactant geometries are schematically shown in Figure 12.1.

Hence, to obtain a certain structure the packing parameter has to be controlled. This can be done by choosing a specific surfactant, which stabilises the desired structure (for example aerosol-OT for oil-continuous so-called L_2 microemulsions) or by mixing the surfactant with a co-surfactant (alcohols; an example is a mixture of SDS and decanol or butanol; see Figure 12.5). Furthermore, the curvature can also be tuned by changing the solvent quality for the hydrophilic or the hydrophobic block. For some surfactants this can be achieved by changing the temperature (C_nE_m surfactants) or in other cases by mixing the solvent with short alcohols or salt.

The salt effect is of special importance with respect to the preparation of decontamination systems since most of the active agents are salts. Their addition might change the phase behavior of the mixture. Moreover, as already mentioned in the introduction, for practical purposes dilution with, for example, sea-water has to be envisaged.

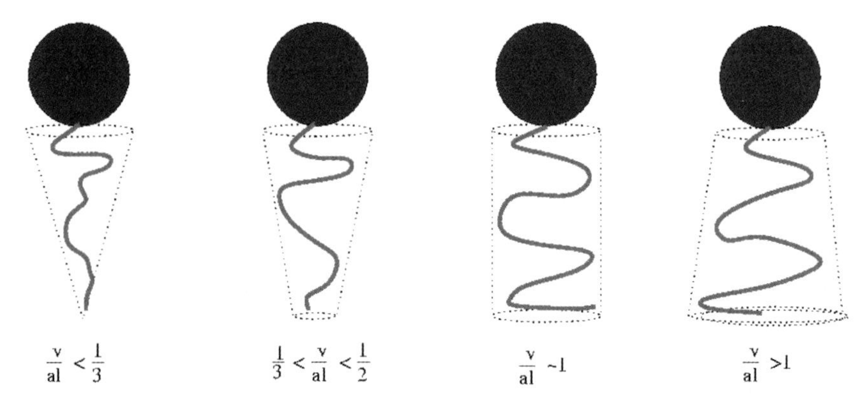

Figure 12.1 The packing parameter is given by the volume occupied by the hydrophobic surfactant tail relative to the headgroup area and hence, determines the curvature of the obtained microstructures in surfactant solutions or microemulsions.

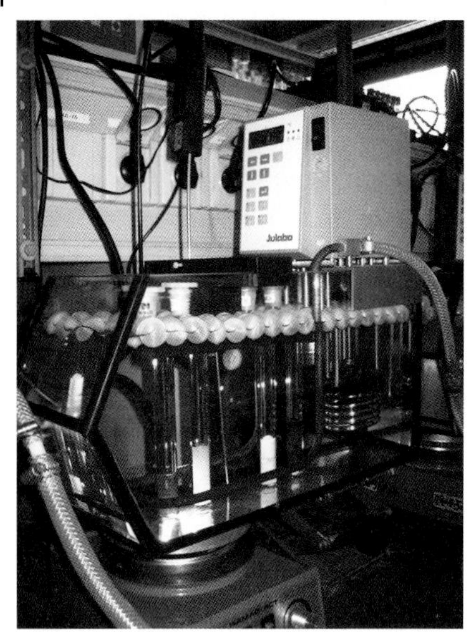

Figure 12.2 Experimental setup for the study of the phase behavior in microemulsion forming systems. The samples are filled into glass cylinders and placed in a thermostated bath equiped with a Pt100 and a magnetic stirrer. The front and the back of the bath are equiped with polaroid filters, which are crossed with respect to each other to allow for the detection of liquid crystalline phases. The samples can be eluminated from the outside.

12.3.2
Phase Behaviour

Most of the studies of the phase behavior in ternary systems still concentrate on non-ionic surfactants of the alkyl oligo-ethyleneoxide type (C_iE_j). This is due to the fact that these surfactants form microemulsions without addition of co-surfactants. Hence, a standard Gibbs triangle is sufficient to represent the phase diagram at a given temperature. Moreover, in these systems the natural curvature of the film can be tuned by simply changing the temperature. This is due to the fact that the hydration of the ethylenoxide headgroups changes with temperature leading to a change of the headgroup area and hence, to a change of p. This allows us to rather easily investigate the different microstructures which can be realised in microemulsion forming liquid mixtures.

Figure 12.2 shows an experimental setup, which is well suited to investigate the phase behavior of mixtures of oil, water, and surfactant. The components are mixed in cylindrical tubes and these are than placed into a thermostated water bath. The front and the back the bath are covered with polaroid filters. The polarization of these two filters is perpendicular to each other. When the samples are illuminated from the outside using a cold light source the filters allow us to identify liquid crystalline phases. Such phases show a strong birefringence in the setup. Often, simple visual inspection of the samples is sufficient to identify

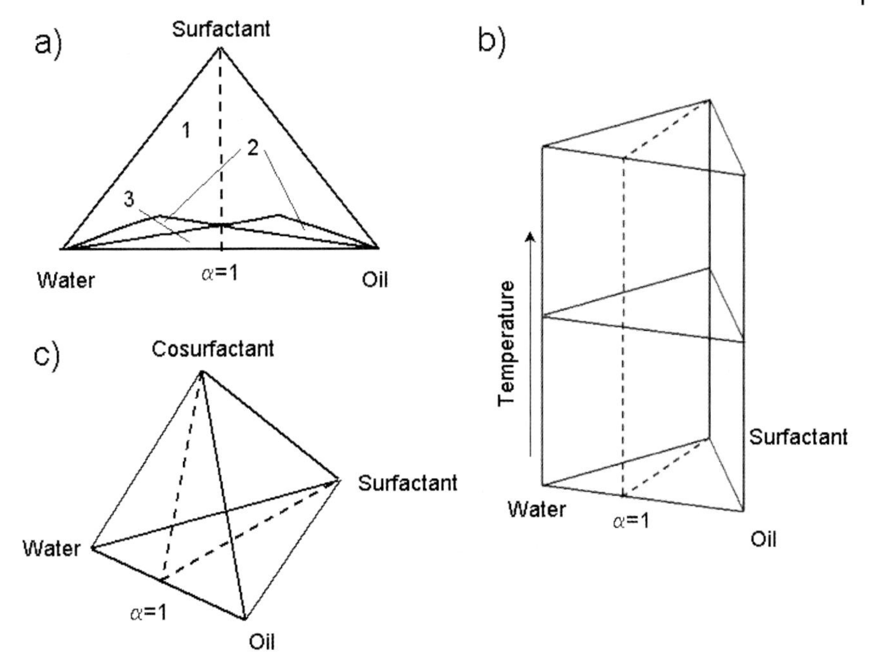

Figure 12.3 (a) The common representation of the phase behavior of a ternary system is the Gibbs triangle. The typical position of the one-, two- and three phase regions is schematically indicated for a constant temperature. (b) In the case of alkyl oligo-ethyleneoxide surfactants (C_iE_j) the temperature can be used to tune the phase behavior and is therefore an additional parameter in the representation of the phase behavior. The Gibbs triangle is expanded to a prism with triangular base. (c) In contrast to non-ionic surfactants of the C_iE_j group ionic surfactants are less temperature sensitive and in this case a fourth component, the so called cosurfactant is the tuning parameter of the phase behavior. The triangle is transformed into an phase tetrahedron at a constant temperature. In all three representations the position of the so called Kahlweit-"fish" at α= oil/(water + oil) = 1 is indicated.

the number of phases for a given set of composition and temperature. When the samples are at rest also the respective phase volumes can be measured.

Now, taking Gibbs triangles for a series of different temperatures leads to a rather complex so-called phase prism (see Figure 12.4b for example, the respective figures in ref. [7]). Often, it is not necessary to study the complete phase behavior of a system. It can be sufficient to look at some specific cuts through the mentioned prism. An especially useful section through the phase prism is obtained at constant water-to-oil ratio (the water-to-oil ratio is expressed as the weight percentage of oil α in the mixture of the two). Usually, approximately equal fractions of water and oil are taken and then the amount of surfactant is varied.

Figure 12.5 shows a scheme of such a section at constant water-to-oil ratio. This section is the so-called Kahlweit-"fish," in which all the single phase microemulsions can be identified in the one phase region inside the fishtail. The commonly used nomenclature for the different phases is as follows. L_1 indicates water-

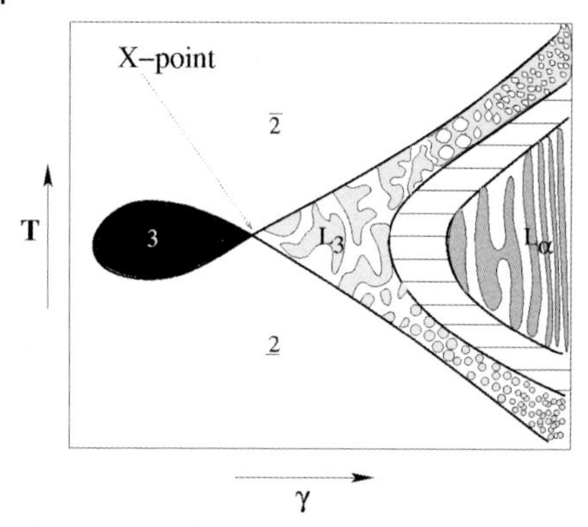

Figure 12.4 Schematic cut through the phase prism of a ternary mixture water-oil-C_iE_j-surfactant at constant ratio water to oil. In the tail of the so called "fish" L_1 and L_2-phases can be found. denotes the surfactant weight fraction. The black area is the 3 phase body.

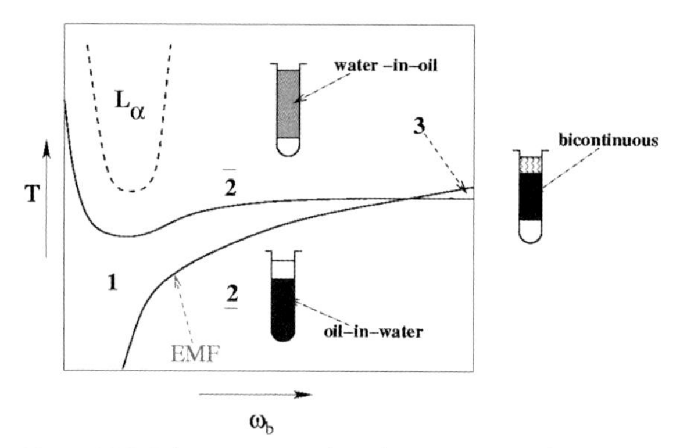

Figure 12.5 Schematic view of another important cut through the phase prism of typical nonionic surfactant (C_iE_j) based microemulsions. Shown is only the left corner (water rich) of the diagram. However, the right corner is usually only a kind of point symmetric to the left one. ω_b denotes the the increasing amount of oil at constant ratio surfactant/(surfactant + water). This cut is especially useful to determine the emulsification failure boundary (EMF). At this boundary the microstructures are spherical. In other systems the curvature tuning parameter can be the co-surfactant or salt concentration instead of temperature.

continuous droplet microemulsions (oil-in-water droplets); L_2 denotes oil-continuous droplet microemulsions (water-in-oil droplets); L_3 specifies the case of a bicontinuous microemulsion with a sponge like microstructure.

Besides the microemulsion phases also other interesting phases occur in mixtures of water oil and surfactant. These are liquid crystalline phases like, for example, the lamellar phase (L_α) or the hexagonal phase (H). These phases are usually of high viscosity and exhibit optical anisotropy.

A second useful cut through the phase prism is shown in Figure 12.2. In this cut, the emulsification failure boundary can be easily identified. Emulsification failure means for example, in the case of an L_1 microemulsion phase, that no more additional oil can be solubilized (incorporated in the droplets) beyond this boundary. Additional oil will simply lead to the formation of an excess oil phase (Figure 12.2; lower schematic test tube). Usually the excess oil forms the upper phase (2 therfore indicates the position of the microemulsion in the test tube). This is true for most oils since they have lower densities then water. Only for perchlorinated oils the excess oil will form the lower phase. Studies of the phase behavior are of course the basis for all investigations of microstructures, which will be described in the following sections.

Microemulsions made of the C_iE_j surfactants are not well suited for the use in decontamination media. This is due to the strong temperature dependence of the phase behavior. As already mentioned in the introduction, for decontamination applications it is desirable to have a system with no temperature dependence over a wide range of temperatures.

However, all the above is also true for other surfactant systems. Especially ionic and sugar surfactants do not exhibit a similar temperature dependence of the packing parameter. In these systems the packing and therefore the natural curvature of the surfactant film can be controlled by co-surfactants. These co-surfactants often are alcohols and due to their slight amphiphilic character they are incorporated into the surfactant film between oil and water. This leads to a change in curvature. Hence, for these systems also fish diagrams can be obtained by plotting the alcohol content instead of the temperature. This type of systems is of special interest with respect to decontamination applications, since temperature dependence is unwanted for this special purpose.

Figure 12.6 shows an experimentally obtained phased diagram for the system perchlorethylen/IHF/isopropanol/water. Marlowet IHF is a technical surfactant mixture containing ionic and non-ionic amphiphiles and is currently used for the preparation of macroscopic decontamination emulsions by the German armed forces ("German emulsion"). However, using the addition of 2-propanol with this rather complex emulsifier it is possible to obtain stable microemulsions.

In Figure 12.6 a "fish"-like diagram is shown, which is obtained using 2-propanol to tune the curvature of the surfactant film. The right corner of the diagram is unfortunately not accessible since IHF already contains some isopropanol. Nevertheless all features which are shown in the schematic diagram in Figure 12.4 are reproduced and all microemulsion phases can be found in the fishtail. The shown diagram was obtained at a constant water to oil ratio of $\alpha = \text{oil}/(\text{oil} + \text{water}) = 0.62$. This value of α corresponds to equal volumes of water and oil.

Some microemulsion phases found in this system have already been tested with respect to their decontamination properties (see last section). The identified L3-phases will serve as a model system to discuss the structural features of these phases in general (see Section 12.4).

Two more examples for microemulsions which are promising candidates for decontamination systems were studied at the DSTL. Initial studies totalling some

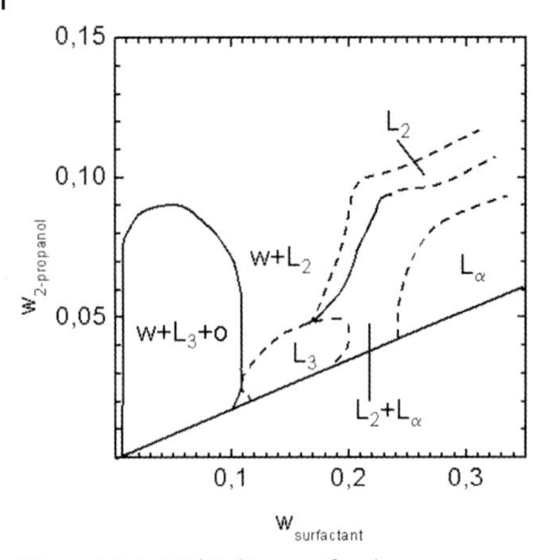

Figure 12.6 "Fish" diagram for the system perchlorethylen/Marlowet IHF/isopropanol/water. The diagram was obtained preparing approximately 200 samples with technical grade IHF but pure perchlorethylen. The lower right corner of the diagram is not accessible since the emulsifier Marlowet IHF already contains 2-propanol. However, further addition of 2-propanol allows to tune the curvature of the surfactant film at the oil water interface even for this technical grade emulsifier and all expected phases including the single phase microemulsions can be identified. L_2 is an oil-continuous droplet microemulsion; L_3 indicates the bicontinuous microemulsion; and L_α is the lamellar phase.

23 single and 97 mixed surfactant O/W microemulsions have identified two particularly promising microemulsion systems:

- The anionic surfactant system sodium dodecylsulfate (SDS), with 1-butanol as a co-surfactant (molecular ratio 1 : 1.68) formulated with toluene or cyclohexane as a disperse oil phase
- The non-ionic surfactant system Triton X100 formulated with toluene as a disperse phase.

(Pseudo)ternary phase triangles for these two microemulsion systems prepared with a toluene oil phase are given as Figures 12.7 and 12.8.

12.4 Investigation of the Microemulsion Structure

12.4.1 Small-Angle Elastic and Quasielastic Scattering: Theory

The microstructures in the different microemulsion phases (L_1, L_2, and L_3) can be studied by employing several experimental techniques like, for example, cryogenic transmission electron microscopy, NMR self diffusion experiments or measurements of conductivity. However, especially in the last 15 years, scattering

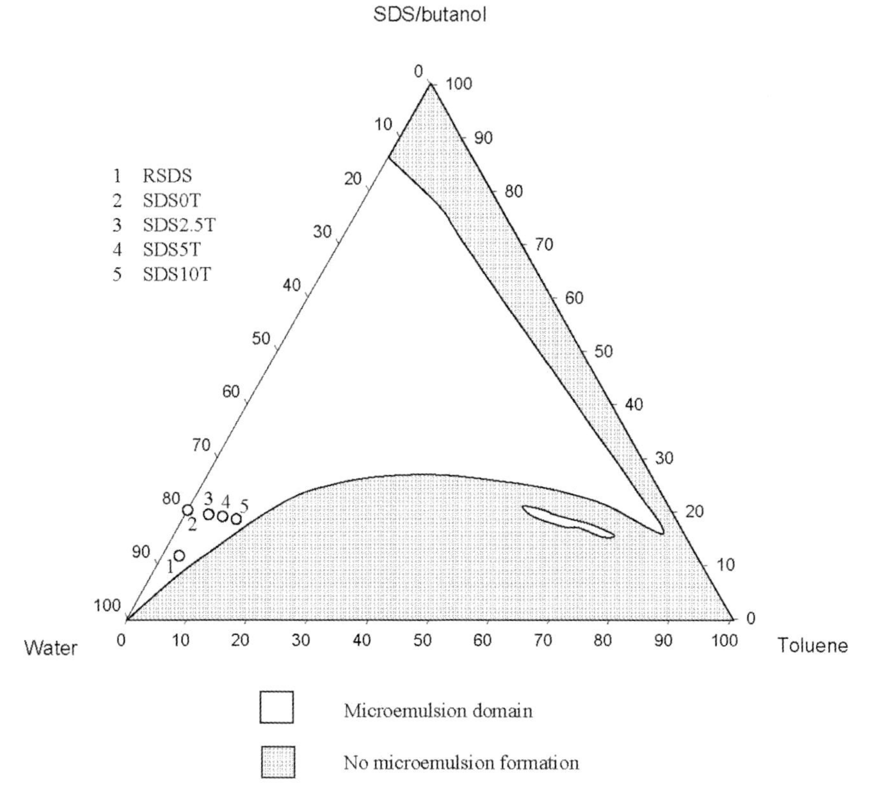

Figure 12.7 Pseudo-ternary phase diagram for water, sodium dodecylsulfate, 1-butanol and toluene at 25°C with ratio of SDS/n-butanol at 1/1.68. Formulations 1–5 were selected for further study.

techniques led to a better understanding of microemulsion and for example, light scattering is available in most laboratories working on colloids. This is why, in the following, only an introduction to the application of neutron scattering techniques for the study of microemulsions will be given.

12.4.1.1 Elastic Small Angle Neutron Scattering (SANS)

The total small angle scattering intensity obtained from a solution of objects with a rather low polydispersity is given by

$$I(q) = NS(q)\int_0^\infty G(R)P(q, R)dR. \tag{12.3}$$

Here, $P(q,R)$ is the particle form factor, $S(q)$ the static inter-particle structure factor and N the number density describing the particle concentration. $G(R)$ is the size distribution of the particles an can be described by a Gaussian [7, 21] or a Schultz distribution function [22].

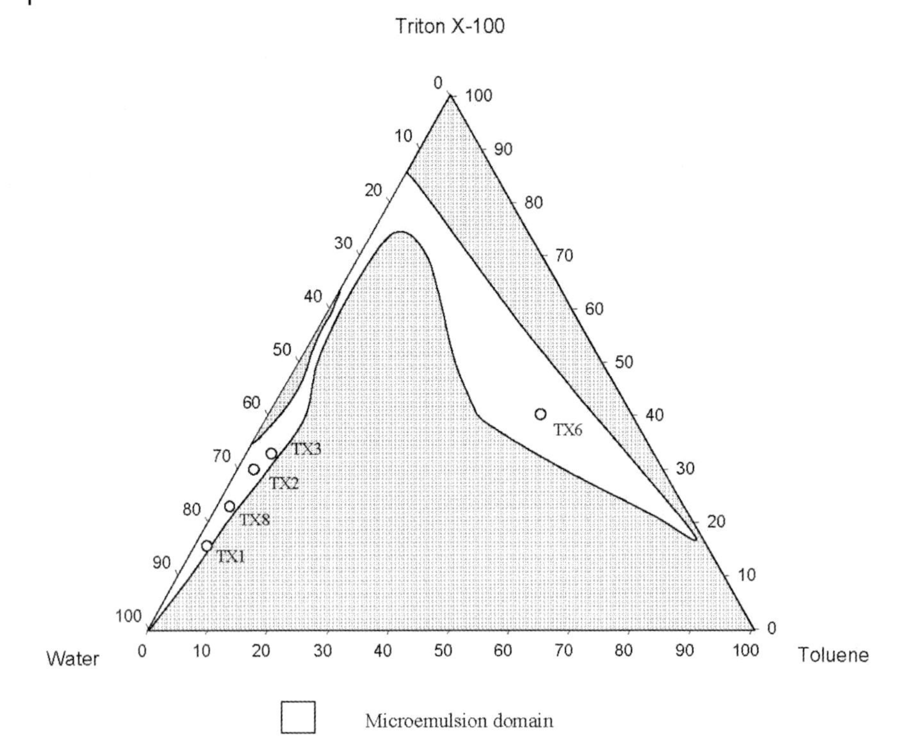

Figure 12.8 Ternary phase diagram for water, Triton X-100 and toluene at 25°C. Formulations TX1–8 were selected for further study.

The particle form factor, $P(q,R)$, for monodisperse thin shells (surfactant film of the microemulsion droplet in shell contrast; oil and water have to have the same scattering length density) can be described by

$$P(q) = 16\pi^2(\rho_{int} - \rho_{ext})^2 \left[R_o^3 p_0(qR_o) - \frac{\rho_{shell} - \rho_{int}}{\rho_{shell} - \rho_{ext}} R_i^3 p_0(qR_i) \right]^2, \qquad (12.4)$$

with the following scattering length densities: ρ_{int} particle core; ρ_{ext} solvent; and ρ_{shell} particle shell. R_i is the inner radius and R_o the outer radius of the shell. q is the magnitude of the scattering vector and equal to $(4\pi/\lambda)\sin(\theta/2)$. The used distribution function, $G(R)$, is given by a Schultz distribution. With this description the R in Equation 12.3 becomes $R = (R_i + R_o)/2$ and the shell thickness, D is simply equal to $R_o - R_i$. For more details see ref. [22]. For low concentrations $S(q) = 1$ is valid, but for higher concentrations this assumption might no longer hold. Knowing the polydispersity index, p, which is obtained as a parameter from the distribution function, $G(R)$, it is possible to derive

$$p^2 = \frac{\langle |u_0|^2 \rangle}{4\pi} = \frac{kT}{8\pi(2\kappa + \bar{\kappa}) + 2kT(\ln\phi - 1)}. \tag{12.5}$$

The term $2kT(\ln\phi - 1)$ accounts for the mixing entropy contribution to the free energy. The quantity $\langle |u_0|^2 \rangle$ arises from the description of the droplet dynamics using spherical harmonics [22–25] and is the amplitude of the mode with $l = 0$. The amplitude of the mode with $l = 2$ is given by

$$\langle |u_2|^2 \rangle = \frac{kT}{16\kappa - 12\bar{\kappa} - \frac{3kT}{4\pi}(\ln\phi - 1)}, \tag{12.6}$$

at emulsification failure.

12.4.1.2 Quasielastic Neutron Scattering: The Intermediate Scattering Function for Shells

Neutron spin-echo (NSE) is a quasielastic neutron scattering technique providing direct knowledge of the intermediate scattering function and therefore, about the dynamics in the scattering sample [26–28]. For scattering shells, as they can be obtained preparing microemulsions using deuterated oil and deuterated water, the intermediate scattering function was derived by Milner and Safran [23, 24, 29] based on the already mentioned spherical harmonics approach and reads like

$$I(q,t) = \left\langle \exp(-\Gamma_1 t) V_s^2 (\Delta\rho)^2 \left[f_0(qR) + \sum_{l \geq 2} \frac{2l+1}{4\pi} f_l(qR) \langle |u_l|^2 \rangle \exp(-\Gamma_l t) \right] \right\rangle_R, \tag{12.7}$$

which is a sum of at least two exponential decays if terms corresponding to $l > 2$ are omitted. The first term represents the translational motion of the particles, the second represents the shape deformations of the particles. Γ_l is the relaxation rate of the corresponding mode. The subscript R indicates an average over the radii distribution of all droplets in the scattering volume. $f_0(qR)$ and $f_l(qR)$ in Equation 12.7 are given by

$$f_0(qR) = j_0(qr)^2 + j_0(qR)((4 - qR) j_0(qR) - 2qR j_1(qR)) \sum_{l>2} \frac{2l+1}{4\pi} \langle |u_l|^2 \rangle \tag{12.8}$$

and

$$f_l(qR) = [(l+2) j_l(qR) - qR j_{l+1}(qR)]^2. \tag{12.9}$$

j_n are the Bessel functions of order n. The peanut-like deformations of the droplets correspond to the modes with $l = 2$ and their relaxation time can be used to calculate the elastic constants applying [23–25, 30]

$$\Gamma_2 = \tau_2^{-1} = \frac{1}{\eta R^3}\left[4\kappa - \bar{\kappa} - \frac{kT(\ln\phi - 1)}{4\pi}\right]\frac{1}{Z(2)}. \tag{12.10}$$

ϕ is the volume fraction of the dispersed phase, η is the viscosity of the continuous phase and $Z(2)$ a hydrodynamic factor that depends on the ratio E of the viscosities of the used oil and the continuous phase (e.g. water). This factor was calculated by Komura and Seki for all values of l [31] and was found to be equal to

$$Z(l) = \frac{[(2l^2 + 4l + 3)E + 2l(l+2)][2(l^2 - 1)E + 2l^2 + 1]}{(l-1)l(l+1)(l+2)(2l+1)(E+1)}. \tag{12.11}$$

Here, E is the ratio of the two viscosities. In previous approaches [23, 24, 30, 32, 33] E was taken to be equal to 1. In practice E is different from 1 and in the case of n-octane for instance $Z(2)$ differs by about 20% from its value for $E = 1$. When the early description is used, the corresponding error on $4\kappa - \bar{\kappa}$ can lead to a significantly different value of $\bar{\kappa}$ at the end.

Applying the equations above it is now possible to calculate a value for the sum $4\kappa - \bar{\kappa}$. Knowing the polydispersity of the particles leads to a second sum $(2\kappa + \bar{\kappa})$ containing the two unknowns (see Equation 12.5) [22, 23] which finally can be used to compute both constants. For more details see reference [34].

12.4.2 "Droplet"-Structures

12.4.2.1 L_1-phases

12.4.2.1.1 Spherical Structures

At present most of the studies on microemulsions still focus on water continuous systems. This is certainly due to the fact that most applications are concerned with the problem of solubilization of oil in water. Also, with respect to decontamination processes, water continuous microemulsions have interesting properties. Water is cheap and available almost everywhere. In these L_1-phases spherical droplet structures are only found close to the so-called emulsification failure boundary (see Figure 12.2). At this phase boundary the microstructures reach the limit of maximum swelling and beyond an excess oil-phase occurs (**2**). The process of oil desorption, which occurs when such a phase separated system is heated above the phase transition temperature was recently investigated [35, 36]. In these studies two different regimes were identified which govern the kinetics of the desorption process. One regime corresponds to diffusion of monomers and the second one to the direct exchange of oil between large and small droplets. The phenomena can be explained in terms of curvature free energy [19].

Concerning the determination of bending elastic constants for surfactant films, which are together with the curvature the most important parameters to describe microemulsions, several interesting works were published recently. One dealing with the determination of κ and $\bar{\kappa}$ by means of measurements of the interfacial tension applying the spinning drop method contains information about a multitude of surfactants of the type C_iE_j [37] is of special interest (besides macroscopic methods like, for example, the spinning drop method, interfacial tension can also be obtained from x-ray [38] or neutron reflectivity). There was also some progress in the determination of the two bending elastic constants from direct measurements of the dynamics of the surfactant film of this type of surfactant measured by means of neutron spin-echo spectroscopy (NSE) [34, 39]. Another NSE study is investigating the influence of charges subsequently incorporated into the film surrounding the droplets. This is done by adding the cationic surfactant tetradecyltrimethylammonium bromide (TTABr) to the non-ionic surfactant tetradecyldimethylamine oxide (TDMAO). The surprising result is that introducing charges does not change the bending elastic constants κ and $\bar{\kappa}$ [40] significantly. Moreover, changes of the ionic strength have almost no influence on the film dynamics. However, the theoretical description of the measured intermediate scattering functions still seems to be controversial [41].

Also a study by Evilevitch and co-workers [42] uses the concept of comicellization of a non-ionic surfactant ($C_{12}E_5$) with a charged one (SDS). In this study, the authors do not investigate the film elasticity as function of charge, but the interaction between charged hard spheres. Comicellization with SDS offers an easy way to control the number of surface charges. The authors report that the obtained droplets (internal phase made of n-decane) do not behave like simple hard spheres, but show the pronounced influence of the charge (quantified by means of dynamic light scattering (DLS)).

A novel approach to analyzing the small angle neutron scattering from droplets and to extracting information about the bending elasticity was also published recently [43]. In this work, AOT (bis(2-ethylhexyl)sulfosuccinate sodium salt) based microemulsions were studied by means of contrast variation using deuterated oil and mixtures of light and heavy water. Also concerned with the analysis of small angle data including influences from the interparticle structure factor, is the work by Brunner-Popela and co-workers [44]. Here, the newly developed GIFT method (based on IFT [45]) is applied to extract size parameters from SANS data. This method is applied to spherical as well as non-spherical structures (for general information on non-spherical strutures in microemulsion L_1-phases see also next sub-section). Besides the already mentioned surfactants, sugar surfactants (C_iG_j) are also of growing interest with respect to microemulsion preparation. This is especially true when thinking about better environmental compatibility. Sugar surfactants can be used to prepare microemulsions, which do not exhibit a strong temperature dependence of the phase behavior. This makes these amphiphiles interesting for deconatmination applications, too. However, articles treating these systems were reviewed not long ago [46] and are therefore not discussed here.

Recently, the so-called "efficency boosting effect" was discovered [47]. This is an effect caused by the addition of small amounts of amphiphilic block-copolymers to the microemulsion forming system, leading to a large increase of the solubilization power of the surfactant. Also, due to this finding, microemulsion phases containing block-copolymers and surfactant are of growing interest [48, 49]. Currently, several groups are trying to investigate the interaction between block-copolymer and surfactant, but most of this work focuses on pseudo-binary systems, not containing oils [50–53]. However, droplet structures containing oil or water will also certainly be investigated in the future, because these structures should allow for a determination of the bending elastic constants as a function of polymer or surfactant concentration. With respect to decontamination applications, the efficiency boosting effect remains to be exploited in the future. Furthermore, the influence of polyelectrolytes on microemulsions is also to be investigated [54].

12.4.2.1.2 Non-spherical Structures

For the sake of completeness it should be mentioned here that the phase structures in a microemulsion are droplets only in very small regions of the phase diagram. Even in the single phase regions non-spherical structures are often found. For example, close to the so called "haze point" a recent theoretical approach predicted the existence of a polymer-like entangled network to explain the inversion of the curvature of the film during the phase transition [55]. Support for these proposed structures came also from Monte Carlo simulations [56] and meanwhile, there is also experimental evidence for their existence. The methods used to prove these structures were cryo transmission electron microscopy (cryo-TEM) [57], dynamic light scattering [58] and NMR relaxation experiments [59].

12.4.2.2 L_2-phases

Most of the chemical warfare agents are hydrophobic and therefore solubilization might be enhanced using oil-continuous microemulsions, the so-called L_2-phases. In addition, oil-continuous microemulsions are also of great interest with respect to the preparation of inorganic nanoparticles [11–14]. Despite this potential with respect to applications, studies of the physical properties of the surfactant film in these structures are rare compared to the studies on L_1-phases. Especially, due to the growing number of groups working in the area of nanoparticle synthesis, the situation seems to change and there is growing interest in the investigation of oil-continuous areas in phase diagrams and of physical properties of these oil-continuous systems. The only study directly investigating the film dynamics of a L_2-phase ($C_{10}E_4$ based) reveals that the bending elastic constants seem to be the same as on the water continuous side of the phase diagram [60]. On the other hand, this study also shows that at least the polydispersity of the droplet structures is apparently higher compared to water continuous microemulsions of the same surfactant.

Most studies on oil-continuous microemulsions use the anionic surfactant AOT, because the preferred curvature of the film is towards the oil for this amphiphile (due to the bulky carbon tails) [61–66]. Nave and co-workers have compared AOT to a series of analog compounds [67]. Based on these works, AOT is not a better microemulsifier compared to the other structures, but combines an appropriate molecular geometry with the appropriate solvent interaction parameters and is, therefore, able to solubilize rather large amounts of water in oil within a temperature range of interest for applications.

Furthermore, pressure dependent phase transitions were investigated by scattering methods and it was shown that L_2-phases made of AOT/n-decane and D_2O can undergo a pressure-induced droplet to lamellar phase transition [68]. In Section 12.2.2 another example for oil continuous microemulsions can be found. Figure 12.6 shows the phase diagram for the technical surfactant Marlowet IHF. An oil continuous L_2-phase can clearly be identified.

12.4.2.3 L_3-Phases ("Sponge"-phases)

The L_3-phase is an isotropic solution with a microstructure consisting of a multiply connected ("sponge"-like) 3D bi-layer [69]. The microstructure of L_3-phases can be studied by elastic small angle neutron scattering. Figure 12.9 shows an example of the SANS intensity distribution I(q) from a sponge structure. As mentioned above q is the magnitude of the scattering vector. The curve was measured with a sample from the quaternary Marlowet IHF based system described in Figure 12.6. The shape of the curve is typical for bulk contrast measurements and already an evidence for the bicontinuous structure. The curve consists of a broad peak and a continuous decrease of the scattered intensity with increasing q values. In bulk contrast measurements a large difference between the scattering length densities of water and oil is usually prepared by using deuterated water. Then, the periodicity of the two subdomains oil and water is measured in the scattering experiment.

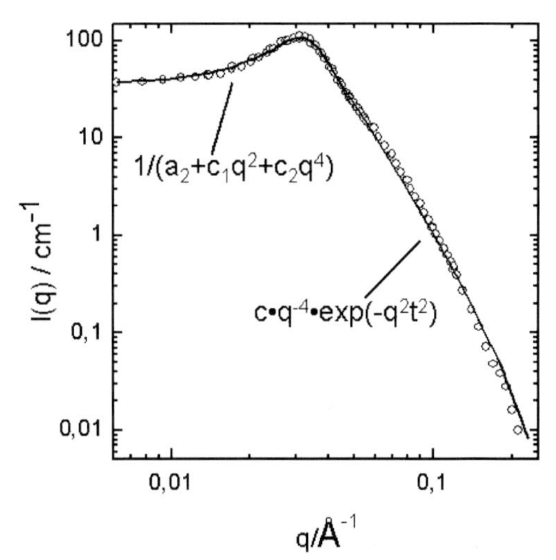

Figure 12.9 Background subtracted bulk contrast SANS intensity distribution from a bicontinuous microemulsion from the PCE/water/IHF/2-propanol system measured at 25 °C. The oil-to-water ratio is 0.5 at a IHF content of 16.6 wt% and a 2-propanol content of 4.6 wt% in the given example. The open circles are data points and the solid lines are results of fitting the data with the Teubner-Strey formula in the low q-range and the modified Porod law in the higher q-range. As a result one finds d = 191 Å, ξ = 96 Å and an effective thickness of the surfactant layer δ = (16 ± 2) Å For details see the text below.

Additionally, using deuterated oils allows for adjusting equal scattering length densities of water and oil which is the so-called interface or film contrast where the periodicity of the surfactant film is measured.

It is very common to analyze the measured spectra by using the Teubner-Strey formula [70]. Here, the scattered intensity I(q) is described by

$$I(q) = \frac{8\pi c_2 \langle \eta^2 \rangle / \xi}{a_2 + c_1 q^2 + c_2 q^4} \tag{12.12}$$

where $\langle \eta^2 \rangle = \phi_a \phi_b (\Delta\rho)^2$ and $(\Delta\rho)$ is the difference of the scattering length densities between media a and b while ϕ is the volume fraction. If the incoherent background is not subtracted an additional baseline has to be taken into account. The parameters a2, c1 and c2 are fitting parameters. The bicontinuous structure can be described by two length scales d and ξ. While d describes the averaged repeat distance of the oil and water domains the correlation length ξ measures the dispersion of d. Both length parameters are given in terms of the Teubner-Strey fitting parameters

$$\xi = \left[\frac{1}{2}\left(\frac{a_2}{c_2}\right)^{1/2} + \frac{c_1}{4c_2}\right]^{-1/2} \quad \text{and} \quad d = 2\pi\left[\frac{1}{2}\left(\frac{a_2}{c_2}\right)^{1/2} - \frac{c_1}{4c_2}\right]^{-1/2}. \tag{12.13}$$

With the Teubner-Strey formula especially the intensity maximum can be described precisely. For the intensity decrease at higher q-values according to the q^{-4} – Porod law other models have to be included. In the case of bulk contrast

$$I(q) = cq^{-4} e^{-q^2 t^2} + b \tag{12.14}$$

gives the large-q part of the curves. From the parameter c the size of the internal surface between oil and water can be extracted and t gives the effective thickness δ of the surfactant interface layer by $\delta = (2\pi t^2)^{1/2}$ [71].

Further methods of analyzing bulk contrast measurements of bicontinuous structures are published in the literature [72]. For example, the modified Berk's random wave model uses an inverse sixth-order polynomial with three parameters to calculate the mean Gaussian curvature of the surfactant film.

Currently, there is growing interest in the local dynamics of bi-layers in L_3-phases. Firstly, this is due to the recent new theoretical description of the dynamics of locally flat bi-layers by Granek and Zilman [73]. Secondly, this is due to the increasing number of available neutron spin-echo spectrometers. Neutron spin-echo allows measurement of the local motions of the layers in the validity limit of the Zilman-Granek theory. As already mentioned, neutron spin-echo spectroscopy is an ideal tool for the investigation of local dynamics in microemulsions. The Zilman-Granek approach leads to a rather simple form for the intermediate scattering function of a "powder" of bi-layers, given by

$$I(q,t)/I(q,t=0) = \exp\left[-(\Gamma t)^{\frac{2}{3}}\right]. \tag{12.15}$$

In this equation Γ is the relaxation rate. For Γ a dispersion relation in q was derived, which predicts an apparent linear q^3 dependence similar to Zimm polymer dynamics.

$$\Gamma = D_{eff} q^3 \tag{12.16}$$

The effective diffusion coefficient D_{eff} is related to the bending modulus κ by the relation

$$D_{eff} = \varepsilon (k_B T)^{\frac{3}{2}} \frac{1}{\kappa^{\frac{1}{2}}} \frac{1}{\eta}, \tag{12.17}$$

were $\varepsilon = 0.025$ for bi-layers. k_b is the Boltzmann constant, T the temperature and η the mean viscosity of oil and water.

Takeda and co-workers were the first using this theoretical description to analyze intermediate scattering functions obtained by NSE experiments [74–76]. Richter and co-workers applied the Zilman-Granek approach to an L_3-phase containing nonionic surfactant and block-copolymers [77]. The method is not only useful to describe NSE data. It has applied to dynamic light scattering data of a sponge phase with large ξ [78]. Also, mixtures of block-copolymers with other polymers may exhibit microemulsion structures [79].

Besides polymers, mixtures of a cationic and non-ionic surfactant also lead to changes in the structure of L_3-phases [80]. Besides the different equilibrium experiments, it is also possible to investigate this type of phases using p- and T-jump techniques. To investigate spinodal lines in L_3-phases for example, T-jump measurements combined with light scattering detection were used recently [81, 82].

12.4.2.4 **Lamellar Phases**

Lamellar phases are not strictly microemulsions, but are strongly related to these. Their formation is of course governed by the same driving forces (bending elastic free energy; curvature) as the other phases mentioned before. These structures were recently reviewed by Dubois and Zemb [83] and also, by Kötz and Kosmella with respect to polyelectrolyte containing systems [84]. Therefore, here, only one publication should be mentioned which indicates that in the $C_{12}E_5/D_2O$/n-octane system the macroscopically lamellar phase might have a different microstructure [85]. Usually, lamellar phases are strongly related to L_3-phases and might also be useful carriers for decontaminating compounds. Compared to the microemulsion phases, they have higher viscosities. This might be advantageous when a longer persistence time of the decontaminant on a vertical surface is required. However, these phases have not yet been investigated with respect to their application as carriers.

12.5 Application Related Results

12.5.1 Decontamination Experiments

12.5.1.1 The Marlowet IHF-System

The German Army used an emulsion system for decontamination purposes under the name German Emulsion (C8). This w/o macroemulsion is composed, by weight, of 15% tetrachloroethylene, 76% water, 1% anionic surfactant, and 8% $Ca(OCl)_2$. It was demonstrated that the macroemulsion is noncorrosive due to the oil contineous phase and a solvent for thickened agents. In addition, another advantage of the German Emulsion is that it can penetrate non-chemical-resistant paint to extract the embedded agent without damaging the paint. After spraying the emulsion on the surface, the thin coherent film formed gives sufficient residence time for the extraction of the agents and their conversion into harmless products by $Ca(OCl)_2$ [87]. However, due to the fact that the German Emulsion has some disadvantages, as the mixture of the different components has been done immediately for the use and is therefore time- and personnel-consuming, and the high water hazard class of the component tetrachlorethylene, investigations were started to improve the formulation. In the first experiments, with the aid of a short chain alcohol it was possible to change the German Emulsion into a German Micro Emulsion under conservation of the decontamination properties (see Figure 12.6). This improvement would reduce the time- and personnel-consuming factor. However, to improve the formulation to meet environmental legal stipulations, further investigations are necessary to replace the harmful components of the German Emulsion with other less harmful components. [86]

12.5.1.2 The SDS-System

The SDS/1-butanol system (see phase diagram in Figure 12.7) in particular shows great promise for use as a microemulsion decontaminant. The large microemulsion phase window gives a wide range of possible formulations. The microemulsion forms rapidly and spontaneously on combination of the components. Formulations within the microemulsion phase window remain stable for many months. These microemulsions exhibit remarkable temperature stability. For example, a variant formulated with a cyclohexane oil phase at 2.5% by volume remained stable throughout the temperature range studied (–12 to 50 °C). This microemulsion has also demonstrated a good capacity for the solubilization of sulfur mustard, incorporating up to 3 weight%. When formulated with toluene as the oil-phase, excellent uptake of thickened sulfur mustard is exhibited. The formulation also remains stable to the addition of militarily realistic quantities of ionic reagents.

The SDS/1-butanol microemulsion system has provided a flexible, well defined classical microemulsion for further study, in which a large number of potential decontamination chemistries have been screened. A detailed fundamental study of the phase behavior of decontamination reaction reagents, products and catalysts within a micro-heterogeneous environment has also been conducted using this model system [to be published].

12.5.1.3 The Triton X-100-System

The Triton X100 system (see Figure 12.8) did not quite exhibit the extreme stability demonstrated by the SDS/1-butanol system. It was also not as effective at solubilizing thickened HD. However, it provided a useful militarily practical non-ionic classical microemulsion for study.

However, the use of hydrocarbon oil phases in these formulations, whilst effective, raises environmental concerns. It has proved possible to reformulate both the SDS/1-butanol and Triton X-100 systems using more environmentally benign oil phases, but these systems have so far failed to match up in terms of solubilization performance or stability when compared to the original versions.

12.5.1.4 Enzymatic Decontamination of Nerve Agents in Microemsulsions

The degradation of soman (GD) was studied in the presence of the enzyme Diisopropylfluorophosphatase (DFPase) in various microemulsions. The aim of these investigations was the identification and characterization of microemulsions as carrier media for enzymes as the active ingredient for decontamination purposes. In complex mixtures like microemulsions quantitative Fourier transform infra-red (FTIR) spectroscopy proved to be a useful tool for in-situ process development studies. The analysis of the decontamination of G-agents was based on real-time monitoring of the characteristic absorption of functional groups present in both the reactants and products (Figure 12.9). The reaction kinetics were monitored by following the absorbance of the P = O stretching band of the G-agents and their resulting acids produced by the hydrolysis reaction. After establishing this method for following the hydrolysis of G-agents by FTIR different microemulsions were tested as possible carriers for enzymes for decontamination purposes. Designed for the physical removal (solubilization) of different chemical warfare agents, one of the microemulsions investigated showed promising results as a carrier medium for enzymes. The microemulsion was buffered with 200 mM (NH_4HCO_3) at pH 8 to neutralize the protons produced during the hydrolysis reaction. In the hydrolysis scheme (18) R means an alkyl group, R′ an alkyl, alkoxy or dialkylamino group and X either fluorine or cyanide group.

$$\mathrm{RO{-}P(=O)(R')\text{-}X} \;\underset{}{\overset{H_2O \;\;\; 2H^+\;X^-}{\rightleftharpoons}}\; \mathrm{RO{-}P(=O)(R')\text{-}O^-} \qquad (12.18)$$

100 units of DFPase were added to the microemulsion. One unit means the conversion of one micromole soman transformed per minute by the enzyme determined in aqueous solution. Without DFPase, soman was stable throughout the whole period of time (P = O stretch in Figure 12.10a). With DFPase, soman was

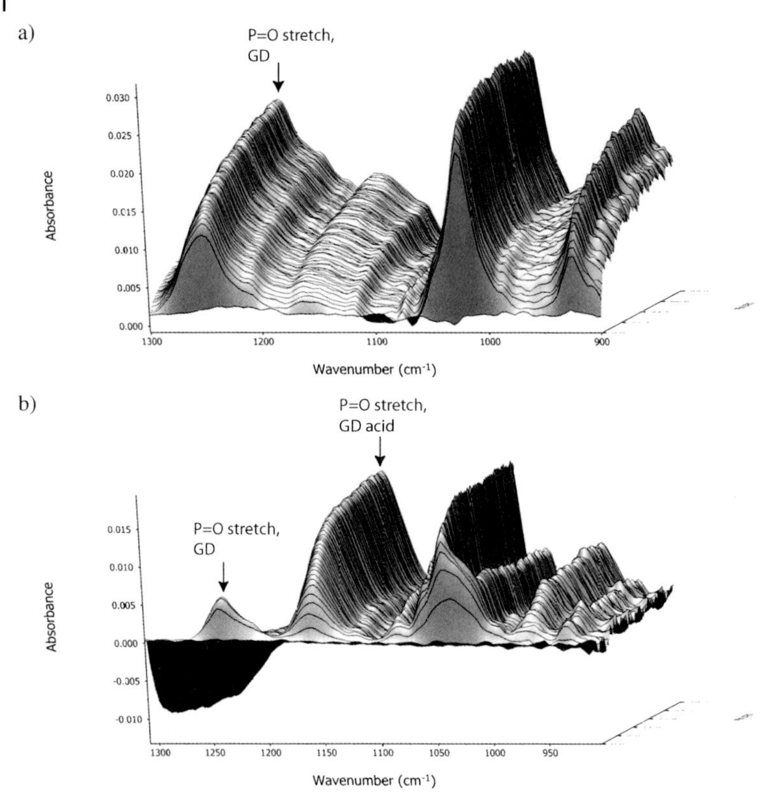

Figure 12.10 Typical absorbance versus wavenumber versus time reaction profile for the hydrolysis of 1% (v/v) Soman in 20 ml microemulsion. Solution buffered with 200 mM (NH_4HCO_3), pH 8.00, 25°C. (a) without enzyme. (b) with 100 units enzyme.

hydrolysed rapidly to the corresponding somanacid (pinacolyl methylphosphonic acid (Figure 12.10b).

In the presence of the enzyme, the whole amount of soman was hydrolyzed in time. One of the next key questions was the comparison of enzyme catalysed hydrolysis and perhydrolysis. The outcome of these investigations could influence the discussion, if an enzyme based technical decontamination system is able to outperform a chemistry based decontamination. Therefore, the same microemulsion as the carrier system was chosen. Phthalamidoperhexanoic acid PAP was identified as an effective G-agent decontaminant, demonstrating rapid conversion of soman to somanacid by a perhydrolysis mechanism and was therefore, the competitor for DFPase. The appearance of somanacid was monitored by FTIR. By increasing the amount of enzyme (2000 units, 10 mg), the rate of hydrolysis of soman increased to become very rapid, and higher than achieved by using a stoichiometric excess of a chemical hydrolytic species such as neutralized peracid (PAP) (Figure 12.11).

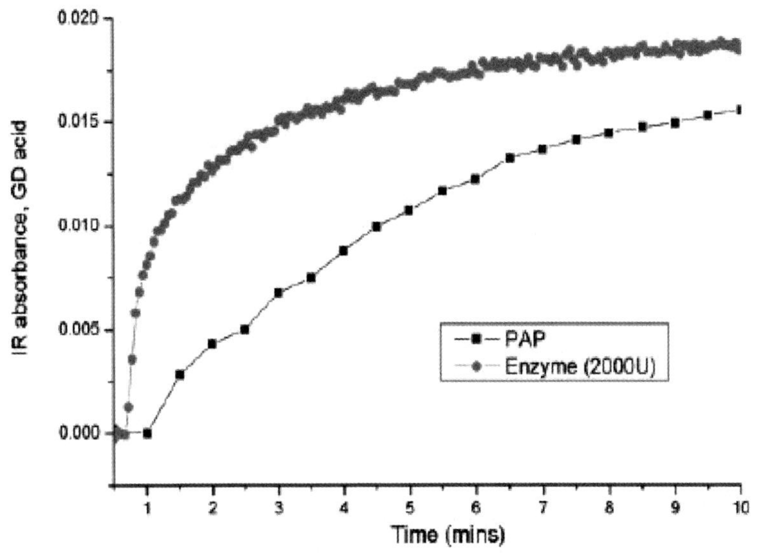

Figure 12.11 Comparison of chemical conversion of Soman in a stoichiometric process by phthalamidoperhexanoic acid (PAP) and by DFPase in a catalytic process. PAP: 1 g in 20 ml microemulsion puffered. Enzyme (DFPase) 2000 units DFPase measured against DFP in an aqueous solution

The results show clearly that an enzyme based conversion of chemical warfare agents can compete with normal chemical based conversion.

12.5.2 Wetting Properties

Beside the appropriate formulation of the microemulsion further aspects like the ability to wet a contaminated surface are of great importance for the choice of a decontamination medium. The maximum contact area between the liquid and the solid is necessary for a rapid diffusional transfer of the CWA into the oil phase of the microemulsion, since real surfaces are heterogeneous in surface structure and chemical composition. Therefore, a decontamination medium has to wet hydrophobic as well as hydrophilic surfaces. A common method to investigate the wetting properties of a liquid uses Young's equation to determine the contact angle θ between liquid and solid surface. The cosine of this angle $\cos\theta = (\sigma_s - \gamma_{sl})/\sigma_l$ is the so called spreading coefficient S, determined by the difference between the surface tension of the solid σ_s and the liquid-solid interfacial tension γ_{sl} divided by the liquids surface tension σ_l. The vector representation of these tensions and the resulting equilibrium situation at the three phase contact point is schematically represented in Figure 12.12a. A positive value of S indicates a situation where the liquid is wetting the solid while at negative values the liquid is not wetting. A video

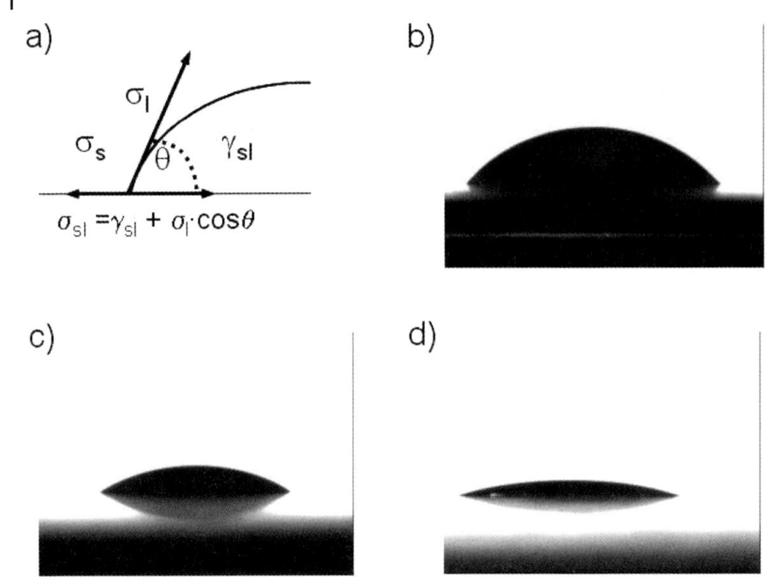

Figure 12.12 (a) The surface and interface tensions at the three phase contact point at the sessile drop. The images (b–d) show the back lightned contour of sessile droplets of water (b) and two microemulsions (c, d) of different water contents. See text for further details.

based contact angle meter can be used to determine contact angles by analysing the contour of homogeneous back lightened droplets deposited on a test surface.

Figure 12.12b–d show an example of three different droplets at a chemically non resistant alkyd paint coated steel sheet surface as it is for example typical for vehicles or technical equipment.

From pure water, picture (b), with a measured surface tension of $\sigma_{l,b} = 71.6\,\text{mJ/m}^2$ a contact angle of $\theta_b = (50 \pm 8)°$ and $S_b = 0.64$ are obtained. The given values are the result of 30 measurements, 3 at different positions of each of 10 test samples. The contact angles of bicontinuous microemulsions of 10 wt% added 2-propanol, an emulsifier-to-PCE ratio of 0.29 and a water content of 40 wt% (picture c) and 32 wt% (picture d) were measured in the same way. The resulting values are $\theta_c = (33 \pm 7)°$ at a surface tension of $\sigma_{l,c} = 27.6\,\text{mJ/m}^2$ and $\theta_d = (12 \pm 3)°$ at a surface tension of $\sigma_{l,d} = 27.0\,\text{mJ/m}^2$. The corresponding spreading coefficients are $S_c = 0.84$ and $S_d = 0.98$. As can be seen from these values, these alkyd painted surfaces are partially hydrophobic. As a result of this, the spreading of liquids of low surface tension is preferred and can be seen from the measured contact angles. Additionally, these values indicate an influence of the water content of the microemulsion on the wetting behavior.

Because real heterogeneous test surfaces influence the observed wetting behavior, a series of measurements at different positions at a single surface and the variation of different test surfaces is necessary to obtain reliable results. Although not all modifications during the lifetime of a surface like weathering, multilayers of paint or the influence of mechanical forces on the local surface structure can be experimentally taken into account, on the basis of the wetting experiments it can be concluded that the use of bicontinuous microemulsions is advantageous for applications where wetting is the crucial point. They are found to wet a large variety of different sample surfaces.

12.6 Conclusions

Microemulsions have shown their great potential as an environment-friendly decontamination media. It seems that micrcoemulsions are able to compromise between the different technical necessities for solvatation of different chemical warfare agents, extraction of entrapped chemicals in porous surfaces and the maintenance of the activity of different active substances like enzymes for a rapid and complete conversion of chemical warfare agents in less harmful substances. However, to achieve the ambitious stipulations for a technical decontamination system and for the use in the field, the existing microemulsions have to be improved and the hydrolysis reaction by enzymes has to be accelerated and stabilized.

References

1 Hoar, T.P. and Schulman, J.H. (1943) Transparent water-in-oil dispersions: the oleopathic hydro-micelle. *Nature (London)*, **102**, 152.

2 Schulman, J.H., Stoeckenius, W. and Prince, L.M. (1959) Mechanism of formation and structure of micro emulsions by electron microscopy. *The Journal of Physical Chemistry*, **63**, 1677–80.

3 Kahlweit, M. and Strey, R. (1985) Phasenverhalten ternärer systeme des typs H_2O-Öl-nichtionisches amphiphil (mikroemulsionen). *Angewandte Chemie*, **97**, 655–69.

4 Kahlweit, M. *et al.* (1987) How to study microemulsions. *Journal of Colloid and Interface Science*, **118** (2), 436–53.

5 Langevin, D. (1992) Micelles and microemulsions. *Annual Review of Physical Chemistry*, **43**, 341–69.

6 Hoffmann, H. (1994) Fascinating phenomena in surfactant chemistry. *Advanced Materials*, **6** (2), 116–29.

7 Strey, R. (1994) Microemulsion microstructure and interfacial curvature. *Progress in Colloid and Polymer Science*, **272**, 1005–19.

8 Strey, R. (1996) Phase behavior and interfacial curvature in water-oil-surfactant systems. *Current Opinion in Colloid and Interface Science*, **1**, 402–10.

9 Hellweg, T. (2002) Phase structures of microemulsions. *Current Opinion in Colloid and Interface Science*, **7**, 50–6. Review.

10 Hao, J. (2000) Effect of the structures of microemulsions on chemical reactions. *Progress in Colloid and Polymer Science*, **278**, 150–4.

11 Schmidt-Winkel, P., Glinka, C.J. and Stucky, D.D. (2000) Microemulsion templates for mesoporous silica. *Langmuir*, **16**, 356–61.

12 Bagwe, R.P. and Khilar, K.C. (2000) Effects of intermicellar exchange rate on the formation of silver nanoparticles in reverse microemulsions of aot. *Langmuir*, **16**, 905–10.

13 Lee, M.-H., Oh, S.-G. and Yi, S.C. (2000) Preparation of Eu-doped Y_2O_3 luminescent nanoparticles in nonionic reverse microemulsions. *Journal of Colloid and Interface Science*, **226** (1), 65–70.

14 Wu, M.-L., Dong, D.-H. and Huang, T.-C. (2001) Preparation of Pd/Part bimetallic

nanoparticles in water/aot/isooctane microemulsions. *Journal of Colloid and Interface Science*, **243** (1), 102–8.

15 Lade, O., Beizai, K., Sottmann, T. and Strey, R. (2000) Polymerizable nonionic microemulsions: Phase behavior of H_2O-n-alkyl methacrylate-n-alkyl poly (ethylene glycol) ether (C_iE_j). *Langmuir*, **16**, 4122–30.

16 Co, C.C., de Vries, R. and Kaler, E.W. (2001) Microemulsion polymerisation. 1. small-angle neutron scattering study of monomer partitioning. *Macromolecules*, **34**, 3224–32.

17 de Vries, R., Co, C.C. and Kaler, E.W. (2001) Microemulsion polymerization. 2. influence of monomer partitioning, termination, and diffusion limitations on polymerization kinetics. *Macromolecules*, **34**, 3233–44.

18 Co, C.C., Cotts, P., Burauer, S., de Vries, R. and Kaler, E.W. (2001) Microemulsion polymerization. 3. molecular weight and particle size distributions. *Macromolecules*, **34**, 3245–54.

19 Helfrich, W. (1973) Elastic properties of lipid bilayers: theory and possible experiments. *Zeitschrift fur Naturforschung* **28c**, 693–703.

20 Israelachvili, J.N. (1991) *Intermolecular and Surface Forces*, 2nd edn, Academic Press, London.

21 Gradzielski, M., Langevin, D., Magid, L. and Strey, R. (1995) Small-angle neutron scattering from diffuse interfaces. 2. polydisperse shells in water-n-alkane-CiEj microemulsions. *The Journal of Physical Chemistry*, **99**, 13232–8.

22 Gradzielski, M., Langevin, D. and Farago, B. (1996) Experimental investigation of the structure of nonionic microemulsions and their relation to the bending elasticity of the amphiphilic film. *Physical Review E*, **53** (4), 3900–19.

23 Milner, S.T. and Safran, S.A. (1987) Dynamical fluctuations of droplet microemulsions and vesicles. *Physical Review A*, **36** (9), 4371–9.

24 Safran, S.A. (1991) Saddle-splay modulus and the stability of spherical microemulsions. *Physical Review A*, **43** (6), 2903–4.

25 Farago, B., Monkenbusch, M., Goecking, K.D., Richter, D. and Huang, J.S. (1995) Dynamics of microemulsions as seen by neutron spin echo. *Physica B*, **213 & 214**, 712–7.

26 Mezei, F. (ed.) (1980) *Neutron Spin Echo, volume 124 of Lecture Notes in Physics*. Springer Verlag, Berlin. Prodeedings from Spin-echo meeting in Grenoble 1979.

27 Hayter, J.B. (1981) Quasielastic neutron spin-echo spectroscopy, in *Scattering Techniques Applied to Supramolecular and Nonequilibrium Systems* (eds S.-H. Chen, B. Chu and R. Nossal), Plenum Press, New York, pp. 3–33.

28 Farago, B. (1994) Dynamics and diffusion in macromelcules, colloids and microemulsions, in *HERCULES: Neutron and Synchrotron Radiation for Condensed Matter Studies. Vol. 3: Applications to Soft Condensed Matter and Biology*, Chapter 7 (eds J. Baruchel, J.-L. Hodeau, M.S. Lehmann, J.-R. Regnard, C. Schlenker), Springer Verlag, Berlin, pp. 93–116.

29 Safran, S.A. (1983) Fluctuations of spherical microemulsions. *The Journal of Chemical Physics*, **78** (4), 2073–6.

30 Huang, J.S., Milner, S.T., Farago, B. and Richter, D. (1987) Study of dynamics of microemulsion droplets by neutron spin-echo spectroscopy. *Physical Review Letters*, **59** (22), 2600–3.

31 Komura, S. and Seki, K. (1993) Dynamical fluctuations of spherically closed fluid membranes. *Physica A*, **192**, 27–46.

32 Farago, B., Richter, D., Huang, J.S., Safran, S.A. and Milner, S.T. (1990) Shape and size fluctuations of microemulsion droplets: the role of cosurfactant. *Physical Review Letters*, **65** (26), 3348–51.

33 Farago, B. (1996) Spin echo studies of microemulsions. *Physica B*, **226**, 51–5.

34 Hellweg, T. and Langevin, D. (1999) The dynamics in dodecane/$C_{10}E_5$/water microemulsions determined by time resolved scattering techniques. *Physica A*, **264**, 370–87.

35 Evilewitch, A., Olsson, U., Jönsson, B. and Wennerström, H. (2000) Kinetics of oil solubilization in microemulsion droplets. mechanism of oil transport. *Langmuir*, **16**, 8755–62.

36 Evilevitch, A., Jönsson, B., Olsson, U. and Wennerström, H. (2001) Molecular transport in a nonequilibrium droplet microemulsion system. *Langmuir*, **17**, 6893–904.

37 Sottmann, T. and Strey, R. (1997) Ultralow interfacial tensions in water-n-alkane-surfactant systems. *The Journal of Chemical Physics*, **106** (20), 8606–15.

38 Mitrinovic, D.M., Williams, S.M. and Schlossman, M.L. (2001) X-ray study of oil-microemulsion and oil–water interfaces in ternary amphiphilic systems. *Physical Review E*, **63** (2), 21601–11.

39 Hellweg, T., Gradzielski, M., Farago, B. and Langevin, D. (2001) Shape fluctuation of microemulsion droplets: a neutron spin-echo study. *Colloids and Surfaces B*, 159–69.183–185

40 Farago, B. and Gradzielski, M. (2001) The effect of the charge density of microemulsion droplets on the bending elasticity of their amphiphilic film. *The Journal of Chemical Physics*, **114** (22), 10105–22.

41 Lisy, V. and Brutovsky, B. (2000) Interpretation of static and dynamic neutron and light scattering from microemulsion droplets: effects of shape fluctuations. *Physical Review E*, **61** (4), 4045–53.

42 Evilevitch, A., Lobaskin, V., Olsson, U., Linse, P. and Schurtenberger, P. (2001) Structure and transport properties of a charged spherical microemulsion system. *Langmuir*, **17**, 1043–53.

43 Arleth, L. and Pedersen, J.S. (2001) Droplet polydispersity and shape fluctuations in aot [bis (2-ethylhexyl) sulfosuccinate sodium salt] microemulsions studied by contrast variation small-angle neutron scatering. *Physical Review E*, **63** (6), 61406–23.

44 Brunner, J., Mittelbach, R.R., Strey, R., Schubert, K.-V., Kaler, E.W. and Glatter, O. (1999) Small-angle scattering of interacting particles. iii. D_2O-$C_{12}E_5$ mixtures and microemulsions with n-octane. *The Journal of Chemical Physics*, **110** (21), 10623–32.

45 Glatter, O. (1977) Data evaluation in small angle scattering: calculation of the radial electron density distribution by means of indirect fourier transformation. *Acta Physica Austriaca*, **47**, 83–102.

46 Stubenrauch, C. (2001) Sugar surfactants–aggregation, interfacial, and adsorption phenomena. *Current Opinion in Colloid and Interface Science*, **6**, 160–70.

47 Jakobs, B., Sottmann, T., Strey, R., Allgaier, J., Willner, L. and Richter, D. (1999) Amphiphilic block copolymers as efficiency boosters for microemulsions. *Langmui*, **15**, 6707–11.

48 Endo, H., Allgaier, J., Gompper, G., Jakobs, B., Monkenbusch, M., Richter, D., Sottmann, T. and Strey, R. (2000) Membrane decoration by amphiphilic block copolymers in bicontinuous microemulsions. *Physical Review Letters*, **85** (1), 102–5.

49 Endo, H., Mihailescu, M., Monkenbusch, M., Allgaier, J., Gompper, G., Richter, D., Jakobs, B., Strey, R. and Grillo, I. (2001) Effect of amphiphilic block copolymers on the structure and phase behavior of oil-water-surfactant mixtures. *The Journal of Chemical Physics*, **115** (1), 580–600.

50 Hecht, E., Mortensen, K., Gradzielski, M. and Hoffmann, H. (1995) Interaction of aba block copolymers with ionic surfactants: influence on micellization and gelation. *The Journal of Physical Chemistry*, **99**, 4866–74.

51 Jean, B., Lee, L.-T. and Cabane, B. (2000) Interactions of sodium dodecyl sulfate with acrylamide–n-isopropyl-acrylamide statistical copolymer. *Colloid and Polymer Science*, **278**, 764–70.

52 Egger, H., Nordskog, A. and Lang, P. (2000) Shape transformation of poly (butadiene)-b-poly (ethyleneoxide) plus dtab compound micelles in aqueous solution. *Macromolecular Symposia*, **162**, 291–306.

53 Bronstein, L.M., Chernyshov, D.M., Vorontsov, E., Timofeeva, G.I., Dubrovina, L.V., Valetsky, P.M., Kazakov, S. and Khokhlov, A.R. (2001) Comicellization of polystyrene-block-poly (ethylene oxide) with cationic and anionic surfactants in aqueous solutions: Indications and limits. *The Journal of Physical Chemistry B*, **105**, 9077–82.

54 Beitz, T., Kötz, J., Wolf, G., Kleinpeter, E. and Friberg, S.E. (2001) Poly (n-vinyl-2-pyrrolidone) and 1-octyl-2-pyrrolidone modified ionic microemulsions. *Journal of

Colloid and Interface Science, **240** (1), 581–9.

55 Tlusty, T., Safran, S.A., Menes, R. and Strey, R. (1997) Scaling laws for microemulsions governed by spontaneous curvature. *Physical Review Letters*, **78** (13), 2616–9.

56 Nelson, P.H., Hatton, T.A. and Rutledge, G.C. (1999) Asymmetric growth in micelles containing oil. *The Journal of Chemical Physics*, **110** (19), 9673–80. Monte Carlo simulation study of linear growth of micelles upon addition of oil.

57 Bernhein, A., Tlusty, T., Safran, S.A. and Talmon, Y. (1999) Direct observation of phase separation in microemulsion networks. *Langmuir*, **15**, 5448–53.

58 Hellweg, T., von Klitzing, R. (2000) Structural changes and complex dynamics in the single phase region of a dodecane/$C_{12}E_5$/water microemulsion: a dynamic light scattering study. *Physica A*, **283** (3–4), 349–58.

59 Nyden, M., Söderman, O. and Hansson, P. (2001) Microemulsions in the didodecyldimethylammonium sulfate (bromide)/hydrocarbon/water system. microstructure and specific counterion effects. *Langmuir*, **17**, 6794–803.

60 Hellweg, T., Brûlet, A. and Sottmann, T. (2000) Dynamics in an oil-continuous droplet microemulsions as seen by quasielastic scattering techniques. *Physical Chemistry Chemical Physics*, **2** (22), 5168–74.

61 Velazquez, M.M., Valero, M. and Ortega, F. (2001) Light scattering and electrical conductivity studies of the aerosol ot toluene water-in-oil microemulsions. *The Journal of Physical Chemistry B*, **105**, 10163–8.

62 Zhou, N., Li, Q., Wu, J., Chen, J., Weng, S. and Xu, G. (2001) Spectroscopic characterization of solubilized water in reversed micelles and microemulsions: Sodium bis (2-ethylhexyl) sulfosuccinate and sodium bis (2-ethylhexyl) phosphate in n-heptane. *Langmuir*, **17**, 4505–9.

63 Laia, C.A.T., Brown, W., Almgren, M. and Costa, S.M.B. (2000) Temperature and composition dependence of the structure of isooctane/aot microemulsion L_2-phases with glycerol and formamide: a light scattering study. *Langmuir*, **16**, 8763–70.

64 Pitzalis, P., Angelico, R., Söderman, O. and Monduzzi, M. (2000) A structural investigation of caaot/water/oil microemulsions. *Langmuir*, **16**, 442–50.

65 Fletcher, P.D.I., Grice, D.D. and Haswell, S.J. (2001) Partitioning of p-nitroaniline between pseudo-phases within a water-in-oil microemulsion. *Physical Chemistry Chemical Physics*, **3**, 1067–72.

66 Hait, S.K., Moulik, S.P., Rodgers, M.P., Burke, S.E. and Palepu, R. (2001) Physicochemical studies of microemulsions. 7. Dynamics of percolation and energetics of clustering in water/aot/isooctane and water/aot/decane w/o microemulsions in presence of hydrotopes (sodium salicylate, α-naphtol, β-naphtol, resorcinol, catechol, hydroquinone, pyrogallol and urea) and bile salt (sodium cholate). *The Journal of Physical Chemistry B*, **105**, 7145–54.

67 Nave, S., Eastoe, J., Heenan, R.K., Steytler, D. and Grillo, I. (2000) What is so special about aerosol-ot? 2. Microemulsion systems. *Langmuir*, **16**, 8741–8.

68 Seto, H., Okuhara, D., Kawabata, Y., Takeda, T., Nagao, M., Suzuki, J., Kamikubo, H. and Amemiya, Y. (2000) Presure and temperature effects on the phase transition from dense droplet to a lamellar structure in a ternary microemulsion. *The Journal of Chemical Physics*, **112** (23), 10608–14.

69 Scriven, L.E. (1976) Equilibrium bicontinuous structure. *Nature*, **263**, 123–5.

70 Teubner, M. and Strey, R. (1987) Origin of the scattering peak in microemulsions. *The Journal of Chemical Physics*, **87** (5), 3195–200.

71 Strey, R.J. and Winkler, L. (1991) Magid: small-angle neutron scattering from diffuse interfaces. 1. Mono- and Bilayer in the Water-Octane-C12E5 System. *Journal of Physical Chemistry*, **95**, 7502–7.

72 Choi, S.-M. and Chen, S.-H. (1997) The relationship between the interfacial curvatures and phase behavior in bicontinuous microemulsions – a SANS study. *Progress in Colloid and Polymer Science*, **106**, 14–23.

73 Zilman, A.G. and Granek, R. (1996) Undulations and dynamic structure factor of membranes. *Physical Review Letters*, **77**, 4788–91.

74 Takeda, T., Kawabata, Y., Seto, H., Komura, S., Gosh, S.K., Nagao, M. and Okuhara, D. (1999) Neutron spin-echo investigation of membrane ondulations in complex fluids involving amphiphiles. *The Journal of Physics and Chemistry of Solids*, **60**, 1375–7.

75 Takeda, T., Kawabata, Y., Seto, H., Gosh, S.K., Komura, S. and Nagao, M. (2000) Membrane undulations in complex fluids involving amphiphiles, in *STATISTICAL PHYSICS. Third Tohwa University International Conference, Volume 519 of AIP Conference Proceedings* (eds M. Tokuyama and H.E. Stanley), American Institute of Physics, Springer Verlag, Berlin, pp. 190–2.

76 Komura, S., Takeda, T., Kawabata, Y., Gosh, S.K., Seto, H. and Nagao, M. (2001) Dynamical fluctuations of the mesoscopic structure in ternary CiEj-water-n-octane amphiphilic systems. *Physical Review E*, **63** (4), 41402–11.

77 Mihailescu, M., Monkenbusch, M., Endo, H., Allgaier, J., Gompper, G., Richter, D., Jakobs, B., Sottmann, T. and Farago, B. (2001) Dynamics of bicontinuous microemulsion phases with and without amphiphilic block-copolymers. *The Journal of Chemical Physics*, **115** (20), 9563–77.

78 Maugey, M. and Bellocq, A.M. (2001) Effect of adsorbed and anchored polymers on membrane flexibility: a light scattering study of sponge phases. *Langmuir*, **17**, 6740–2.

79 Morkved, T.L., Stepanek, P., Krishnan, K., Bates, F.S. and Lodge, T.P. (2001) Static and dynamic scattering from ternary polymer blends: bicontinuous microemulsions, lifshitz lines, and amphiphilicity. *The Journal of Chemical Physics*, **114** (16), 7247–59.

80 Silas, J.A. and Kaler, E.W. (2001) The phase behavior and microstructure of efficient cationic-nonionic microemulsions. *Journal of Colloid and Interface Science*, **243** (1), 248–54.

81 Schwarz, B., Mönch, G., Ilgenfritz, G. and Strey R. (2000) Dynamics of the "sponge" (L_3) phase. *Langmuir*, **6**, 8643–52.

82 Le, T.D., Olsson, U., Wennerström, H., Uhrmeister, P., Rathke, B. and Strey, R. (2001) Binodal and spinodal curves of an L3 (sponge) phase. *Physical Chemistry Chemical Physics*, **3**, 4346–54.

83 Dubois, M. and Zemb, T. (2000). Swelling limits for bilayer microstrutures: the implosion of lamellar structure versus disordered lamellae. *Current Opinion in Colloid and Interface Science*, **5**, 27–37. Review.

84 Kötz, J. and Kosmella, S. (1999) Polymers in lyotropic liquid crystals. *Current Opinion in Colloid and Interface Science*, **4**, 348–53.

85 Choy, D. and Chen, S.-H. (2001) Clipped random wave analysis of anisometric lamellar microemulsions. *Physical Review E*, **63** (2), 21401–10.

86 Richardt, A. and Mitchell, S. (2006) Enzymes for environmentally friendly decontamination of sensitive equipment. *Journal of Defence Science, GBR* **10**, 261–5.

87 Yang, Y.-C., Baker, J.A. and Ward, J.R. (1992) Decontamination of chemical warfare agents. *Chemical Reviews*, **92**, 1729–43.

13
Immobilization of Enzymes

Birgit Hülseweh, André Richardt and Bernd Niemeyer

13.1
Introduction

Biological and chemical processes that can effectively detect, detoxify and demilitarize chemical warfare agents, pesticide waste and chemical stockpiles are in great demand.

The "biologization" of surfaces is a major task employing biological functions for technical and medical purposes. This technology could be used to achieve the stipulations not only for a fast and complete conversion of harmful substances into less harmful products, but could also lead to very sensitive detection of these compounds by immobilized enzymes. This chapter gives not only an introduction into the basics of immobilization of enzymes, but also an overview regarding the immobilization of enzymes for decontamination and detection of chemical warfare agents and other toxic compounds like pesticides. For all of these applications the immobilization procedure is crucial for the resulting features of the functionalized "bio-"carrier and thus its proceeding usage. Consequently, one of the earliest efforts in biotechnology focused on the development of these methods. A deep understanding of these different methods is necessary for achieving a new biotechnology based system for decontamination and detection, for chemical warfare agents and other toxic compounds like pesticides.

13.2
History of Immobilization Technologies

In the 1960s supports on the base of natural glycol-products, such as starch, cotton and agarose were the earliest matrices used for the binding of enzymes and later on for antibodies [1, 2]. The main reasons for their early application might have been their advantageous characteristics such as

- good binding characteristics, for example, a high overall number of anchoring (hydroxyl-) groups on the matrix onto

Decontamination of Warefare Agents.
Edited by André Richardt and Marc-Michael Blum

ISBN: 978-3-527-31756-1

which the enzymes can finally be bound as ligands, and hydroxyl groups, which are easily convertible for the immobilization procedure as well as
- favorable processing features based on a loose network structure of the bio-carrier, providing large macropores serving high diffusion rates, and thus minimizing the required processing time.

The drawbacks of low physical, chemical and biological stability significantly restrict their industrial utilization.

Organic polymers were developed in the 1970s to deliver tailor-made supports, employing sophisticated techniques to generate porous structures. Thus, the pore geometry, size and distribution and finally, the particle size of the matrix were intended to be well designed, according to the requirements of the biological and technical systems. The choice of different polymeric materials, including their various preactivated forms, deliver additional variable chemical characteristics. Polymer supports generally provide high chemical stability [3–5] and highly adaptable features for process optimization. The development of organic matrices for chemical analytical purposes such as liquid chromatography systems and their medical and biotechnological applications discussed above, proved mutually sustaining during this period.

Inorganic supports are largely based on the oxides of titanium and zirconium, on silica, ceramic and glass matrices. Their preparation with porous structures requires greater technological effort. These new matrices deliver high stability against shear forces and physical parameters, such as thermo- and pressure-stability, as well as of the texture under elevated physico-chemical conditions, for example, at high pressure utilizations or in rigorous regeneration procedures [6–10].

As innumerable different matrices are available now, new technological requirements and new materials such as membranes and nano-structured particles are generating new carriers. Recent developments include nano based magnetic supports for use in expanded bed adsorption separation [11, 12].

The recovery of the biological activity of the ligand after its immobilization is most important and strongly depends on the method applied [13–15]. Thus intensive research has been undertaken to develop numerous techniques [16–18]. Comprehensive overviews of the state of the art are summarized elsewhere [19, 20]. Important parameters besides the activity preservation during the immobilization procedure are the optimal ligand density, a high degree of accessible enzymes, the long-term stability, and the whole design of the bio-carrier according to the physico-chemical requirements (e.g. hydrophilic/hydrophobic properties) of the technical and biological systems which are being optimized in ongoing processes [21–24].

It is obvious that a large variety of features delivered from bio-carriers have to comply with the requirements of the biological and technical systems where they are employed, resulting in sophisticated optimization procedures [13, 14, 25, 26].

13.3 Heterogeneous Bio-systems – Benefits and Drawbacks

Prior to using bio-molecules for technical purposes, they have to be produced in an organism or in a biotechnological process, followed by various separation steps to remove unrequired matter of the organism, and by-products, respectively. These multi-step separation processes are involved in order to deliver the protein in its purified form, to realize highly optimized and reproducible conditions during its technical application. Technical processes might include biocatalytic conversion to generate new products [27]. Similar systems are also relevant to the highly selective bio-separation in downstream processing [28, 29]. Aside from the bio-product processing, the elucidation of structures of bio-molecules and their interactions can be realized with these types of systems [30–32]. Analysis via bio-sensors is also vitally based on the heterogeneous systems of immobilised proteins [33, 34]. These measurement techniques may be utilized as stand-alone systems or integrated into processing plants for process control. Thus, heterogenization is of fundamental interest in modern biotechnology and biochemical analyses. The basic difference of homogeneous versus heterogeneous systems will be discussed later.

The proteins in their purified state are ready to be dissolved directly into liquids to generate homogeneous reaction systems. After their usage for reaction purposes the enzymes are either wasted after one single conversion step, or they have to be recovered from the reaction broths in similar multi-step separation processes as for their original isolation. These procedures more or less decrease their activities, are time consuming and finally cost intensive. The immobilization of enzymes enables their multi-fold use with little or no effort for recovery and regeneration. Consequently, homogeneous systems are widely switched towards heterogeneous ones. Important advantages involved are:

- increased stability of the immobilised enzyme, especially under rigid process conditions. The shear stability of the protein is improved, and the micro-environment of the bio-carrier “buffers” extreme process parameter values of, for example, the temperature or pH-value;
- processing is easier with immobilised proteins, as their handling is less difficult in a heterogenised state. Also, they do not interfere negatively during the separation of the products in downstream processing;
- high catalytic productivity of the heterogeneous systems, since a high protein concentration can be realized at the surface of the carrier;
- repeatedly applicable to the process and widely maintaining their bio-activity as no further protein isolation from the reaction broth is necessary;
- employment of further types of reactors such as fixed bed, circulating bed, and fluidised bed reactors, respectively. Reactors regularly used for dissolved enzymes are stirred

tank vessels and bubble columns. These additional reactor types reveal possibilities for process optimization and
- above all, from the technical point of view, the possibility for continuous processing in heterogeneous systems is of major importance.

All of the discussed advantages and parameters result in reduced overall costs for processing and analyzing with heterogenised systems. For more details refer to textbooks and review papers discussing the different items [14, 16, 19, 35].

Heterogenization also has disadvantages that are mostly connected to the development of the whole process. These shortcomings are:
- More or less reduction of the bio-activity of the enzyme in its bound state. This is attributed to the immobilization procedure, where some deactivate the proteins bio-activity to a higher degree;
- enhanced mass diffusion resistance. The mass transport through the stagnant film around the solid surface from and to the liquid bulk phase as well as in the pores of the carrier are based on the slow molecular diffusion. These are often the time limiting steps in the sequence of paces of substrate transport to the enzyme, bio-conversion and transport of the product from the enzyme back into the bulk-phase, which finally results in longer processing times. This may be minimized with large-pore supports (as seen from Figure 13.1), consequently resulting in a reduced available surface for the conversion, ending in an optimization routine of the space-time-yield for the bioreactor and

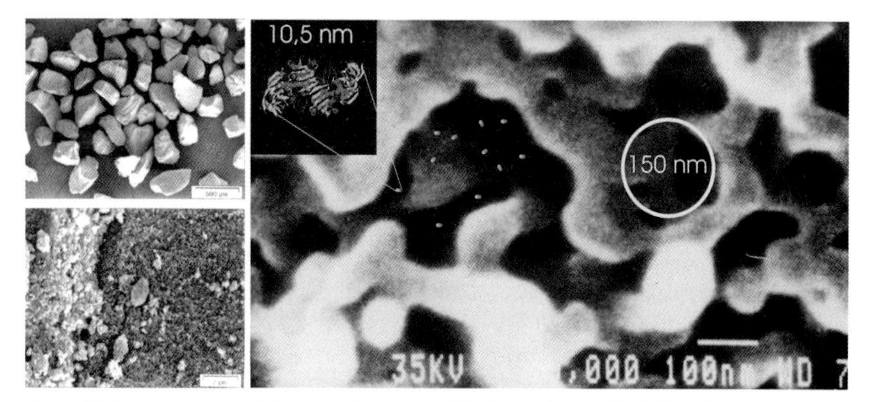

Figure 13.1 Comparison of the pore size (150 nm) of the open porous support Silicagel XWP-P005 (Grace, Worms, FRG) versus the size of the Alcohol dehydrogenase from Baker's Yeast (Y-ADH) (molar weight: around 150 kDa; main length: 10.5 nm) proves a significant pore width for minimized protein interaction wirth the solid matter, even if the support is functionalized.

- time consuming efforts for the selection and the determination of process design parameters of an optimal heterogeneous system.

It becomes obvious that the selection of the support as well as the fixation technologies employed are the crucial aspects for success in bio-processing and for the application in analytical systems, respectively.

13.4 Basic Technologies for the Immobilization and Methods for the Characterization of Heterogeneous Bio-systems

The modification of the supports' surfaces with organic reagents offers a multiplicity of advantages. They do not only affect the physical or chemical attributes of the modified materials like wettability, adhesion and contamination [36–38], but also deliver the possibility of introducing functional groups together with an organic layer, onto which, in a second step, proteins may be chemically attached. Many different protein coupling techniques have been developed and patented [39, 40], several hundred enzymes have been immobilised in different forms, but only a few biocatalysts are so far used in industrial applications [39]. Each method has its advantages and drawbacks and there is no generally appropriate method of protein binding. Instead, immobilization is always a question of the biocatalyst itself, its stability, nature, source and purity. Non specific interactions have to be evaluated versus site-directed fixation approaches.

For a comprehensive overview of protein and enzyme heterogenization refer to textbooks [19, 35] and review articles [40–43]. These methods are presented in Figure 13.2.

The most widely used method of immobilization is the covalent binding onto carrier surfaces or solid supports. Adsorption techniques represent early attempts of surface modification and protein heterogenization. Classical adsorbents are ion-exchange matrices, porous carbon, clays, metal oxides, glasses, silica gels as well as polymer resins.

Immobilization by adsorption is a relatively mild procedure and causes little or no conformational change of the enzyme or its active site. The technique is nearly universally applicable, simple, cheap and usually, no reagents and only a minimum of activation steps are required. However, the method strongly depends on parameters like pH, temperature and solvent and has the disadvantage that the adsorbed biocatalyst may leak from the solid support during utilization due to weak binding forces. Another disadvantage is non-specificity of the binding, this means further adsorption of other proteins or substances onto the bio-carrier, other than the immobilised enzyme during the processing. One of the earliest examples of enzyme fixation by physical adsorption was the coupling of beta-D-fructo-furanosidase onto aluminum hydroxide. Other classical examples are alpha-amylase adsorbed to activated carbon or clay, AMP deaminase onto silica and

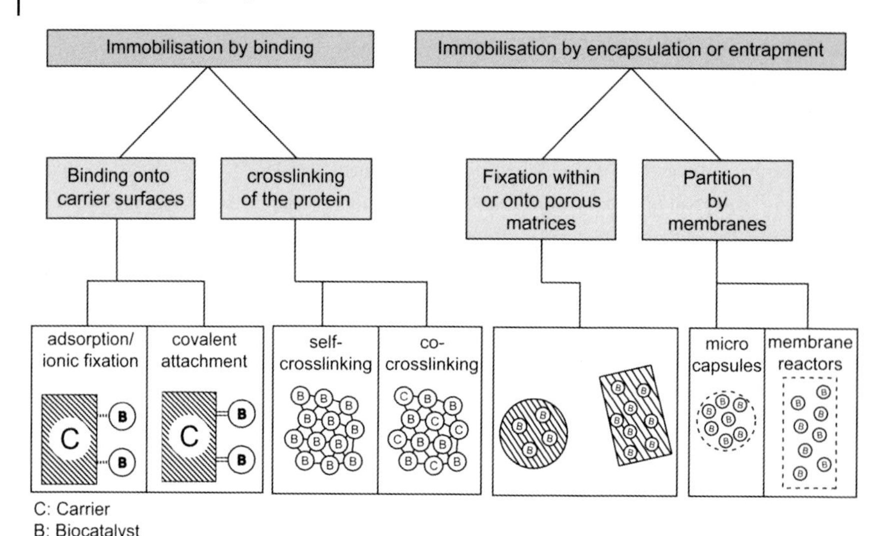

Figure 13.2 Overview of the most widely applied protein and enzyme immobilization methods.

chymotrypsin onto kaolinite [39]. In parallel chemical bound organic layers, the so-called Self-Assembled Monolayers (SAMs) have also been investigated. Organic molecules are fixed on the surface through an interface group and are proved to be thermodynamically stable and mechanically robust [44, 45]. Particular head-groups of special interests are the following systems:

- alkyl-trichlorosilane and alkoxy-alkylsilane on hydroxylated surfaces (SiO_2, Al_2O_3, glass, TiO_2, etc.) [46]
- alkanethiolates on gold, silver, copper and platinum [47]
- alcohols and amines on platinum [47]
- organosulfur compounds, for example, dialkyl sulfide, dialkyl disulfide, xanthate and thiocarbamate coordinate strongly to silver, copper, platinum, mercury, iron, nanosize γ-Fe_2O_3 particles, gold and colloidal gold particles, GaAs and InP surfaces [48, 49]
- fatty acids on silver, copper, aluminium and aluminium oxide [50].

The organic layers are formed by immersing a surface-active metal in a solution of organic reactants. This way, for example, a monolayer of trichlorosilane dissolved in toluene, precipitates on the surface of metal oxides, such as titanium dioxide [51]. For the attachment of large ligands like enzymes, the appropriate bi-functional molecules are chosen (e.g. peptides possessing free –SH-groups, in analogy to the alkane thiols) to bind bio-active ligands on their outer end and thereby to equip the outer surface of the carrier with selective bio-functional attributes, which can be used particularly to immobilise enzymes.

Covalent coupling of bio-molecules to an insoluble matrix delivers multiple advantages and best suits the purpose where the long-term stability of the biocatalyst is particularly important. Consequently, it is an extensively researched technique and, at present, most widely employed for protein immobilization. Frequently, covalent immobilization of an enzyme results in better activity and reduced non-specific adsorption [23]. In general, very little leakage of enzyme from the support occurs. Functional groups that may take part in the covalent binding of enzymes to water-insoluble carriers or organic layers are amino, carboxyl, sulfhydryl, hydroxyl, imidazole, phenolic, tresyl, and indol groups. The usefulness of various groups found in enzymes for covalent link formation depends upon their availability, reactivity, and nucleophilicity. Nowadays, for the covalent binding of large proteins, such as enzymes, onto carrier surfaces, defined structures are applied onto the bio-carrier [12–14, 16, 19, 20, 52, 53]. These are:

1. carrier or support based on organic (polymer based) or inorganic (silica, titanium dioxide, etc.) matter onto which a
2. bi-functional reagent is bound to introduce a reactive anchoring group onto the surface for improved reactivity for
3. the introduction of a so-called spacer. It should realize a sufficient distance [16–18, 54] from the bio-molecule from the carrier backbone, maintaining its high steric flexibility and thus providing a high bio-activity of
4. the finally fixed ligand onto the spacer. Large sized ligands in the range of a diameter of up to 12 nm comprehend the enzymes.

These different reaction steps have to be optimized individually, and according to the whole system. Parameters characterising the overall result are high

- coupling yield of the ligand,
- surface coverage of the carrier and
- accessibility of the ligand, in static as well as dynamic processes.

Additionally, parameters like

- low unspecific interaction, and adsorption of substances from the reaction media, respectively and
- high long term stability of the heterogenised ligand, maintaining its bio-activity under process conditions

are also of vital significance for the application of heterogeneous systems, which can be influenced by the above outlined procedure no. 1 to 4.

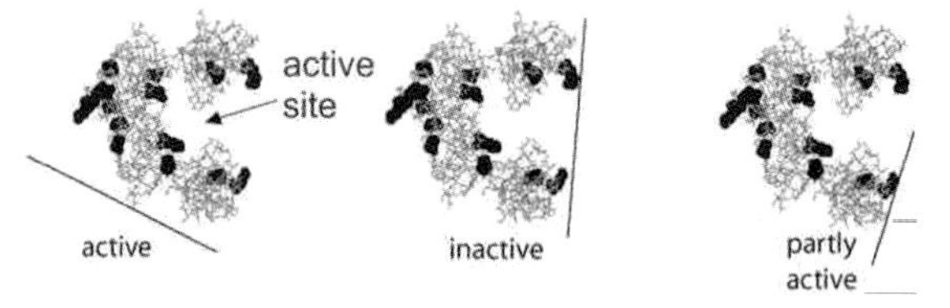

Figure 13.3 Schematic presentation of different orientations of immobilised proteins onto solid supports.

13.4.1
Random Versus Oriented Immobilization

An aspect in science that should be addressed in the fixation of biocatalysts is the orientation of the enzymes that can be either random or oriented, which is schematically drawn in Figure 13.3. A highly ordered coupling of the lignad can be achieved by the terminal biotinylation of genetically engineered proteins with subsequent end-specific attachment to a streptavidin functionalized surface.

Because of the ease of biotinylation, successful immobilization has been highlighted in many scientific reports [55, 56] and streptavidin–biotin capturing has been employed for bio-sensors as well as for microarrays [56, 57]. Further very versatile procedures to capture biocatalysts in an oriented way are the amino- or carboxy-terminal addition of protein-TAGs and the utilization of Protein A-coated surfaces to bind IgG antibodies.

13.4.2
Characterization of Immobilised Enzymes

For reasons of quality control during the synthesis of the bio-active support with the bound enzyme, the different steps of the immobilization procedure have to be examined by specific analytical methods as presented, for example, in Chapter 4 by Hermanson *et al.* and elsewhere [5, 53].

The basis for the characterization of heterogenised proteins is in kinetic studies, investigating the concentration-time course of the substrate conversion by the enzyme bound. The chemical analytical principle most often employed is the spectrometric detection of the conversion progress [58, 59]. Modern analytical methods are retained in order to understand and to steer the binding and finally, the conversion process. For this, it is vital to examine the interaction of the protein immobilised with the compound binding onto it in more detail. Therefore, modern analytical methods such as AFM [60], NMR [61], IR [62, 63], SPR [64] and others have been used.

The experimental results have to be modeled for

1. comparison with homogeneous systems, particularly regarding the activity (which is equivalent the reaction rate of the enzyme) and

2. obtaining process engineering parameters such as ligand density and accessibility onto the carrier surface, rate and the Arrhenius constants.

Diverse models are available in the literature, contributing to the elucidation of species interaction [31, 65–68]. Many of these also cover a wide range of applicability.

13.5 Applications of Immobilized Enzymes for Detoxification, Decontamination and Detection of Chemical Warfare Agents

Chemical warfare agents, pesticide waste and chemical stockpiles pose a potential threat to human health and the environment. Therefore, biological and chemical processes that can effectively detect, detoxify and demilitarize chemicals compounds are in great demand.

Today, biocatalytic methods for detoxification are recommended because, compared to chemical decontamination processes, enzymatic methods are not corrosive and harmful [69]. Enzymatic procedures are generally safe for the user and environment and limitations like their short life-time and their non-reusability can be overcome by the utilization of heterogenised enzymes. The main enzymatic classes that are employed for detoxification and decontamination procedures are oxidoreductases and hydrolases.

Immobilised laccase (oxygen oxidoreductase, EC 1.10.3.2) for example exhibits broad substrate specificity and holds great promise for the textile industry in terms of decolorization and detoxification of textile dyes [70, 71]. In addition, immobilised laccases of various origins may be useful for the removal of phenolic compounds that are present in wastes from industrial processes such as coal conversion, petroleum refining and olive oil production [72–74]. Furthermore, chemically bound laccases have demonstrated their potential to oxidise alkenes and to convert herbicides into an inactive form [75].

Further heterogenised enzymes that have been investigated in great detail for their degradation capacities and potential employment for detoxification and decontamination are organophosphorus hydrolase (OPH) from *Pseudomonas diminuta*, organophosphorus acid anhydrolase (OPAA) from *Alteromonas sp.*, acetylcholinesterase (AChE), mammalian paraoxonase and squid *O,O*-di-isopropyl fluorophosphatase (DFPase) from *Loligo vulgaris*. The first reports of an immobilised pesticide-hydrolyzing enzyme preparation were published by Munnecke in 1977 and 1979 [76, 77]. The coupled enzyme maintained good stability for over 180 days, but was susceptible to deactivation by solvents and unknown inhibitors in the aqueous waste stream. Since then, detoxification of soil and water by heterogenised OP-degrading enzymes has become the focus of many studies. For example Caldwell and Raushel investigated the degradation of OP-pesticides by *Pseudomonas diminuta* phosphotriesterase and bound the biocatalyst by adsorption onto trityl agarose and by covalent coupling using glutaraldehyde onto various nylon supports [78]. Other work focused on the use of cellulose-immobilised OPH [79–81] and examined OPAA and OPH covalently-coupled to solid supports like azlactone polyacrylamide, glyoxal agarose and glyoxal-aminopropyl for the treatment of contaminated water supplies. At least chemically bound OPH-fusion enzymes proved to be a promising tool for the remediation of coumaphos contaminated water [79].

A further appealing approach for fixed OP degrading enzymes is the enzymatic decontamination and demilitarization of chemical warfare agents. Similar to OP pesticides, G-agents like soman, sarin and VX are organophosphorus esters and display a relatively high toxicity and persistence. Worldwide stores of OP nerve agents are estimated to exceed 200000 tons that have to be eliminated and destructed within stipulated deadlines according to the OPCW treaty (*Organization for the Prohibition of Chemical Weapons*).

Decontamination of organophosphate nerve agents by immobilised enzymes has been studied in great detail by Russell and coworkers [82, 83] and recent advances in enzyme immobilization illustrate cross linking of OP degrading enzymes within polyurethane matrices [84–86]. By this method, free amino-acid groups of the enzyme are covalently incorporated into a polyurethane prepolymer that contains multiple isocyanate functionalities. Depending on the reactivity of the isocyanate, either foams or gels can be prepared. Havens and Rase were the first to investigate the incorporation of OPH into a polyurethane sponge [87].

Further enzymes that were efficiently coupled to polyurethane foams were parathion hydrolase, OPH and AChE [83, 85]. For OPH more than 50% specific activity, and for AChE approximately 90% specific activity was retained after optimization of the polymer synthesis. The bound enzymes displayed no altered enzyme function and were extremely resistant to proteolysis [83].

Today, the capability to effectively incorporate OP degrading enzymes within polyurethane foams is commercialized by ICx-Agentase and numerous benefits such as improved stability, reusability, and environmental resistance are incurred when these enzymes are co-polymerised within polyurethane polymers.

However, copolymerization is not restricted to G-agent degrading enzymes, and thus ICx-Agentase has developed a suite of enzyme-based sensors (CAD-Kit) for the detection of a variety of chemical weapons that is shown in Figure 13.4.

The enzyme-sensors work on a positive response mechanism, are easy to use and provide first responders with the ability to conduct surface, solid and liquid interrogation of nerve, blood and blister agents. As described in Table 13.1 a positive detection result is indicated by a color change.

Furthermore, OP detoxifying biocatalysts have been incorporated into conventional formulations of fire fighting foams, decreasing the volatilization of contaminants and enhancing the surface wettability. The foams allow control of the rate of enzyme delivery to large areas and make them valuable for the decontamination of electronic devices, equipment, machineries, vehicles and aircraft. Further novel developments and recent advances in material synthesis have allowed the prepara-

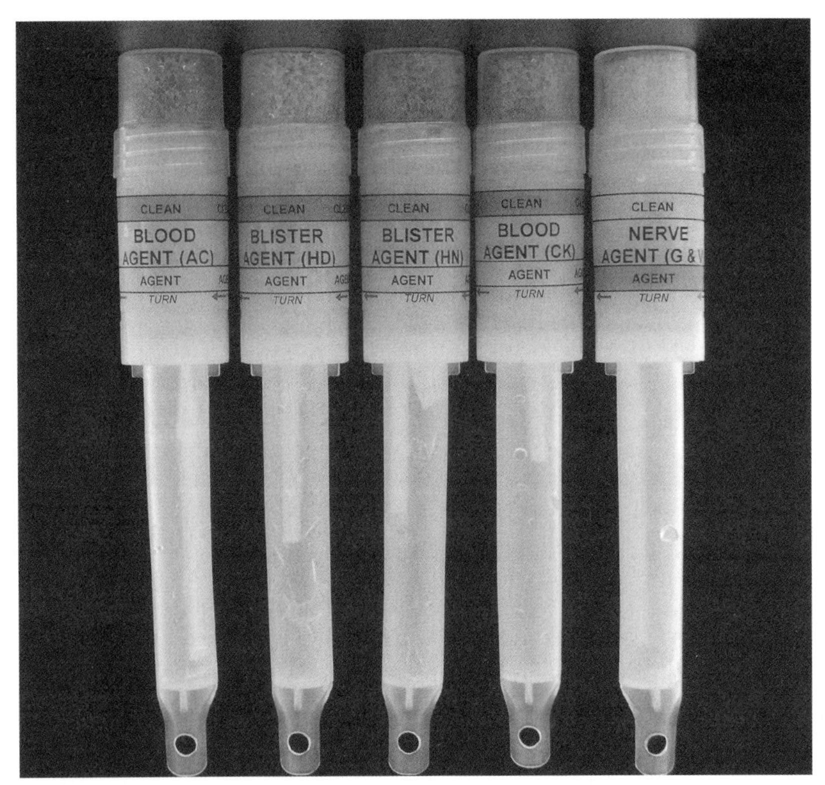

Figure 13.4 ICx-Agentase suite of enzyme-based sensors (CAD-Kit) for the detection of nerve, blood and blister agents. The sensors are built with an easy to use construct with separate sections that include the enzyme polymer, a carrier fluid and solid reagents mixed together upon activation of the enzyme-sensor. The figure was reproduced with permission from J. Berberich, ICx-Agentase.

Table 13.1 Color change of the CAD-Kit.

	Sensor	*Agent absent*	*Agent present*
Nerve	Nerve Agent (G & V)	Yellow	orange
Blister	Sulfur Mustard (HD)	Pink	yellow
	Nitrogen Mustard (HN)	Orange	yellow
Blood	Cyanogens (CK)	Light blue	yellow
	Hydrogen Cyanide (AC)	Orange	yellow

tion of wipes, gas filters and column packings as well as coatings and paints with self-decontaminating potential [82, 88].

With the constant threat of chemical warfare, significant research has also gone in protective clothing. Current technology makes use of adsorptive polyurethane foam embedded between several layers of polyester fabric. However, the layer adds extra weight to the clothing and this sort of clothing has to be disposed after decontamination since it cannot be detoxified.

Therefore, future developments that highlight enzyme-containing polymeric materials might be self-decontaminating clothes. Early investigations in this direction date back to 1991 to K.S. Rajan, who studied squid-type DFPase-immobilised matrices and focused his binding studies on fabrics like nylon, polyester and cotton [89].

Since the ability to monitor contaminants and toxic chemicals in the environment is instrumental for understanding and managing risks, immobilised proteins or enzymes have also emerged as an important and accepted tool for bio-sensors. Bio-sensors for analytical monitoring act as early warning systems. In addition, they can contribute important information on the progress of remediation treatments. Key elements of a bio-sensor are the biological component or biocatalyst, its transducer and amplifier. Detection can be either based on optical, electrical electrochemical and piezoelectrical principles, respectively [90].

Although the history of bio-sensors dates back to the early sixties, it took further 30 years for them to become a mature technology. In 1962, Clark published work on the oxygen electrode. This invention and its subsequent modification with glucose oxidase entrapped at the electrode, identifies him as the father of the bio-sensor. Since then, there have been numerous variations on the basic design and many different enzymes have been attached by workers for various purposes.

Outstanding progress has been made in recent years on bio-sensors for laboratory and field analysis of water and soil [91–93]. Today, biocatalysis-based bio-sensors for environmental applications include the employment of tyrosinase and laccase for the detection of phenols [94–96] and the utilization of organophosphate hydrolase (OPH) for the detection of organophosphorus pesticides [97–100].

Affinity bio-sensors as well as catalytic bio-sensors also attracted the field of chemical defense and risk evaluation where an early and definite identification is as important as sensitivity and specificity.

Early work for screening OP neurotoxins concentrated on enzyme-inhibition sensors based on acetylcholine esterase (AChE) or butyryl cholinesterase (BChE) [101–103]. However, these bio-sensors suffered from a number of limitations, including that environmental factors can cause loss of enzyme acticivity that will result in false positive signals. Also due to the irreversible nature of enzyme-analyte interaction, a continous application or reuse of an affinity-bio-sensor is excluded.

One of the ways to overcome this disadvantage of affinity-bio-sensors is to replace the inhibition receptor enzyme with a catalytic receptor enzyme.

Therefore, prominent examples for the screening of organophosphate nerve agents today employ either organophosphorus hydrolase (OPH) or organophosphorus acid anhydrolase (OPAA) as the immobilised biological recognition element [104]. Bio-sensors that can discriminate between organophosphate neurotoxins and chemically similar agriculture compounds offer great promise. Simonian and coworkers, for example, identified a multi-enzyme strategy to discriminate between different classes of neurotoxins. Wang *et al.* [93] described an enzymatic microchip assay that allowed the differentiation between individual OP substrates. Compared to conventional OPH-based bio-sensors, the OPH-biochip combines a pre-column reaction of OPH with electrophoretic separation and conductivity detection of the phosphonic acid reaction products.

13.6 Conclusions and Outlook

The basis for the utilization of biological macromolecules and functions is their successful heterogenization. Thus, focus must be put on the protein immobilization procedure, which has to provide a highly bio-active support, carrying a maximal number of bio-molecules, ensuring their accessibility for the interacting substrate, and guaranteeing long-term high stability under process conditions. Therefore, the chemical, biological and technical sub-systems have to be adapted to each other and optimised for the required employment. From the chemical point of view, the choice of the matrix, its functionalization and the attachment of the ligand is decisive. Efforts towards the optimal heterogenization of biological systems have been sustained for a long time.

Further investigations are required due to

- the improvement of analytical methods detecting the behavior of the protein immobilised and the interacting substrate, disclosing basic insights and opening new optimization strategies,
- new biological sources for the protein utilized, for example, derived from protein engineering procedures, and from special biological habitats, such as microorganisms living under extreme conditions, respectively,
- the growing number and fields using heterogenised proteins in biotechnology, and analyses. Modern biotechnological methods have started to "take over" conventional chemical process routes, so it is forseen that, even for the production of bulk chemicals, in the near future biotechnology will be applied to utilize renewable sources ("Green Chemistry") [105]. In chemical warfare detoxification as well as analyses, new horizons were highlighted in this contribution. Modern medical analyses already widely employ heterogenised proteins in assays for disease identification. This strong trend will be ongoing. One widely uncovered area is the effective analyses of the immunsystem.

It seems that breathtaking new results from basic research in the field of immobilization cannot be expected, although strong competencies are required to introduce new scientific results in medicine into daily life, and employing biotechnology to cope with actual challenges, for example, and changing the industrial source bases from oil era towards post-oil industry, respectively. In this way, immobilization contributes to the successful widespread application of biotechnology and thus contributes to the maintenance of high living standards, even in times of transition.

References

1 Axen, R., Porath, J. and Ernback, S. (1967) Chemical coupling of peptides and protein to polysaccharides by means of cyanogen halides. *Nature*, **214**, 1302–4.
2 Unger, K., Berg, K. and Gallei, E. (1969) Herstellung oberflächenmodifizierter Adsorbentien. *Kolloid-Zeitung, Zeitschrift für Polymere*, **243**, 1108–14.
3 Engelhardt, H. and Mathes, D. (1979) Chemically bonded stationary phases for aqueous high-performance exclusion chromatography. *Journal of Chromatography*, **185**, 311–20.
4 Messing, R.A. and Oppermann, R.A. (1979) Pore dimensions for accumulating biomass. *Biotechnology and Bioengineering*, **21**, 49–58.
5 Hermanson, G.T., Mallia, A.K. and Smith, P.K. (1992) *Immobilized Affinity Ligand Techniques*, Academic Press, San Diego, CA, USA.
6 Eaton, D.L. (1974) The optimization of porous materials for Immobilized Enzymes (IME) systems, in *Immobilized Biochemicals And Affinity Chromatography* (ed. R.B. Dunlap), Plenum, New York, NY, USA.
7 Jervis, L. (1978) Affinity chromatography on porous inorganic supports in chromatography of synthetic and biological polymers, in *Chromatography of Synthetic and Biological Polymers* (ed. R. Epton), Horwood, Chichester.
8 Unger, K.K. (1979) *Porous Silica*, Elsevier, Amsterdam, Netherlands.
9 Janowski, F. and Heyer, W. (1982) *Poröse Gläser*. VEB Deutscher Verlag für Grundstoffindustrie, Leipzig, Germany.
10 Robinson, P.J., Dunnill, P. and Lily, M.D. (1971) Porous glass as a solid support for immobilization of affinity chromatography of enzymes. *Biochimica et Biophysica Acta*, **242**, 659–61.
11 Khedmati, M. (2005) Herstellung und charakterisierung superparamagnetischer polystyrol-mikropartikel, in *Fortschritte der Polymerisationstechnik* (ed. H.-U. Moritz), Wissenschaft & Technik Verlag, Berlin, Germany.
12 Gebauer, K.H., Thömmes, J. and Kula, M.R. (1997) Breakthrough performance of high-capacity membrane adsorbers in protein chromatography. *Chemical Engineering Science*, **52**, 405–19.
13 Monsan, P. (1977) Optimization of glutaraldehyde activation of a support for enzyme immobilization. *Journal of Molecular Catalysis*, **3**, 371–84.
14 Groman, E.V. and Wilchek, M. (1987) Recent developments in affinity chromatography supports. *Trends in Biotechnology*, **5**, 220–4.
15 Lasch, J. and Koelsch, R. (1978) Enzyme leakage and multipoint attachment of agarose-bound enzyme preparations. *European Journal of Biochemistry*, **82**, 181–6.
16 Lowe, C.R. (1979) An introduction to affinity chromatography, in *Laboratory Techniques in Biochemistry and Molecular Biology*, Vol. 7, Part II (eds T.S. Work and E. Work), Elsevier, North-Holland, Netherlands.
17 Turkova, J. (1978) *Affinity Chromatography*, Elsevier, Amsterdam, Netherlands.
18 Steers, E., Cuatrecasas, P. and Pollard, H.B. (1971) The purification of beta-Galactosidase from Escherichia coli by affinity chromatography. *The Journal of Biological Chemistry*, **246**, 196–200.
19 Hartmeier, W. (1986) *Immobilisierte Biokatalysatoren*, Springer, Berlin, Germany.

20 Narayanan, S. and Crane, L. (1990) Affinity chromatography supports: a look at performance requirements. *Trends in Biotechnology*, **8**, 12–6.

21 Weetall, H.H. (1970) Storage stability of water-insoluble enzymes covalently coupled to organic and inorganic carriers. *Biochimica et Biophysica Acta*, **212**, 1–7.

22 Jones, K. (1987) Optimization procedure for the silanisation of silicas for reversed-phase high-performance liquid chromatography II. Detailed examination of significant variables. *Journal of Chromatography*, **392**, 11–6.

23 Williams, G.J., Nelson, A.S. and Berry, A. (2004) Directed evolution of enzymes for biocatalysis and the life sciences. *Cellular and Molecular Life Sciences*, **61**, 3034–46.

24 Rosenfeld, H., Aniulyte, J., Helmholz, H., Liesiene, J., Thiesen, P.H., Niemeyer, B. and Prange, A. (2005) Comparison of modified supports on the base of glycoprotein interaction studies and of adsorption investigations. *Journal of Chromatography A*, **1092**, 76–88.

25 Fowell, S.L. and Chase, H.A. (1986) A comparison of some activated matrices for preparation of immunoadsorbents. *Journal of Biotechnology*, **4**, 355–68.

26 Cartellieri, S., Helmholz, H. and Niemeyer, B. (2001) Preparation and evaluation of Ricinus communis agglutinin affinity adsorbents using polymeric supports. *Analytical Biochemistry*, **295**, 66–76.

27 Bommarus, A.S. and Polizzi, K.M. (2006) Novel biocatalysts: recent developments. *Chemical Engineering Science*, **61**, 1004–16.

28 Schmidt-Kastner, G. (1984) Biospezifische adsorption hochmolekularer substanzen, in *Aufarbeitung biologischer Medien, Physikalisch-chemische Grundlagen* (eds K. Schügerl, M.-R. Kula and U. Onken), VCH, Weinheim, Germany.

29 Cartellieri, S., Hamer, O., Helmholz, H. and Niemeyer, B. (2002) One-step affinity purification of fetuin from fetal bovine serum. *Biotechnology and Applied Biochemistry*, **35**, 83–9.

30 Björklund, M. and Hearn, M.T.W. (1997) Characterisation of silica-based heparin affinity sorbents from equilibrium binding studies on plasma fractions containing thrombin. *Journal of Chromatography*, **762**, 113–33.

31 Hirabayashi, J., Arata, Y. and Kasai, K. (2000) Reinforcement of frontal affinity chromatography for effective analysis of lectin-oligosaccharide interactions. *Journal of Chromatography A*, **890**, 261–71.

32 Green, E. and Baenziger, J. (1989) Characterization of oligosaccharides by lectin affinity high-performance liquid chromatography. Trends in Biochemistry. *Science*, **14**, 168–72.

33 Xu, J.-J., Zhang, X.-Q., Fang, Z.-H., Yu, H.-Q. and Chen, H.-Y. (2001) A stable glucose biosensor prepared by co-immobilizing glucose oxidase into poly(p-chlorophenol) at a platinum electrode. *Fresenius' Journal of Analytical Chemistry*, **369**, 486–90.

34 Ligler, F.S., Taitt, C.R., Shriver-Lake, L.C., Sapsford, K.E., Shubin, Y. and Golden, J.P. (2003) Array biosensor for detection of toxins. *Analytical and Bioanalytical Chemistry*, **377**, 469–77.

35 Schmidt, R. (1991) *Herstellung und Charakterisierung formselektiver Biokatalysatoren mit immobilisierten Mikroorganismen*. Dr. Thesis, University of Erlangen, Nuremberg, Germany.

36 Hautman, J. and Klein, M.L. (1991) Microscopic wetting phenomena. *Physical Review Letters*, **67**, 1763–6.

37 Whitesides, G.M. and Laibinis, P.E. (1990) Wet chemical approaches to the characterization of organic surfaces: self-assembled monolayers wetting and the physical-organic chemistry of the solid-liquid interface. *Langmuir*, **6**, 87–96.

38 Lele, B.S., Papworth, G., Katsemi, V., Ruterjans, H., Martyano, I., Klabunde, K. J. and Russell, A.J. (2004) Enhancing bioplastic-substrate interaction via pore induction and directed migration of enzyme location. *Biotechnology and Bioengineering*, **86**, 628–36.

39 Cao, L. (2005) Immobilized enzymes: past, present and prospects, in *Carrier-bound Enzymes: Principles, Applications and Design*, Wiley-VCH, Weinheim, Germany, pp. 1–52.

40 Tischer, W. and Wedekind, F. (1999) Immobilized enzymes: methods and

applications. *Topics in Current Chemistry*, **200**, 95–126.

41 Bornscheuer, U.T. (2005) Trends and challenges in enzyme technology. *Advances in Biochemical Engineering/Biotechnology*, **100**, 181–203.

42 Cao, L. (2005) Immobilised enzymes: science or art? *Current Opinion in Chemical Biology*, **9**, 217–26.

43 Cao, L., Langen, L. and Sheldon, R.A. (2003) Immobilised enzymes: carrier-bound or carier-free? *Current Opinion in Biotechnology*, **14**, 387–94.

44 Schreiber, F. (2000) Structure and growth of self-assembling monolayers. *Progress in Surface Science*, **65**, 151–256.

45 Ulman, A. (1991) *An Introduction to Ultrathin Organic Films*, Academic Press, San Diego, CA, USA.

46 Wasserman, S.R., Tao, Y.-T. and Whitesides, G.M. (1989) Structure and reactivity of alkylsiloxane monolayer formed by reactions of alkyltrichlorsilanes on silicon substrates. *Langmuir*, **5**, 1074–87.

47 Troughton, E.B., Bain, C.D., Whitesides, G.M., Nuzzo, R.G., Allara, D.L. and Porter, M.D. (1988) Monolayer films prepared by spontaneous self-assembly of symmetrical and unsymmetrical dialkyl sulfides from solution onto gold substrates: structure, properties, and reactivity of constituent functional groups. *Langmuir*, **4**, 365–85.

48 Liu, Q. and Xu, Z. (1995) Self-assembled monolayer coatings on nanosized magnetic particles using 16-mercaptohexadecanoic acid. *Langmuir*, **11**, 4617–22.

49 Stratmann, M. (1990) Chemically modified metal surfaces – a new class of composite materials. *Advanced Materials*, **2**, 191–5.

50 Tao, Y.-T. (1993) Structural comparison of self-assembled monolayers of n-alkanoic acids on the surfaces of silver, copper, and aluminium. *Journal of the American Chemical Society*, **115**, 4350–8.

51 Rye, R.R., Nelson, G.C. and Dugger, M.T. (1997) Mechanistic aspects of alkychlorosilane coupling reactions. *Langmuir*, **13**, 2965–72.

52 Aniulyte, J., Liesiene, J. and Niemeyer, B. (2006) Evaluation of cellulose-based biospecific adsorbents as a stationary phase for lectin affinity chromatography. *Journal of Chromatography B*, **831**, 24–30.

53 Helmholz, H., Cartellieri, S., Thiesen, L.-Z., He, P.H. and Niemeyer, B. (2003) Process development in affinity separation of glycoconjugates employing lectins as ligands. *Journal of Chromatography A*, **1006**, 127–35.

54 Thiesen, P.H. and Niemeyer, B. (2005) Maßgeschneiderte Adsorbentien im Anwendungsspektrum Bio-, Medizin- und Umwelttechnik. *Chemie Ingenieur Technik*, **77**, 373–83.

55 Schaeferling, M. and Kambhampati, D. (2003) Protein microarray surface chemistry and coupling schemes, in *Protein Microarry Technology*, Wiley-VCH, Weinheim, Germany.

56 Schaeferling, M., Schiller, S., Paul, H., Kruschina, M., Pavlickova, P., Meerkamp, M., Giammasi, C. and Kambhampati, D. (2002) Application of self-assembly techniques in the design of biocompatible protein microarray surfaces. *Electrophoresis*, **23**, 3097–105.

57 Pavlickova, P., Knappik, A., Kambhampati, D., Ortigao, F. and Hug, H. (2003) Microarray of recombinant antibodies using a streptavidin sensor surface self-assembled onto a gold layer. *Biotechniques*, **34**, 124–30.

58 Bisswanger, H. (1994) *Enzymkinetik*, 2nd edn, VCH, Weinheim, Germany.

59 Finette, G.M., Mao, Q.-M. and Hearn, M.T.W. (1997) Comparative studies on the isothermal characterisation of proteins adsorbed under batch equilibrium conditions to ion-exchange, immobilised metal ion affinity and dye affinity matrices with different ionic strength and temperature conditions. *Journal of Chromatography*, **763**, 71–90.

60 Feldman, K., Hahner, G., Spencer, N.D., Harder, P. and Grunze, M. (1999) Probing resistance to protein adsorption of oligo(ethylene glycol)-terminated self-assembled monolayers by scanning force microscopy. *Journal of the American Chemical Society*, **121**, 10134–41.

61 Caravajal, G.S., Leyden, D.E., Quinting, G.R. and Maciel, G.E. (1988) Structural characterisation of (3-aminopropyl)triethoxysilane-modified silicas by silicon-29 and carbon-13 nuclear magnetic resonance. *Analytical Chemistry*, **60**, 1776–86.

62 Ihs, A. and Liedberg, B. (1994) Infrared study of ethyl and octyl xanthate ions adsorbed on metallic and sulfidized copper and silver surfaces. *Langmuir*, **10**, 734–40.

63 Richardt, A. (2006) *Biotechnologie-basierte Dekontamination biologischer und chemischer Kampfstoffe – Grundlagen der Dekontamination und die Notwendigkeit von umweltfreundlichen, zukunftsweisenden Dekontaminationsmitteln*, Chapter 2.2. Habilitation, helmut-Schmidt-University/University of the Federal Armed Forces Hamburg, Germany.

64 Thiesen, P.H., Rosenfeld, H., Konidala, P., Garamus, V.M., Prange, L.-Z., He, A. and Niemeyer, B. (2006) Glycolipids from a colloid point of view. *Journal of Biotechnology*, **124**, 284–301.

65 Yang, C. and Tsao, G.T. (1982) Affinity chromatography, in *Advances in Biochemical Engineering/Biotechnology*, Vol. 27 (ed. A. Fiechter), Springer, Berlin, Germany.

66 Arnold, F.H., Blanch, H.W. and Wilke, C.R. (1985) Analysis of affinity separations. *Journal of Chemical Engineering*, **30**, 189–223.

67 Yang, C.M., Goto, M. and Goto, S. (1989) Enzyme purification by affinity chromatography combined with batchwise adsorption and columnwise elution. *Journal of Chemical Engineering of Japan*, **22**, 532–7.

68 Yeh, Y.-L., Liao, C.-M., Chen, J.-S. and Chen, J.-W. (2001) Modelling lumped parameter sorption kinetics and diffusion dynamics of odour causing VOCs to dust particles. *Applied Mathematical Modelling*, **25**, 593–611.

69 Richardt, A. and Mitchell, S.J. (2006) Enzymes for environmentally friendly decontamination of sensitive equipment. *Journal of Defence Science*, **10**, 261–5.

70 Abadulla, E., Tzanov, T., Costa, S. and Gübitz, G. (2000) Decolorization and detoxification of textile dyes with a laccase from Trametes hirsute. *Applied and Environmental Microbiology*, **66**, 3357–62.

71 Zilles, A. (2005) *Laccase reactions for textile applications*. Ph.D thesis, University of Minho, Portugal.

72 Aggelis, G., Iconomou, D., Christou, M., Bokas, D., Kotzailias, S., Christou, G., Tsagou, V. and Papanikolaou, S. (2003) Phenolic removal in a model olive oil mill wastewater using Pleurotus ostreatus in bioreactor cultures and biological evaluation of the process. *Water Research*, **37**, 3897–904.

73 Hublik, G. and Schinner, F. (2000) Characterization and immobilization of the laccase from Pleurotus ostreatus and its use for the continuous elimination of phenolic pollutants. *Enzyme and Microbial Technology*, **27**, 330–6.

74 Ahn, M.Y., Dec, J., Kim, J.E. and Bollag, J.M. (2002) Treatment of 2,4-dichlorophenol polluted soil with free and immobilized laccase. *Journal of Environmental Quality*, **31**, 1509–15.

75 Mougin, C., Boyer, F.D., Caminade, E. and Rama, R. (2000) Cleavage of the diketonitrile derivate of the herbicide isoxaflutole by extracellular fungal oxidases. *Journal of Agricultural and Food Chemistry*, **48**, 4529–34.

76 Munnecke, D.M. (1977) Properties of an immobilized pesticide-hydrolyzing enzyme. *Applied and Environmental Microbiology*, **33**, 503–7.

77 Munnecke, D.M. (1979) Hydrolysis of organophosphate insecticides by an immobilized-enzyme system. *Biotechnology and Bioengineering*, **21**, 2247–61.

78 Caldwell, S.R. and Raushel, F.M. (1991) Detoxification of organophosphate pesticides using a nylon based immobilized phosphotriesterase from Pseudomonas diminuta. *Applied Biochemistry and Biotechnology*, **31**, 59–73.

79 Mansee, A.H., Chen, W. and Mulchandani, A. (2005) Detoxification of the organophosphate nerve agent coumaphos using organophosphorus hydrolase immobilized on cellulose materials. *Journal of Industrial Microbiology and Biotechnology*, **32**, 554–60.

80 Wang, A.A., Chen, W. and Mulchandani, A. (2005) Detoxification of organophosphate nerve agents by immobilized dual functional biocatalysts in a cellulose hollow fiber bioreactor. *Biotechnology and Bioengineering*, **91**, 379–86.

81 Wang, A.A., Mulchandani, A. and Chen, W. (2002) Specific adhesion to cellulose and hydrolysis of organophosphate nerve agents by a genetically engineered Escherichia coli strain with a surface-expressed cellulose-binding domain and organophosphorus hydrolase. *Applied Biochemistry and Biotechnology*, **68**, 1684–90.

82 Russell, A.J., Berberich, J.A., Drevon, G.F. and Koepsel, R.R. (2003) Biomaterials for mediation of chemical and biological warfare agents. *Annual Review of Biomedical Engineering*, **5**, 1–27.

83 Lejeune, K.E., Dravis, B.C., Yang, F., Hetro, A.D., Doctor, B.D. and Russel, A.J. (1998) Fighting nerve agent chemical weapons with enzyme technology. *Annals of the New York Academy of Sciences*, **864**, 153–70.

84 Drevon, G.F. and Russel, A.J. (2000) Irreversible immobilization of diisopropylfluorophosphtase in polyurethane polymers. *Biomacromolecules*, **1**, 571–6.

85 Gordon, R.K., Gunduz, A., Doctor, B.P., Skvorak, J.P., Maxwell, D.M., Ross, M. and Lenz, D. (2000) OP Nerve Agent Decontamination, Detoxification and Detection Using Polyurethane Immobilized Enzymes. Conference Proceedings, Chemical and Biological Medical Treatment (CBMTS) III, 1–5, Spiez, Switzerland.

86 Lejeune, K.E. and Russel, A.J. (1996) Covalent binding of a neve agent hydrolyzing enzyme within polyurethane foams. *Biotechnology and Bioengineering*, **51**, 450–7.

87 Havens, P.L. and Rase, H.F. (1993) Reusable immobilized enzyme/ polyurethane sponge for removal and detoxification of localized organophosphate pesticide spills. *Industrial and Engineering Chemistry Research*, **32**, 2254–8.

88 Drevon, G.F. (2002) *Enzyme immobilisation into polymers and coatings*. Ph.D thesis, University of Pitsburgh, USA.

89 Rajan, K.S. (1991) Enzymatic detoxification of chemical warfare agents: Immobilization of the enzyme on material surfaces. Progress Report of the Naval Research Office, 1–5.

90 Andreescu, S. and Sadik, O.A. (2004) Trends and challenges in biochemical sensors for clinical and environmental monitoring. *Pure and Applied Chemistry*, **76**, 861–78.

91 Rogers, K.R. and Mascini, M. (1998) Biosensors for field analytical monitoring. *Field Analytical Chemistry and Technology*, **2**, 317–31.

92 Andreescu, S. and Marty, J.L. (2006) Twenty years research in cholinesterase biosensors: from basic research to practical applications. *Biomolecular Engineering*, **23**, 1–15.

93 Wang, J., Chen, G., Muck, A., Chatrathi, M., Mulchandani, A. and Chen, W. (2004) Microchip enzymatic assay of organophosphate nerve agents. *Analytica Chimica Acta*, **505**, 183–7.

94 Freire, R.S., Thongngamdee, S., Durán, N., Wang, J. and Kubota, L.T. (2002) Mixed enzyme (laccase/tyrosinase)-based remote electrochemical biosensor for monitoring phenolic compounds. *The Analyst*, **127**, 258–61.

95 Vianello, F., Ragusa, S., Cambria, A. and Rigo, A. (2006) A high sensitivity amperometric biosensor using laccase as biorecognition element. *Biosensors and Bioelectronics*, **21**, 2155–60.

96 Ghindilis, A.L., Gavrilova, V.P. and Yaropolov, A.I. (1992) Laccase-based biosensor for determination of polyphenols: determination of catechols in tea. *Biosensors and Bioelectronics*, **7**, 127–31.

97 Mulchandani, A., Kaneva, I. and Chen, W. (1998) Biosensor for direct determination of organophosphate nerve agents using recombinant Escherichia coli with surface-expressed organophosphorus hydrolase. 2. Fiber-optic microbial biosensor. *Analytical Chemistry*, **70**, 5042–6.

98 Mulchandani, A., MulchAndani, P., Kaneva, I. and Chen, W. (1998) Biosensor for direct determination of organophosphate nerve agents using recombinant Escherichia coli with surface-expressed organophosphorus hydrolase. 1. Potentiometric microbial electrode. *Analytical Chemistry*, **70**, 4140–5.

99 Mulchandani, P., MulchAndani, A., Kaneva, I. and Chen, W. (1999) Biosensor for direct determination of organophosphate nerve agents. 1. Potentiometric enzyme electrode. *Biosensors and Bioelectronics*, **14**, 77–85.

100 Mulchandani, A., Mulchandani, P. and Chen, W. (1998) Enzyme biosensor for determination of organophosphates. *Field Analytical Chemistry and Technology*, 2, 363–9.

101 La Rosa, C., Pariente, F., Hernandez, L. and Lorenzo, E. (1994) Determination of organophosphorus and carbamic pesticides with an acetylcholinesterase amperometric biosensor using 4-aminophenyl acetate as substrate. *Analytica Chimica Acta*, **295**, 273–82.

102 Mionetto, N., Marty, J.L. and Karube, I. (1994) Acetylcholinesterase in organic solvents for the detection of pesticides: biosensor application. *Biosensors and Bioelectronics*, **9**, 463–70.

103 Imato, T. and Ishibashi, N. (1995) Potentiometric butyrylcholine sensor for organophosphate pesticides. *Biosensors and Bioelectronics*, **10**, 435–41.

104 Simonian, A.L., Flounders, A.W. and Wild, J.R. (2004) A FET-based biosensor for the direct detection of organophosphate neurotoxins. *Electroanalysis*, **16**, 1896–906.

105 Braun, M., Teichert, O. and Zweck, A. (2006) Biokatalyse in der industriellen produktion, fakten und potenziale zur weißen biotechnologie, in Zukünftige Technologien Nr. 57, (ed. VDI-Technologiezentrum), VDI, Düsseldorf.

14
Road Ahead

Marc-Michael Blum, Heiko Russmann and André Richardt

14.1
Introduction

The current threat posed by biological and chemical weapons and effective counter measures have been discussed in previous chapters. In this chapter, we want to speculate about future developments in biological and chemical warfare and how to compete with such a threat.

The rapid development in life science and the enormous growth in the worldwide traffic could lead to new dangerous developments. If a better strategy of risk management for both military and civilian sites is to be developed, more has to be considered than only the "old" biological and chemical warfare agents. For example, the increase of information from the Human Genome Project could not only be used for medical applications and for improvement of human life, but

it is also conceivable that this information could be abused for genetic warfare. Also, the developments in the lab-on-the-chip technology could lead to easier production of highly toxic compounds, which then can also be abused for criminal intentions. Furthermore, the increase of traffic worldwide could lead to easier spreading of diseases (natural or unnatural) over the whole world in a couple of days. The outbreak of "Bird flu" in Asia indicates the potential risks of such an outbreak [1–3]. However, perhaps the greatest danger is that we will follow this path without intensive public debate. Therefore, an improved risk management strategy and a close co-operation between politicians, civilian decision makers and military sites are necessary.

14.2
Developments in Science

To maintain the ability to respond to biological and chemical incidents correctly, it is also necessary to follow the developments in science constantly [4]. In the last decades, new and incredible advances have driven technological change in the basic sciences. For over 50 years, since the genetic code was identified by James

Decontamination of Warfare Agents.
Edited by André Richardt and Marc-Michael Blum

ISBN: 978-3-527-31756-1

D. Watson and Francis Crick in 1953, discoveries in molecular genetics and the development of genetic engineering have offered new possibilities to manipulate organisms for different applications and to understand possible key targets for selective highly toxic compounds, and to start genetic warfare in the future.

For a long time, the purposeful genetic manipulation of microorganisms to change their properties was lengthy, often unsuccessful and could only be done by specialists. With the progress in chemical science, lab-on-a-chip technology, automated high-through-put synthesis/screening technology, genetic engineering, and the advances in biochemical structure-response relationships, development of new toxic molecules and manipulation of pathogens has acceded to a level where biochemical work is now commercially available and easy to perform. Genetic tailoring of microorganisms for pharmaceutical purposes is a rapidly growing field of biotech research. The deeper understanding of physiological properties of pathogens makes it more and more easy to enhance certain, sometimes deadly, properties. Initial attempts to alter pathogens during the Soviet bioweapon program were previously described by Ken Alibek in 1999. The USSR used the signing of the Biological Weapons Convention by their potential enemies as an opportunity to gain advantage in the field of biological weapons and supported a huge program until the beginning of the 1990s. More than 30000 workers were involved in research, development and production of biological weapons. The R&D activities included recombinant research studies with viral haemorrhagic fevers and smallpox. Even the creation of genetic chimera, new hybrid microorganisms created by joining DNA or RNA fragments from two or more different microorganisms, had been tested and was, according to Ken Alibek, successful [5, 6].

Since then, knowledge, technology and instrumentation have spread worldwide to a level where most countries and even terrorist groups might have the means or at least the potential to alter microorganisms for bio weapon purposes. States like Libya, Iran, Syria and Iraq actively sought to recruit former scientists from the Soviet bio weapon program.

The thread from bio and chemical warfare agents has gradually changed from a chemical via a biotechnology phase to a genetic phase, in which the door is open to new potential applications with two general tendencies: improvement of known pathogens to enhance military use, and improvement of the specificity for a target. Within the next decade, some possible developments could include:

- Genetically engineered pathogens that evade diagnosis and treatment
- Pathogens that are exceptionally lethal
- Pathogens intended to disable permanently
- Pathogens with enhanced contagiousness
- Pathogens with enhanced environmental stability
- New toxic compounds with new docking targets [7, 8].

The genetic modification of microorganisms to evade therapy was part of the Soviet R&D program in the 1980s and 1990s, and has been described by Ken Alibek. The goal was to alter known pathogens to make them resistant to Western drugs and to develop new and more powerful strains of pathogens [9].

In nature, the virulence of pathogens is a result of selection. Pathogens that kill their host quickly have a limited time to infect other hosts. This selection is not valid for strains cultivated in laboratory. In theory, a non-selective mechanism is limited in virulence, but, by genetic engineering, organisms with an exceptionally high and rapid lethality could be developed. With a better understanding of human bioregulators, for example, microorganisms designed to release pharmaceutically active substances causing long-lasting physical or mental disabilities could be engineered [10].

One main limitation of biological agents is their stability during dissemination and in aerosol form. Most pathogens have very limited transmissibility due to short-time persistency. With a better knowledge of why some pathogens are more stable than others, comes the possibility to direct genetic modifications [10].

Recent publications in scientific literature demonstrate that with modern molecular genetics it is even possible to reawaken past diseases, like the Spanish flu, responsible for the influenza pandemia from 1918 to 1920. The Spanish flu killed between 50 and 100 million people worldwide and had a higher death toll than World War I. Unlike other influenza pathogens, the Spanish flu strain H1N1 was exceptionally deadly for young adults, a property most interesting for developers of military agents [11–13].

Even the building of synthetic viruses has proved possible. In 2002 the group of Eckhard Wimmer chemically synthesized the genetic sequence of a polio strain with a length of 4.741 base pairs by assembling smaller DNA chains. He proved that the synthetic virus was infectious. The genetic sequence was accessible via the Internet. The synthesis of the genetic fragments is available from commercial companies specializing in DNA-synthesis. Although this virus has a relatively short DNA sequence, with further advances in DNA synthesis other viruses like Ebola or smallpox could be synthesized [14].

Still more frightening was the outcome of mousepox experiments in Australia. Ron Jackson and Ian Ramshaw performed experiments with different viruses to make rodents sterile. They planned to develop a contraceptive vaccine for mice. This material should cause an immune reaction of the mouse against fertilized egg cells to ensure that they are repelled. For this purpose the scientists genetically inserted a gene which produces large quantities of Interleukin-4 (IL-4) in mousepox virus. The protein IL-4 plays an important role in the development of T-cells. However, to their surprise, the scientists found that parts of the mouse immune system were completely suppressed. While mousepox normally causes only weak symptoms, all animals infected with the new virus died within nine days. And even with mice which were vaccinated against mousepox, the mortality rate still was about 50%. Although mousepox is not seen as dangerous for humans, many scientists are concerned that this method could be used for the development of biological weapons [15].

> "It would be safe to assume that if some idiot did put human Il-4 into human smallpox, they'd increase the lethality quite dramatically," Ron Jackson said, "We wanted to warn the general population that this potentially dangerous technol-

> ogy is available. We wanted to make it clear to the scientific community that they should be careful, that it is not too difficult to create severe organisms."

In the long term, even more frightening developments and manipulations might be possible, like synthetic viruses and prions, stealth pathogens, a microbial equivalent to sleeper cells, and ethnic-specific human pathogens or autoimmune diseases [10].

In awareness of this new threat, the Conference of the Biological and Toxin Weapons Convention (BTWC) stated in 1986:

> The Conference, conscious of apprehensions arising from relevant scientific and technological developments, inter alia, in the fields of microbiology, genetic engineering and biotechnology, and the possibilities of their use for purposes inconsistent with the objectives and the provisions of the Convention, reaffirms that the undertaking given by the States Parties in Article I applies to all such developments
>
> The Conference reaffirms that the Convention unequivocally applies to all natural or artificially created microbial or other biological agents or toxins whatever their origin or method of production. Consequently, toxins (both proteinaceous and non-proteinaceous) of a microbial, animal or vegetable nature and their synthetically produced analogues are covered.

14.3 Bioterrorism

In the past, terrorist groups have avoided using biological weapons and only very few attempts have been verified. The reason for this might be, in part that they did not want to alienate their supporters, and in part because of a lack of the expertise required to produce the agent and to build an effective weapon. But with increasing knowledge and technical skills this might change in future. Furthermore, of the seven countries listed by the U.S. Department of State as sponsoring international terrorism, at least five are suspected to have biological warfare programs.

There are several reasons why the extent of bioterrorism might increase in future [16]:

1. The number of militant religious groups, like Al Qaeda, and their supporters is constantly increasing. These extremists might label their victims as heretics or infidels and thus, too unworthy to live. The incentive of extremists to kill as many people as possible may not be constrained by any moral scruples.

2. Information about building biological weapons is increasingly available. Recipes for the preparation of biological agents and technical data for dissemination devices are readily available via the Internet. Through globalization and increased worldwide travel, more people have inconspicuous access to biological material and pathogens.
3. Religiously motivated international terrorism leads to increased commonality of groups in disparate areas. Religious or ideological ties are outweighing former national or ethnical bonds. As a result, groups are less worried about harming homelands or affecting potential constituents.
4. In the past, several terrorist groups have demonstrated interest in biological weapons and even prepared agents. In particular, Al Qaeda has a long-standing interest in acquiring capabilities. Traces of anthrax and ricin were found at several locations in Afghanistan and there is evidence for their interest in Botulinum toxin and plague [17].

But there are still arguments as to why the extent of biological and chemical terrorism might not increase [16]:

1. There are more effective weapons far more accessible than biological and chemical weapons. During the rare incidents of their use, biological and chemical weapons have demonstrated that the number of casualties was relatively low, compared to conventional weapons and explosives.
2. The use of biological weapons will presumably put the handler at risk of exposure to biological agents and might lead to long-lasting suffering. Spectacular and instant death in an explosion might be more desirable for any terrorist.
3. Terrorist groups tend to prefer weapons that are approved and have been demonstrated to be successful in the past.
4. The preparation and use of biological weapons remains difficult and could lead to a failure. For non-state actors, several problems like getting access to pathogens or toxins, cultivation of the biological material, transforming the agent into a weapon and effective dissemination, have to be solved. Furthermore, the production of the agent puts the producer at risk of being infected during the preparation.

The unsuccessful attempts of the Japanese Aum Shinrikyo sect might stand as an example of the difficulties involved in developing effective biological and chemical weapons. Aum Shinrikyo's abilities made world-wide news after the Tokyo subway attack with sarin in 1995, but the quest for a biological weapons capability actually predated the chemical program. Despite intensive efforts to acquire the means to develop and disseminate biological agents, actual attempts to distribute botulinum toxin and anthrax in and around Tokyo from 1990 to 1993 failed. At one incident more than 1000 liters of anthrax slurry was released from the laboratory roof in Tokyo and was obviously aimed at killing tens of thousands of people. The failure was probably caused by Aum Shinrikyo's scientists' unintentional use of a vaccination strain of anthrax, and the botulinum toxin was either inactivated during the release by the sun light or it was a non-toxigenic strain of *Clostridium botulinum*, the toxin-producing bacteria, that was cultivated. The anecdotal allegation that Aum Shinrikyo also sought to acquire Ebola was never proven. Whether this might prove the impossibility to effective use of biological weapons by terrorist groups in future is questionable, especially after the 2001 anthrax letters.

In 1984 the Bagwan Shree Rajneesh sect spread salmonella cultures at restaurant salad bars in Dallas, USA. Health authorities identified over 750 reported cases of salmonellosis, of which 45 had to be hospitalized, with a much higher number of unreported cases. Although no one died in the attack, it was startlingly successful. The attack was a trial run to test the possibility of disturbing a local election with the intention of affecting the outcome. The intended attack at the election was never performed because, even with this interference, no encouraging results could have been expected [8].

The anthrax letter attack of 2001, directed at politicians and news media that killed five people and infected 16 more, can also be seen as bioterrorism, even though no perpetrator has been identified yet. The attack consisted of two waves of letters. During the first wave in September 2001, it is presumed that five letters were mailed to different news media offices. Only two letters were actually identified, the existence of three more letters was reasoned from the pattern of anthrax infections. All letters contained dried anthrax powder that was partly released during the postal distribution. A second wave of two letters, addressed to US senators appeared three weeks later. Early reports described the powder in the latest letters as highly refined dry powder consisting of approximately one gram of nearly pure spores. An analysis by Dany Shoham and Stuart Jacobsen demonstrated that sophisticated additives and processing was used to create the powder [18]. The quality of preparation of the anthrax powder was far more sophisticated than experts had expected from non-state actors and was sometimes characterized as "weaponized" or "weapons grade" anthrax.

All preparations were derived from one potent anthrax strain, Ames, which was first researched at the U.S. Army Medical Research Institute of Infectious Diseases [19]. This and the quality of the preparation led to the never proven assumption that the perpetrator was either working for a US military laboratory or had direct access to military knowledge and material. Still the discussion about if any terrorist group might be able to reproduce an attack of that sophistication, is highly controversial.

Nevertheless, the successful and clandestine preparation and use of anthrax led to widespread fear of biological warfare agents and raised concerns about public safety and potential counter measures against bioterrorism attacks.

To predict future use of bio weapons by terrorists or individuals by retrospectively looking at past incidents is a questionable method because shock is still an effect that terrorism aspires to. But the public attention given to the anthrax letters and their economic impact, with financial spending of several billion US $ during recent years for the USA alone, demonstrate the profound effects of bioterrorism, and thus why it could be very attractive for the next generation of criminal individuals and terrorists.

14.4
Agricultural Biowarfare

Although most potential bio warfare agents are primarily animal pathogens, the effect of biological attacks on agriculture is grossly underestimated [20–22]. In the past, interest in biowarfare and bioterrorism focused on attacks against humans and increased in perception after the anthrax letters in 2001. Nevertheless, for several years, many experts have pointed out that the social and economic consequences of an attack against livestock or crops could be equally disastrous [21, 22]. Agricultural warfare is effective, because most countries have eradicated diseases that were endemic for centuries and that targeted animals or plants, and they have maintained this situation for decades by systematic and thorough hygiene measures. Confidence in the effectiveness of these measures led to extensive and wide-ranging transportation of animals, crops and agricultural material, and leaves modern economies vulnerable against biological attacks.

Their low perception by the public suggests that agricultural biowarfare and bioterrorism are relatively new threats. But biological attacks directed at a nation's agriculture have a long historical record and have proven to be very effective.

To illustrate the possible and serious consequences of a possible attack on a country's agriculture, the epidemic of foot-and-mouth disease (FMD) in Taiwan 1997 is a good example. The outbreak deeply affected the Taiwanese economy and had long-lasting effect on the country's agricultural sector. At the time of the first reported cases in one swine farm, 27 other farms were already contaminated. Within one week, over 700 farms were affected by the disease and within one month FMD was spread across the whole country. 4 million pigs were slaughtered and more than 500 000 tons of pork had to be destroyed. This and the following loss of export markets caused a drop in the year's gross domestic product by 2%. Within one week of the outbreak, pork prices dropped by 60%, about 50 000 people became unemployed and 6.9 billion US $ were lost in export revenue. In the following year, more than 4 billion US $ were spent in an unsuccessful effort to eradicate the disease. The overall losses amounted to 15 billion US $. Even three years later Taiwan had not fully regained its export markets. Interestingly, during this outbreak Taiwan lost its most lucrative pork market in Japan, which it gained in1982 from Denmark after an FMD outbreak. Even after Denmark declared free of FMD, it never fully regained the Japanese market [22, 23].

A similar outbreak of FMD in 2001 in Great Britain resulted in similar consequences. More than 4 million farm animals had to be slaughtered and the epidemic caused an economic loss of several billion US $. The following export ban imposed on British beef by the European Union caused additional political problems and tense relations between Great Britain and the EU states [23].

The spread of the pathogen can often be traced back to contaminated seed or material. An outbreak of FMD in 2000 in Japan was most likely caused by infected straw imported from China [24].

The examples above demonstrated "natural" outbreaks that started at one or very few locations and where the containment of the disease was difficult or unsuccessful. Spreading contaminated material with a criminal intent at multiple locations would undoubtedly result in a hardly containable epidemic. The greatest advantages of this are that the agents are typically not harmful for perpetrator, no technically challenging "weaponization" or effective dissemination is necessary, and suitable pathogens are easily accessible in endemic regions.

Military, terrorist groups and criminal individuals may have different motivations to use biological agents against agriculture (Table 14.1).

According to Mark Wheelis [25], states most vulnerable to economic an attack on their agricultural sector are those with several or most of the following attributes:

- High-density, large area agriculture
- Heavy reliance on monoculture of a restricted range of genotypes
- Free of specific serious animal and plant pathogens or pests
- Major agricultural exporter, or heavily dependent on a few domestic agricultural products

Table 14.1 Motivations to use biological agents against agriculture.

Military	Terrorism /Crime
• Attack the food supply of enemy troops	• Weakening economy by forcing export restriction
• Hamper transportation by targeting draft animals (for example Afghanistan)	• Destabilize the government
• Weakening economy by forcing export restriction	• Destabilize financial markets and chang market shares, even to realize profit
• Undermining civilian support and destabilize the government in the enemies homeland	• Support ideological ideas by interfering markets (for example disturb the meat market by animal rights activists)
• Force different supply and demand patterns for a single commodity	• Control of undesirable plants or animals (biocontrol)
• Control of undesirable plants or animals (biocontrol)	• Blackmailing
• Retaliation	

- Suffering serious domestic unrest, or the target of international terrorism, or unfriendly neighbor of states likely to be developing BW programs
- Weak plant and animal epidemiological infrastructure.

14.5 Public Health Response to Biological and Chemical Weapons

Different technologies and strategies are available to protect individuals against biological and chemical weapons [4, 26]. It is important to know the different technologies and to include them in an overarching concept [4, 26, 27]. A concept is only as strong as the weakest point in the whole concept. Responding to a biological or chemical threat is a complex task and therefore, it is necessary to take actions to eliminate, reduce and control the risks posed by biological and chemical weapons. Decision makers can use some basic principles to structure their concept and their ability to respond to biological and chemical incidents depends on

1. Preparedness: before an incident takes place
2. Response: action after a warning or a release [26].

The chapter will provide an outline of the matters that will need to be considered. Further detailed information can be found in the literature [26, 28–30].

14.5.1 Preparedness

Preparedness can be divided into

- Threat analysis [26, 31]
- Pre-emption of attack [26, 32, 33]
- Preparing to respond [26, 34]
- Preparing communication [35, 36]
- Validation of response capabilities [37].

Decision makers have to know that most of civilian health-care providers and first responders have little or no experience with chemical or biological weapons. Therefore, it is important to arouse public interest and to de-mystify these weapons by education and training. Such intensive and time-consuming training will also reduce or prevent fear and anxiety in healthcare workers and first-time responders. Also, realistic training simulations can be used to identify areas that can be improved and would help national and international authorities to make maximum use of existing respond capabilities [26].

14.5.2 Response

After a warning or a release of chemical or biological warfare agents is it necessary to respond immediately. The kind of response that takes place depends on the kind of information obtained.

- Response after warning, before any release, would lead to an early identification of the nature of the warning like an unusual package or device. After identification, an effective decontamination or neutralizing is possible.
- Distinguishing between biological and chemical agents leads to the right activities to decrease the number of ill or wounded individuals (Table 14.2).
- Response to biological attack leads to some major activities (Table 14.3). It is important to know that responses to

Table 14.2 differentiation of biological and chemical attack.

Indicator	Chemical Attack	Biological Attack
Epidemiological features	Patients with similar symptoms seeking care Clusters of patients arriving from a single locality Pattern of symptoms evident	Increasing disease evidence over hours or days in a healthy population Unusual increase in people seeking care Unusual numbers of patients with fatal illness and relatively uncommon disease
Animal indicators	Sick or dying animals	Sick or dying animals
Devices, liquids, liquid spray or vapour	Suspicious devices and/or packages, Un-normal clouds or fog, liquids with an unusual odour	Suspicious devices and/or packages

Source: adapted from references [26, 38, 39].

Table 14.3 Response activities for biological attack.

Sequence	Activities
Assess the risks	Determine the release or the outbreak Warfare agent identification [41, 42] Risk characterization
Manage the risk	Protect responders and health-care workersInfection-prevention and control procedures Conduct case triage Medical care of infected persons
Monitoring activities	Local and national resources and/or international assistance Active surveillance to monitor a) effectiveness of the prevention and control procedures, b) to follow up distribution of cases, c) to adjust response activities
Communication	Risk communication program for the affected population

Source: adapted from reference [26].

Table 14.4 Response activities for chemical attack.

Sequence	Activities
Assess the risks	Hazard identification
	Specialists for definitive identification needed for forensic and legal purposes
Manage the risk	Protect responders
	Control contamination
	Conduct casualty triage
	Medical care of casualties
Monitoring activities	Local and national resources and/or international assistance
	Continue monitoring the hazard level on the site
	Implement activities of for example, long-term injuries and rehabilitation
Communication	Risk communication program for the affected population

Source: adapted from reference [26].

natural and intentionally caused outbreaks follow the same line [40].

- Response to chemical incidents follows a similar pattern to that as used for biological outbreaks (Table 14.4).

Response to biological or chemical incidents can only happen if the time- and cost-consuming preparedness takes place. Political decision makers should have in mind that savings in civilian health care could cost more then 10-fold more, especially in the context of stunning developments in life science. In an overarching concept, knowledge of different decontamination technologies is necessary to choose the right option in the case of a biological or chemical incident.

14.6 State of the Current Decontamination Technology

Different decontamination technologies exist worldwide to decontaminate equipment, personnel and infrastructure. All of the existing technologies have their advantages and disadvantages. In previous chapters, the current conditions for an effective decontamination of biological and chemical warfare agents were shown. The most critical point regarding the existing technology can be divided in two sub-points:

- Many of the decontaminants are harmful to the environment and most of them cannot be used for decontamination of sensitive equipment, personnel, or for wounded persons. Also, for decontamination, experienced personnel are required and the decontamination is time- and cost-consuming.

- The decontamination can only take place after a positive detection and the determination of the kind of contamination. Particularly in the case of chemical warfare agents, this could lead to a very short time window for an effective decontamination.

However, with the existing decontamination equipment, the public health infrastructure would be stretched to its limits and maybe beyond the capability of the system to cope with the threat, if a biological warfare agent could not destroyed in time. The reader should also bear in mind that, today, over a quarter of the world's million deaths were due to infectious disease. Especially in developing countries without decontamination technology for chemical and biological warfare agents, this threat could lead to mass destruction of life or mass casualties.

14.7 Road ahead for Enzymatic Decontamination

Enzymatic decontamination is not the answer to all questions in the area of decontamination. To fulfill the requirements for field decontamination, enzymatic decontamination has to meet the ambitious stipulations necessary for uses under conditions fairly different from those in the laboratory. These stipulations are

- Cheap large scale production of enzymes
- Long term storage of enzymes under realistic conditions
- Design of carrier systems for decontamination purposes
- Thorough decontamination of chemical warfare agents like nerve agents in a decontamination media under realistic conditions [43–45].

For hydrolytic enzymes and with designed carrier systems (micro emulsions) the NATO standard criteria for thorough decontamination for one class of chemical warfare agents (G-type nerve agents) can be achieved [44, 45]. Also, there is evidence that other hydrolytic or oxidative enzymes can be used for the decontamination of mustard and VX. Current results suggest that beside hydrolytic enzymes, oxidoreductases will play a significant part, especially in the decontamination of spores, in a future benign decontaminant [43, 46]. To open the door for this advanced decontamination technology the German Army will introduce an enzyme module for the decontamination of sensitive equipment [49]. This enzyme module can be expanded by other enzyme systems to be prepared for future challenges [43, 47, 48].

However, the easiest way to eliminate or to reduce the possible threat of biological or chemical warfare is to improve living conditions not only in the developed world, but also in the developing countries. Improved living conditions and better terms of trade would lead to better chances for everybody and the ground for recruiting new terrorists and religious extremists would be desiccated.

References

1 Mactariane, J.T. and Lim, W.S. (2005) Bird flu and pandemic flu. *BMJ*, **331**, 975–6.

2 Chen, H., Smith, J.D., Zhamg, S.Y., Qin, J., et al. (2005) Avian flu: H5N1 virus outbreak in migratory waterfowl. *Nature*, **436**, 191–2.

3 Capua, I. and Alexander, D.J. (2006) The challenge of avian influenza to the veterinary community. *Avian Pathology: Journal of The WVPA*, **35** (*3*), 189–205.

4 Demirev, P.A., Feldman, A.B. and Lin, J.S. (2005) Chemical and biological weapons: current concepts for future defenses. *Johns Hopkins APL Technical Digest*, **26**, 321–33.

5 Alibek, K. and Handelman, S. (1999) *Biohazard: The Chilling True Story of the Largest Covert Biological Weapons Program in the World–Told from Inside by the Man Who Ran it*, Hutchinson, Great Britain, ISBN: 0091801354.

6 Alibek, K. (2001) *Mighty Microbe, Autumn, Defense Review*, p. 44.

7 Whitby, S., Millett, P. and Dando, M. (2002) The potential for abuse of genetics in militarily significant biological weapons. *Medicine, Conflict, And Survival*, **18**, 138–56.

8 Wheelis, M. (2002) Biotechnology and biochemical weapons. *Nonproliferation Review*, **9** (*1*), 48–53.

9 Bailey, K.C. (2001) *The Biological and Toxin Weapons Threat to the United States*. National Institute for Public Policy.

10 Wheelis, M. (2006) *Will New Biology Lead to New Weapons? Biological Weapons Convention Reader*, pp. 46–52.

11 Tumpey, T.M., Basler, C.F., Aguilar, P.V., Zeng, H., Solorzano, A., Swayne, D.E., Cox, N.J., Katz, J.M., Taubenberger, J.K., Palese, P. and Garcia-Sastre, A. (2005) Characterization of the reconstructed 1918 Spanish influenza pandemic virus. *Science*, **310** (*5745*), 77–80.

12 Taubenberger, J.K., Reid, A.H., Lourens, R.M., Wang, R., Jin, G. and Fanning, T.G. (2005) Characterization of the 1918 influenza virus polymerase genes. *Nature*, **437** (*7060*), 889–93.

13 Kobasa, D. et al. (2007) Aberrant innate immune response in lethal infection of macaques with the 1918 influenza virus. *Nature*, **445**, 319.

14 Cello, J., Paul, A.V. and Wimmer, E. (2002) Chemical synthesis of poliovirus cDNA: generation of infectious virus in the absence of natural template. *Science*, **297**, 1016–8.

15 Jackson, R.J., Ramsay, A.J., Christensen, C.D., Beaton, S., Hall, D.F. and Ramshaw, I.A. (2001) Expression of mouse interleukin-4 by a recombinant ectromelia virus suppresses cytolytic lymphocyte responses and overcomes genetic resistance to mousepox. *Journal of Virology*, **75** (*3*), 1205–10.

16 Cronin, A.K. (2003) *Terrorist Motivations for Chemical and Biological Weapons Use: Placing the Threat in Context*. Report for Congress, March 28, 2003.

17 Miller, J. (2002) Labs suggest Qaeda planned to build arms, officials say. *New York Times*, September 14, 2002.

18 Shoham, D. and Jacobsen, S.M. (2007) Technical intelligence in retrospect: the 2001 anthrax letters powder. *International Journal of Intelligence and CounterIntelligence*, **20**, 79–105.

19 US Army Medical Research Institute of Infectious Diseases (2001) *USAMRIID's Medical Management of Biological Casualties Handbook*, 4th edn, US Army Medical Research Institute of Infectious Diseases, Fort Detric, MD, USA.

20 Koblentz, G. (2003/2004) Pathogens as weapons: the international security implications of biological warfare. *International Security*, **28**, 84–122.

21 Frazier, T.W. and Richardson, D.C. (eds) (1999) *Food and Agricultural Security: Guarding against Natural Threats and Terrorist Attacks Affecting Health, National Food Supplies, and Agricultural Economics*, New York Academy of Sciences, New York.

22 Wheelis, M., Casagrande, R. and Madden, L.V. (2002) Biological attack on agriculture: low-tech, high-impact bioterrorism. *Bioscience*, **52** (*7*), 569–76(8).

23 Koczura, R., Kaznowski, A. and Mickiewicz, A. (2004) The potential impact of using biological weapons against livestock and crops. *ASA-Newsletter*, 4–6.

24 Matsubara, K. (2000) *Final Eradication of Foot and Mouth Disease in Japan*, Ministry of Agriculture, Tokyo.

25 Wheelis, M. (2000) *Agricultural Biowarfare and Bioterrorism*, Occasional Paper of The Edmonds Institut, ISBN 1-930169-14-0.

26 WHO guidance (2004) *Public Health Response to Biological and Chemical Weapons, Second Edition of Health Aspects of Chemical and Biological Weapons: Report of a WHO Group of Consultants*, World Health Organization, Geneva.

27 Richardt, A. and Grabowski, A. (2006) Dekontaminationstechnologien – Notwendigkeit eines Gesamtkonzeptes für den militärischen und zivilen Schutz. *Bevölkerungsschutzzeitung*, **4**, 18–22.

28 United Nations International Strategy for Disaster Reduction, Living with risk: a global review of disaster reduction initiatives, www.unisdr.org (July 2002).

29 Leonard, B. (ed.) (2000) *Emergency Response Guidebook: A guidebook for first Responders During the Initial Phase of a Dangerous Goods/Hazardous Materials Incident*, Diane Publishing Co., Ottowa, ON, ISBN: 0788183990.

30 Health Council of the Netherlands (2001) *Defense Against Bioterrorism*, Health Council of the Netherlands, The Hague (publication no. 2001/16).

31 Pan American Health Organization (2000) *Natural Disasters: Protecting the Public's Health*, Pan American Health Organization, Washington, DC (Scientific Publication No. 575).

32 Roberts, A. and Guelf, R. (2000) *Documents on the Laws of War*, 3rd edn, Oxford University Press, Oxford.

33 Krutsch, W. and Trapp, R. (1999) *Verification Practise under the Chemical Weapons Convention*, Kluwer Law International, The Hague.

34 Cosivi, O. (2003) Health Preparedness for the deliberate use of biological agents to cause harm: WHO's activities (eds J. Kocik, M.K. Janiak and M. Negut), Proceedings of the NATO Advanced Research Workshop on Preparedness against Bioterrorism and Re-emerging Infectious Diseases, 15–18 January 2003, Warsaw, Poland, pp. 9–12.

35 Norlander, L. et al. (eds) (1995) *A FOA Briefing Book on Biological Weapons*, National Defence Research Establishment, Umea.

36 Ivarsson, U., Nilsson, H. and Santesson, J. (eds) (1992) *A FOA Briefing Book on Chemical Weapons: Threat, Effects, and Protection*, National Defence Research Establishment, Umea.

37 Henderson, D.A., Inglesby, T.V. and O'Toole, T. (2002) Shining light on dark winter. *Clinical Infectious Diseases*, **34**, 972–83.

38 Sidell, F.R., Patrick, W.C. and Dashiell, T.R. (1998) *Jane's chem.-bio Handbook*, Jane's Information Group, Alexandria, VA.

39 APIC Bioterrorism Task Force, Centers for Disease Control Hospital Infections Program Bioterrorism Working Group (2002) *Interim Bioterrorism Readiness Planning Suggestions*, Washington DC, http://www.apic.org/Content/NavigationMenu/PracticeGuidance/Topics/Bioterrorism/APIC_BTWG_BTRSugg.pdf (12.10.07).

40 Bres, P. (1986) *Public Health Action in Emergencies Caused by Epidemics: A Practical Guide*, World Health Organization, Geneva.

41 Franz, D.R., et al. (1997) Clinical recognition and management of patients exposed to biological warfare agents. *Journal Of The American Medical Association*, **278**, 399–411.

42 Green, M. and Kaufmann, Z. (2002) Surveillance for early detection and monitoring of infectious disease outbreaks associated with bioterrorism. *The Israel Medical Association Journal*, **4**, 503–6.

43 Richardt, A., Blum, M.M., Mitchell, S. and Danielsen, S. (2007) Enzymatic Decontamination of Nerve-Agents by DFPase – From Concept to Product, In: *Ninth International Symposium on Protection against Chemical and Biological Warfare Agents*, Gothenburg, Sweden, May, 2007, 61, ISSN 1650-942.

44 Richardt, A., Niederwöhrmeier, B. and Danielsen, S. (2006) Novel, improved catalytic enzymes – a new road to CB-

decontamination. ESW 2006, Proceedings of the 3rd European Survivability Workshop, 16–19 Mai, P 10: 1–9.

45 Richardt, A., Blum, M.M. and Mitchell, S. (2006) Was wissen Calamari über Sarin?–Enzymatische Dekontamination von Nervenkampfstoffen. *Chemie in unserer Zeit*, **40**, 252–9.

46 Whited, G. (2007) Perhydrolase–Safe, Effective in Situ Peracetic Acid Generation by a Biocatalyst–New Concepts, Procedures, and Planning, In: *Ninth International Symposium on Protection against Chemical and Biological Warfare Agents*, Gothenburg, Sweden, May 2007, 68, ISSN 1650-942.

47 Richardt, A. (2005) Enzymatische Dekontamination -Vom Konzept zum kommerziellen Produkt. *Wehrtechnik*, **5**, 129–31.

48 Richardt, A. and Mitchell, S. (2006) Enzymes for environmentally friendly decontamination of sensitive equipment. *Journal of Defence Science*, **10**, 261–5.

49 Bradschett, C. and Richardt, A. (2007) Dekontaminations vertahren und–systemes. *Strategie und Technik*, **Sept**, 60–62.

Index

Decontamination of Warefare Agents.
Edited by André Richardt and Marc-Michael Blum

ISBN: 978-3-527-31756-1